A Book of Biotechnology

A BOOK
OF
BIOTECHNOLOGY

By

Dr. Syed Mohammad Ahmad
Rtd. Professor HOD Botany
Saifia Science College
Bhopal (M.P.)

Rehana Khan
Asstt. Prof. Zoology and Environment
Cambridge Girls P.G. College
Bhopal (M.P.)

UNIVERSITY SCIENCE PRESS
(An Imprint of Laxmi Publications Pvt. Ltd.)
BANGALORE • CHENNAI • COCHIN • GUWAHATI • HYDERABAD
JALANDHAR • KOLKATA • LUCKNOW • MUMBAI • RANCHI
NEW DELHI

Published by :
UNIVERSITY SCIENCE PRESS
(An Imprint of Laxmi Publications Pvt. Ltd.)
113, Golden House, Daryaganj,
New Delhi-110002

Phone : 011-43 53 25 00
Fax : 011-43 53 25 28

www.laxmipublications.com
info@laxmipublications.com

First Edition : 2010

OFFICES

✆ **Bangalore**	080-26 61 15 61	✆ **Chennai**	044-24 34 47 26
✆ **Cochin**	0484-237 70 04, 405 13 03	✆ **Guwahati**	0361-254 36 69, 251 38 81
✆ **Hyderabad**	040-24 65 23 33	✆ **Jalandhar**	0181-222 12 72
✆ **Kolkata**	033-222 74 384	✆ **Lucknow**	0522-220 95 78
✆ **Mumbai**	022-24 91 54 15, 24 92 78 69	✆ **Ranchi**	0651-221 47 64

UBI-9472-160-A BOOK OF BIOTECHNOLOGY-AHM **C—**

Typeset at : Sukuvisa Enterprises, New Delhi.

Printed at : Repro Knowledgecast Limited, Thane

CONTENTS

PREFACE

The aim of this book is to present the subject of Biotechnology to the Graduate and Post graduate students of Biotechnology of various universities of India.

The authors have been extra cautious in presenting the matter in easy and lucid language. The diagrams are especially designed for clarity and simplicity. Thus, in this book many topics which are not easily available have been incorporated for the convenience of the students. So, the non-availability of literature, we started a sincere effort to search, collate, and edit the various topics. This was a challenging job. The completion of the book has not at all easy task. It has been completed with the co-operation of so many hands.

Suggestions for the development of this book shall be gratefully acknowledged.

— *Authors*

Acknowledgement

The authors wish to express their gratitude to Prof. Shakoor Khan, Principal Cambridge Girls P.G. College, Bhopal for his guidance and active participation in going through the manuscript.

Our thanks to other authors whose material has been put and utilized for promotion of the student knowledge of the subject through this book.

We further express our special thanks to our family members who were the source of inspiration in completing this academic effort.

— *Authors*

Chapter 1

MICROSCOPY

INTRODUCTION

Almost all cells are too small to be examined directly with the human eye and so, our knowledge of cells has depended very much on microscopic techniques for magnifying them. The history of cell biology is an excellent example of the effect of one scientific discipline on another. The improvements in microscopy produced by developments in physics have been closely correlated with the expansion of cell biology. It is interesting to compare the appearance of the microscope used by Robert Hook in the seventeenth century with that of modern light and electron microscope. The magnifications attainable by these microscopes range from X 100 to X 400,000. In addition, several different kinds of microscopy are available, and many techniques have been developed by means of which specimens can be prepared for examination. Each type of microscopy and each method of preparing specimens for examination offers advantages for demonstration of specific morphological features.

THE MICROSCOPE

Microscope is one of the most important instruments used in the biological sciences; its function being to produce an enlarged image of the object. For making "enlarged" images, simple lenses like "magnifiers" or "reading glasses" have been in use for centuries, especially in Ancient China. These are forms of simple microscope. Such simple magnifiers had definite limitations due to their crude construction. About 1590, a Dutch spectacle maker, Zacharias Janssen, used a second lens to magnify the image produced by a primary lens. This is the basic principle of the compound microscope used even today. Galileo invented an improved compound microscope in 1610.

Antony Von Leeuwenhoek (1632-1723) was the first to record observations of bacteria, yeasts, and protozoa with the help of microscope invented by himself. A cloth maker and tailor by trade as well as the official winetaster of Delft, Holland, Leeuwenhoek's interest in

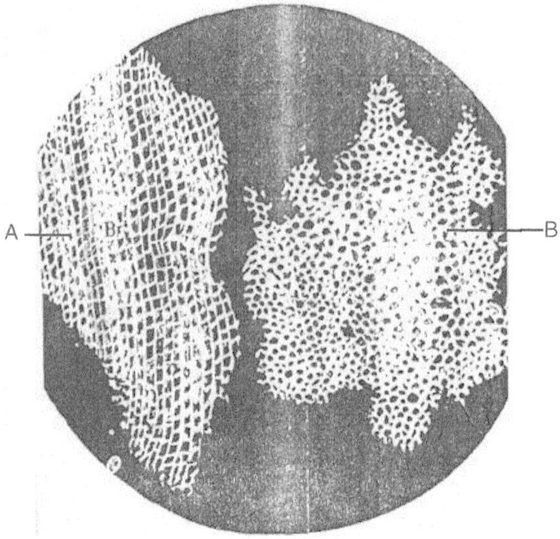

Left: In 1665 Robert Hooke made these drawing of thin slices cut from a piece of cork and viewed through his microscope. *A* is a cross-section, and *B* is a longitudinal section

Right: Robert Hooke's microscope

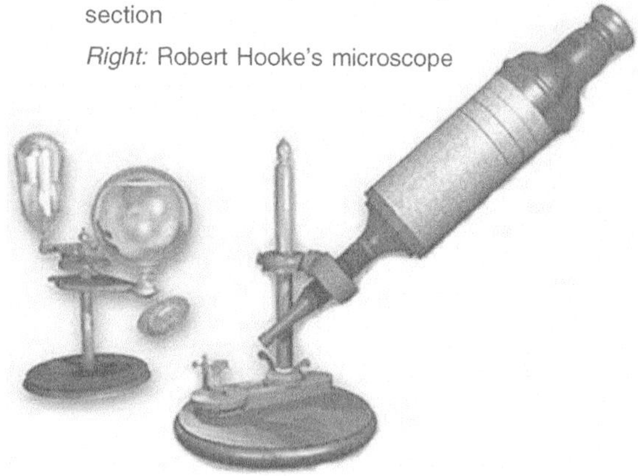

Fig. 1.1. (Beckett, p.II)

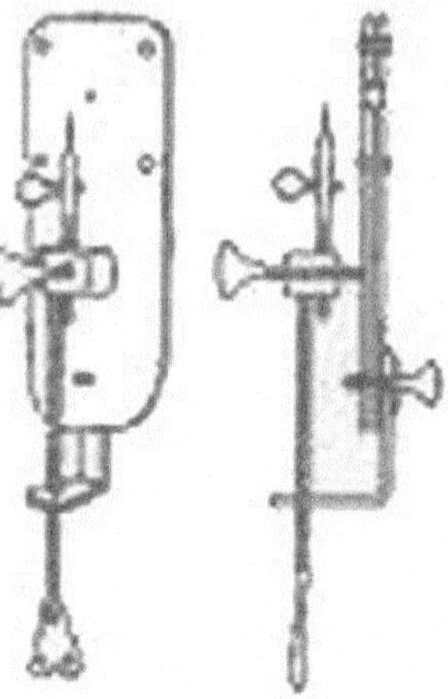

(A)

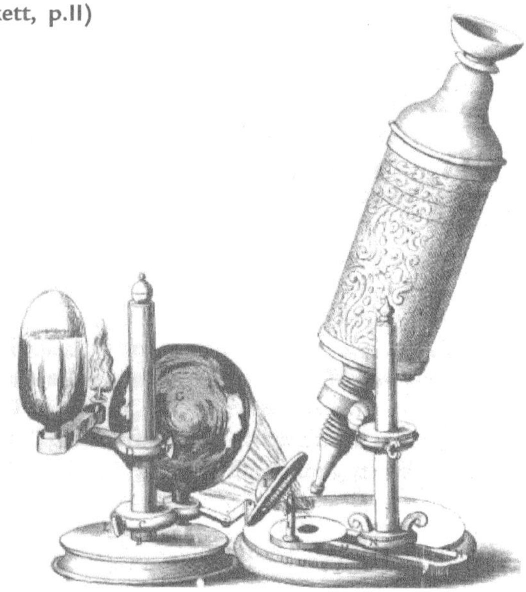

(B)

Fig. 1.2. Some early microscopes, used in the late seventeenth century :

(A) This microscope used by Leeuwenhoek consists of a single biconvex lens (I) inserted into a small hole on the left side of the leather base. The specimen was placed on the small pointed wire attached to the screw, which moved the specimen back and forth to bring it into the focus of the fixed glass.

(B) Hooke's compound microscope of the same period more closely resembles the appearance of a modern microscope. Note the elaborate decoration of the body of the microscope, the candle illumination source and the separation of the eyepiece from the objective lens.

(Atlas, p. 5)

microscopes was probably related to the use of magnifying glasses by drapers to examine fabrics. Leeuwenhoek had the means and opportunity to pursue his hobby of lens grinding and microscope making. During his lifetime he made more than 250 microscopes consisting of home-ground lenses mounted in brass and silver.

The earliest microscope used by Leeuwenhoek was basically a very tiny, perfect lens that was used essentially like a magnifying glass. The difference between it and a conventional magnifying glass lies in the amount of magnification. A high-quality magnifying glass magnifies 10 to 20 fold, whereas Leeuwenhoek's best microscope magnified objects 200 fold. Leeuwenhoek mounted a specimen on the tip of an adjustable screw that he could position to focus the specimen. This is an example of a simple microscope. As early as 1847 Carl Zeiss Jena (Germany) built the first simple types of microscopes. The Olympus Optical Company was established in 1919 and began production of microscopes.

Fig. 1.3. Antony van Leeuwenhoek (1632-1723) here seen holding one of his microscopes, opened the doors to the hidden world of microbes when he described bacteria. Although only a amateur scientist, Leeuwenhoek has a keen interest in optics, and his diligence allowed him to make this important discovery. (Atlas, p.5)

CARL ZEISS—A SYNONYM FOR QUALITY

Carl Zeiss (*1816–1888*)

The brandname Zeiss is a synonym for precision, perfection, and state-of-the-art design. This has been true of the company throughout its 130 years history. In the early days the small but exclusive Zeiss company in Jena had a standard answer to fastidious customers requesting products of special quality. "The Carl Zeiss Company guarantees identical high standards of quality for every product. No individual product is superior to the others. If this does happen to occur, then from that point on the entire production of that item is rigorously made to comply with those standards.

NO COMPROMISE ON QUALITY

This guarantee was backed up by the name of the company owner: Carl Zeiss (1816-1888), a man who would countenance no compromises where his microscopes were concerned. If the slightest drop in quality was discovered in the products lining the shelves of his small shop, he would personally smash the entire batch with a hammer.

Fig. 1.4. This large-scale planetarium projector, developed by Zeiss in Oberkochen, is a prime example of precision engineering.

(German News) Vol. XXVII, No. 3, March 21,1985

Hard work and exacting standards were a family tradition. Carl Zeiss came from a long line of skilled craftsmen, and his father had been a master wood-turner.

His son had loftier ambition. At an early age Carl Zeiss developed an inquiring mind and later attended university lectures.

RANKING AMONG PROBLEM-SOLVERS

Having becomes a qualified mechanic, Carl Zeiss joined the ranks of the problem-solvers. A hit- and -miss technique played a large part in the manufacture of his microscopes, and it was a time consuming process trying out hundreds of lenses until he had the right one. Often he had to rely on sheer good luck.

Carl Zeiss was not discouraged, however at the age of 50, he made the acquaintance of the 26-year-old Ernst Abbe, a private scholar from Jena. The two men complemented each other perfectly: the one a gifted but extremely sensitive physicist who suffered from recurrent bouts of migraine; the other an adventurous entrepreneur who never gave up.

It was nine years before the breakthrough finally came: the first comprehensive and exact scientific method of constructing systems of lenses.

WORLDWIDE REPUTATION

By 1886, the Zeiss factory in Jena had already sold its 10,000th microscope. The company had long since established a worldwide reputation–not least thanks to the advent of a further expert in his field: Dr. Otto Schott, a specialist in the chemistry of glass.

After the Second World War the division of Germany also meant that the Carl Zeiss Foundation split into two parts. The special glass is today produced both by the state-owned company Carl Zeiss Jena in the GDR and by the Schott glassworks in Mainz. And the high-precision optical equipment bearing the brandname Zeiss is manufactured in Jena and in Oberkochen, a small town near Stuttgart, Federal Republic of Germany.

Before the many facets of microscopy are considered, it is important to understand a fundamental property of microscopes: *resolution* or *resolving power*.

RESOLVING POWER

The problem of the magnification of an object is best considered in relation to the property known as *resolution*. The ability of lens system to show fine detail is termed its resolving power and this may be defined as the ability of an optical system to show as distinct and separate to points that are close together. Mere increase in size (greater magnification) without the ability to distinguish structural details (greater resolutions) is not beneficial.

The simplest illustration of resolving power is the visibility of double stars. Although the two stars may be separated by a vast distance, the visual angle reaching the eye is very small, and the stars appear to be close together. Many individuals can see but one star. Other persons, whose eyes have better resolving power can see the two stars distinctly. Applying this principle to the microscope, a lens of poor resolving power will show a slender chromosome as a single thread, whereas a lens of good resolving power will show the chromosome as two interwound threads.

The human eye has a fundamental limitation in that it cannot distinguish clearly points closer together than about 0.1 mm; it has a resolving power of about 0.1 μm. Points closer together than this will be seen as a single image. The resolving power of the eye is, in fact, less than that which is theoretically possible, due to the diffraction or scattering of light.

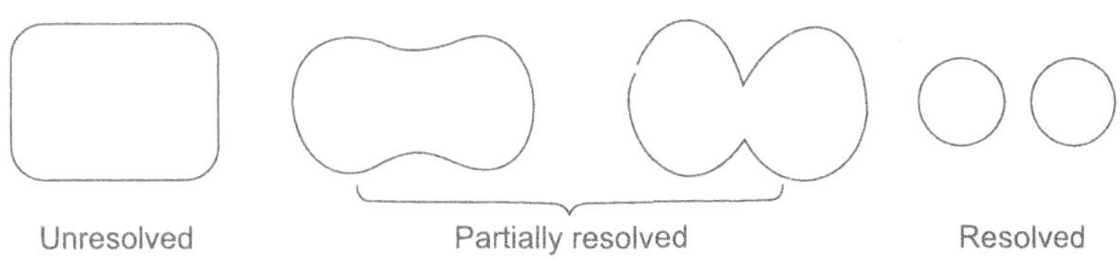

Fig. 1.5. Resolution of two points under a microscope. At low resolution structures blur together; the greater the resolution, the more detail can be observed. (Atlas, p.43)

Consider two points and the image that they produce at the retina of the eye. Ideally, in the absence of diffraction, these would produce two sharp peaks of light intensity. However, because diffraction occurs, a more diffuse area of light will be produced and the two images will overlap. The diffuse area of light is known as Airy disc and if the centres of the two discs are closer together than the radius of each, the eye will not resolve the two points and will record only a single region of light.

The greater the resolving power, the greater the definition of an object, hence microscopes with high resolving power are especially good for viewing small structures. Resolving power in a compound microscope depends on the wavelength of light used and an optical property of the objective lens known as its numerical aperture. Since the wavelength of light is usually fixed, the resolution of an object is in practice a function of the numerical aperture; the larger the numerical aperture, the smaller the object resolved. There is a rough correspondence between the magnification of an objective lens and its numerical aperture; lenses with higher magnification usually have higher numerical apertures. (The value of the numerical aperture is printed on the side of the lens). But the medium through which the light passes also affects numerical aperture.

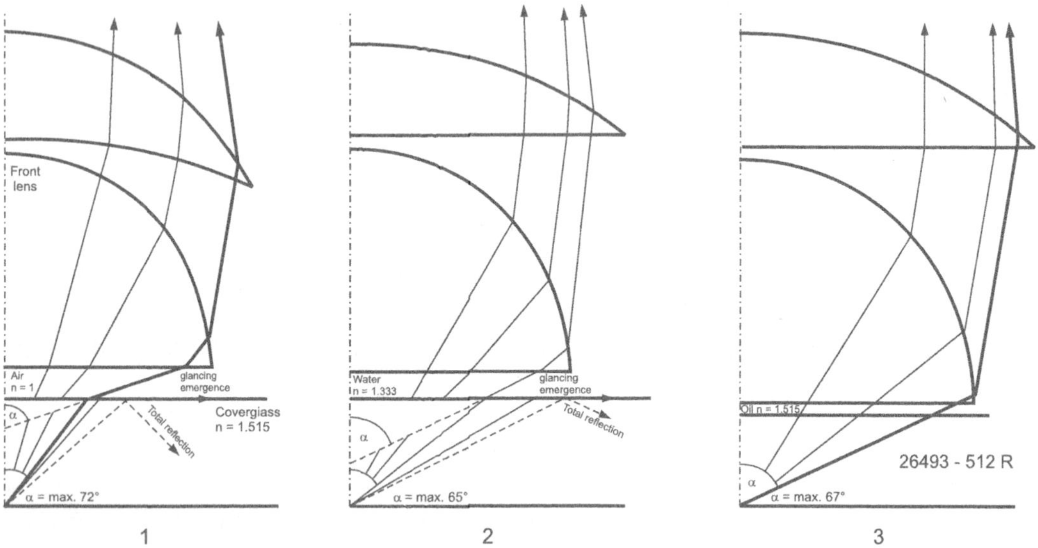

Fig. 1.6. Diagrammatic representation of the aperture values of–

1. a dry system,

2. a water immersion objective, and

3. an oil immersion objective.

The light microscope is able to increase the resolution of the eye because, essentially, it increases the aperture of the eye. This reduces the radius of the Airy discs and so allows the eye to distinguish points closer together.

By the end of the last century, great improvement had been made in light microscopy, and it was possible to obtain resolving power of about 0.2 μm, which is considered better than that of the unaided eye. However, the resolution of any microscope is fundamentally limited by the wavelength of the illumination employed. The average wavelength of visible light is about 550 nm, and so to improve on the resolution of a normal light microscope shorter wavelength illumination must be used. In some microscopes, ultra-violet light with a wavelength of about 250 nm has been utilized to give a resolving power required to see many cell structures. This problem has now been overcome with the development of the electron microscope which makes use of the wave-like properties of a beam of electrons. The electron beam in an electron microscope normally has a wavelength of about 4 nm, some sixty times less than that of ultra-violet light, and this has enabled resolving powers of about 0.2 nm to be obtained, about 1000 times better than that of light microscope.

LIMIT OF RESOLUTION

The limit of resolution is the smallest distance by which two objects can be separated and still be distinguishable as two separate objects. The greatest resolution in light microscopy is obtained with the shortest wavelength of visible light and an objective with the maximum NA. The relationship between NA and resolution can be expressed as follows :

$$d = \frac{\lambda}{2\text{NA}}$$

where d = resolution and λ = wavelength of light. Using the values 1.3 for NA and 0.55 μm, the wavelength of green light, for λ resolution can be calculated as :

$$d = \frac{0.55}{2 \times 1.30} = 0.2\ \mu\text{m}$$

From these calculations we may conclude that the smallest details that can be seen by the typical light microscope are those having dimensions of approximately 0.2 μm.

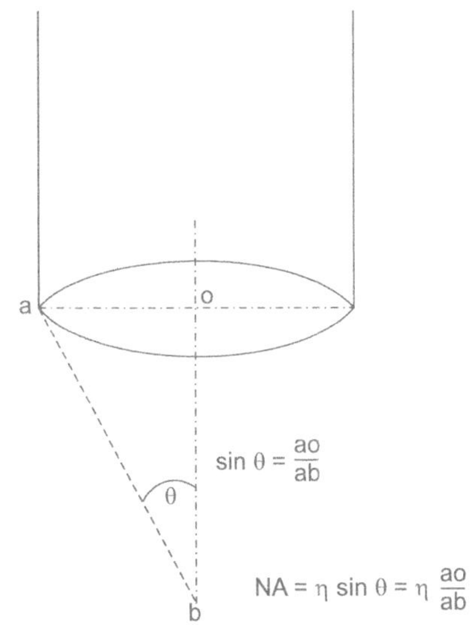

A diagrammatic representation of numerical aperture.

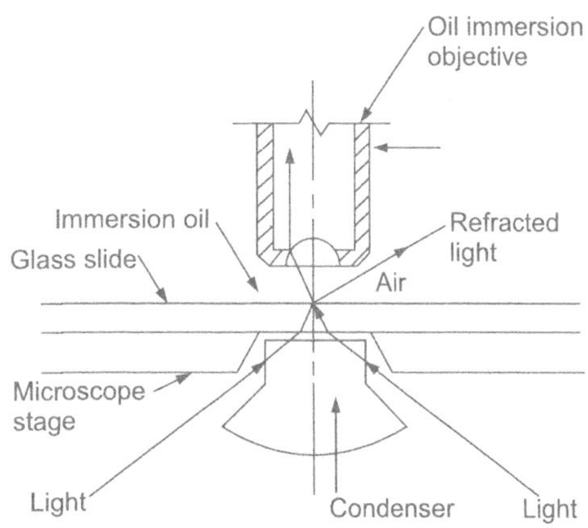

Fig. 1.7. Comparison between the effects of oil and air on the passage of light rays from a specimen to the front lens of the oil-immersion objective (Wistreich & Lechtman, p.56)

NUMERICAL APERTURE (NA)

Numerical Aperture is a number that is indicative of relative resolving power. The value of numerical aperture is engraved on most objectives. This number is 0.25 in a 10x objective, for example and increase through progressively higher magnification, attaining the value 1.4 in a 90x objective.

Knowing the numerical aperture, we can make a simple computation and arrive at resolving power of the lens. Assume that we are using an objective of NA 1.0 and using light having a wavelength, in round numbers of 0.0005 mm. The formula is :

$$\frac{\text{Wavelength}}{2\,\text{NA}} = \frac{0.0005}{2} = 0.00025 \text{ mm}$$

This means that if two bacterial cells or two chromomeres on a chromosome are separated by a space of 0.00025 mm, they can be seen as two distinct bodies. As the numerical aperture increases, the resolving power increases and the depth of field decreases.

The practicable upper limit of NA 0.95 is obtainable with dry lenses, used with an air space between the objective and the cover glass. To achieve values greater than this, the objective must be immersed in a medium of higher refractive index than that of air.

Using immersion oil between the specimen and the objective also improves the resolving power of the microscope because more light can enter the lens. The observation of several microorganisms (algae, fungi, and protozoa) can be achieved with dry objectives, that is, where air occupies the space between the specimen and the objective, but the viewing of bacteria in sufficient detail to determine the shape and arrangement of cells normally requires the use of an *OIL IMMERSION LENS*.

ADVANTAGES OF USING IMMERSION OIL

There are at least two advantages in using immersion oil with very high-powered objectives :

1. In order to have the details of an object clear-cut and sharply defined, the light rays passing through each part of it must not be exactly parallel but must diverge from it before entering the lens. If the space between the cover glass and the lens is filled with air, the most divergent rays will be refracted so far that they will miss the lens, and there will be consequent loss of definition. With oil filling this space, these divergent rays will reach the lens and better definition will be obtained; also, there will be greater illumination of the microscopic field from the same source of light.
2. The front lens of the objective is plano-convex flat at the bottom and so convex at the top that if it were used as a dry lens the plane of focus would be almost at its bottom surface, since the bending of the rays as they pass from the air beneath the lens into the glass would result in bringing the plane of focus closer. The immersion oil, however, has the effect of making the lens deeper from top to bottom, as the oil and glass form one continuous system.

The rays are not bent on entering the lens from below and the plane of focus is farther away from it.

The liquids best suited for this purpose are specially prepared cedarwood oil, or a mineral non-drying immersion oil. The latter has the advantage of not hardening on the lens or the slide when it is not promptly cleaned off. To preserve the microscope components, however, all oil should be removed from the objectives and all other parts after use. Lens paper is used for this purpose.

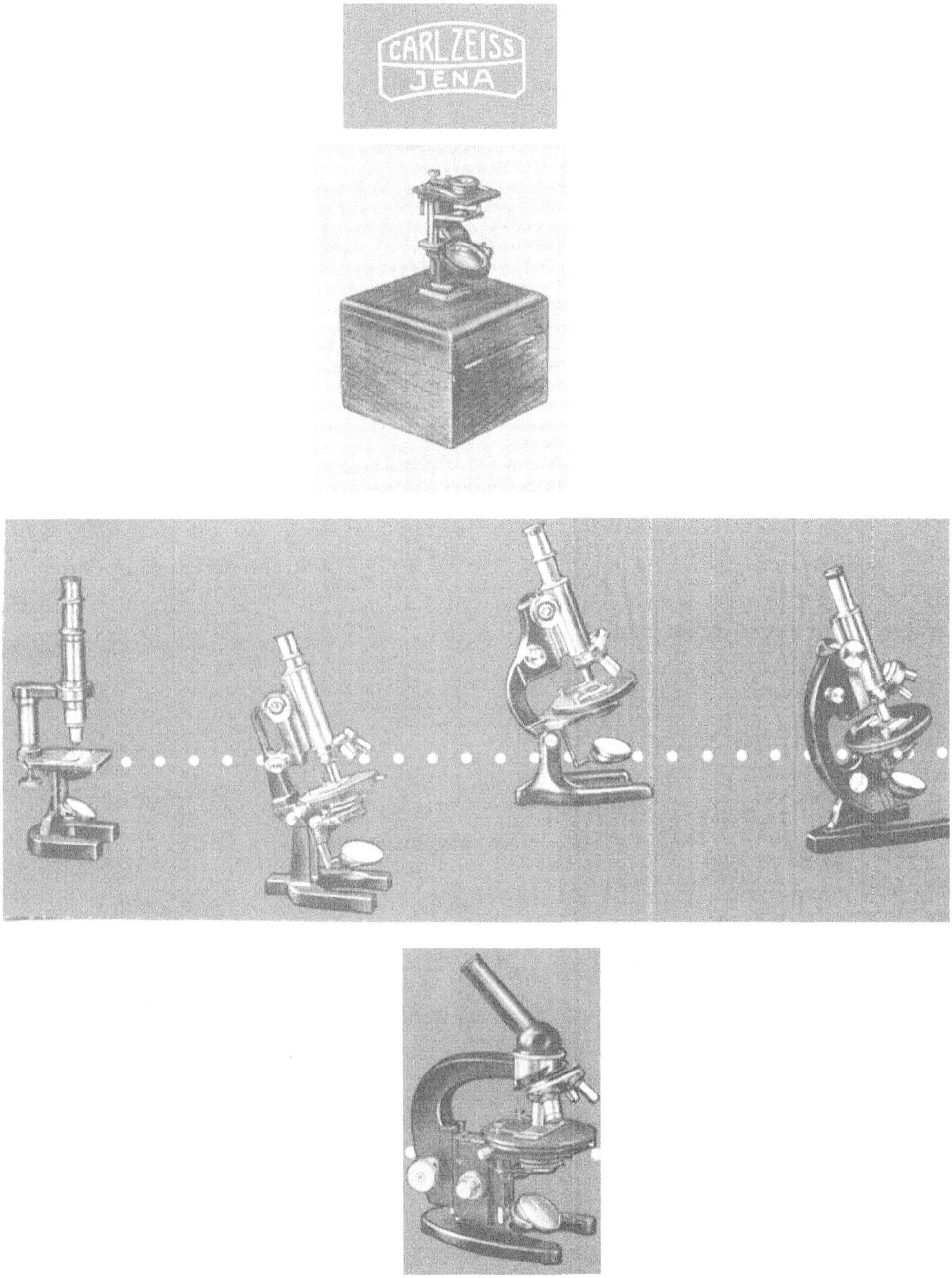

Fig. 1.8. Evolution in microscope building.

If a drop of cedarwood oil or synthetic silicon immersion fluid is used to connect the immersion objective with the slide, the resolving power is increased. An NA of 1.10 can be obtained with a water immersion lens, and NA 1.40 using cedar oil. An oil immersion lens used in air presents a very unsatisfactory image, however; the lens should never be employed without the use of oil.

THE NEED FOR A COVER GLASS

If the upper surface of the object is curved or irregular, as is usually the case, the rays of light passing from it through the air above to a dry objective will be so bent that the image will be distorted. This does not happen, however, if a cover glass is used and the space between it and the object is filled with transparent embedding material. If, however, an oil immersion lens is used the oil fills the space between it and the object, thus excluding the air. Under these conditions, the rays pass through without bending, and the cover glass is unnecessary, unless one desires to preserve the specimen.

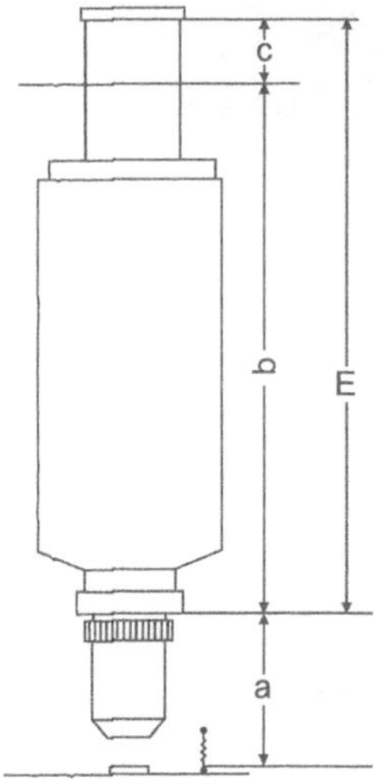

Fig. 1.9. Object without coverglass.

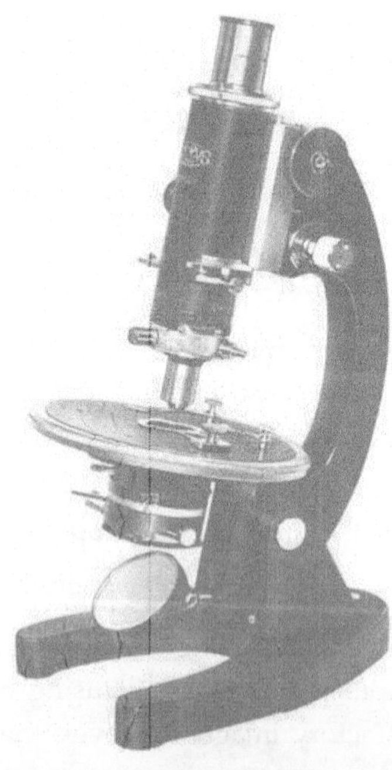

Fig. 1.10. Polarizing microscope.

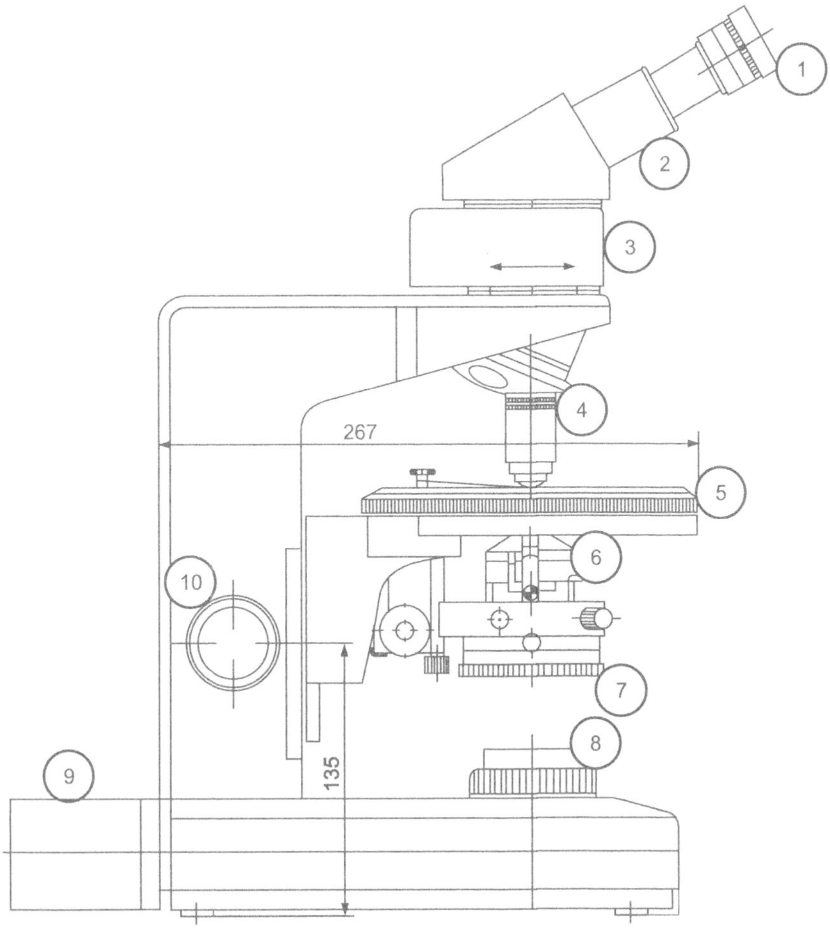

Fig. 1.11. Polarizing microscope

1. Eye piece
2. Tube, pinhole lens
3. Intermediate tube with analyzer
4. Centerable objective nosepiece with tube slot and objectives
5. Stage (specimen)
6. Condenser with top and aperture diaphragm
7. Polarizer
8. Field diaphragm
9. Light source
10. Coarse and fine focus controls

PARFOCALIZATION

Two or more objectives are parfocal with each other when it is possible to focus one objective on an object, turn to the next objective without focusing, and see the object in more or less sharp focus. Therefore, if the microscope is parfocal, once the specimen is in focus with one lens, it remains in focus when switching to other objective lenses. This permits focusing with a low power objective, and changing to a higher power objective with some minor focusing adjustment. Excessive breakage of slides and scratching of objectives occur if the objectives are not parfocal.

MAGNIFICATION

The most obvious property of objectives is magnification, which is a fixed value. The objective magnifications used most commonly on standard monobjective microscopes range from 3.2x to 100x. Magnifications below this range are used on paired objective stereoscopic prism binocular dissecting microscopes. The conventional low-power objective is 10x.

Most laboratory microscopes are equipped with three objectives, each capable of a different degree of magnification. These are referred to as the *oil immersion, high -dry* and *low power objectives*. The primary magnification provided by each objective is engraved on its barrel. The total magnification of the system is determined by multiplying the magnifying power of the objective by that of the eyepiece.

The total magnification of a compound microscope is the product of the magnifications of its objective lens (the lens nearest the specimen) and the ocular lens (the one nearest the eye). With an objective magnifying 40x and an ocular magnifying 10x, the total magnification is 400x. The compound microscope is therefore able to achieve considerably greater powers of magnification than is a microscope constructed of only a single lens. The latter, called a *simple microscope,* is used mainly in hand lenses and magnifying glasses.

With the compound microscope, magnification greater than 1,000 are attainable by choice of appropriate oculars, but increased magnification does not lead to increased resolution, although it may make an object easier to see. High magnification that increases the size of an object but does not increase its detail is called *empty magnification.*

TABLE 1.1. Optical Data for Commonly Used Objective Lenses

		Objective Magnification		
		10x	**40x**	**90x (Oil immersion)**
1.	Magnification with 10x eyepiece	100x	400 x	900x
2.	Numerical aperture	0.25	0.76	1.30
3.	Resolving power	2.0	0.45	0.27
4.	Approximate depth of field	7.0	1.3	0.5
5.	Approximate area of field with 10x Eyepieces, mm	1.5	0.35	0.17

(Brock, P. 18)

DEPTH OF FIELD (DEPTH OF FOCUS)

A minute body or a very thin section has thickness or depth. If a deep cell is being viewed with a 10x objective and the lens is focused on the upper wall of the cell, the bottom wall, may also be in focus. If a 45x objective is focused on the top wall, the bottom wall may be completely out of focus and practically invisible. If the lens is focused on the bottom wall, the top wall becomes obscured. The vertical extent of the zone of sharp focus is the depth of field. The term "depth of focus" is in common use. Depth decreases as the magnification increases. Under oil immersion, the depth of field is very shallow, usually less than 1 μm.

The Area of Field is represented by the diameter of the specimen that is within view. Area of field is greater at low power than at high power, and it is for this reason that low-power lenses are useful for scanning a slide.

To understand microscopy, one must be aware of certain fundamental properties of light and optics. With this foundation, many of the problems encountered in using a microscope will be more readily understood.

PROPERTIES OF LIGHT

Light travels with a velocity of approximately 186,300 miles, or 3×10^{10} centimeters, per second. According to the wave theory, light is propagated through space as waves of varying electric and magnetic fields.

Amplitude

Light wave are described in terms of their amplitude, frequency, and wavelength. An analogy can be made between a light wave and a jump rope. When two people pull on the rope and stretch it tight, a position of equilibrium is established. Now if the rope is swung, a wavelike effect is produced. The maximum displacement of the rope from the equilibrium position as represented by the crests and troughs of the curves produced, is the amplitude of the wave.

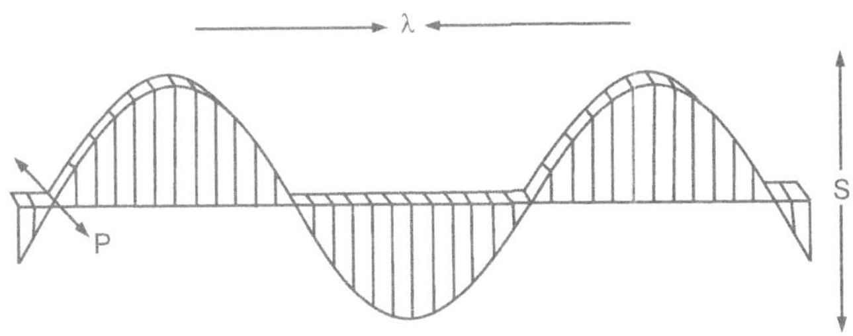

Fig. 1.12. The wave nature of light : Wavelength λ, vibration direction s, and polarizing direction p of a transverse wave.

Frequency

This property of a light wave refers to the number of vibrations that occur in one second. Specifically, *frequency* is the number of times a wave crest or trough passes a particular point per second. If two people take a rope and shake it, the frequency of the wave can be regulated by the speed with which the rope is shaken.

Wavelength and Frequency

A *wavelength* is the distance between two corresponding points on a wave, such as the distance between two successive peaks or crests. Because the frequency of light is the number of wave crests or troughs that move past of a point in one second and because the speed of light in a medium is constant, frequency is inversely related to wavelength :

$$\text{Frequency} = \frac{\text{velocity}}{\text{wavelength}}$$

The wavelength of light rays that make up the visible spectrum range from approximately 4000 to 7000 Å (400 to 700 nm). The colours we see result from a combination of factors, including amplitude and wavelength. Wavelengths either less than or greater than the limits of the visible spectrum also exist. Ultraviolet light rays, for example have wavelengths ranging from approximately 1000 to 3850 Å (100 to 385 nm). Infrared light has wavelengths greater than those of visible light.

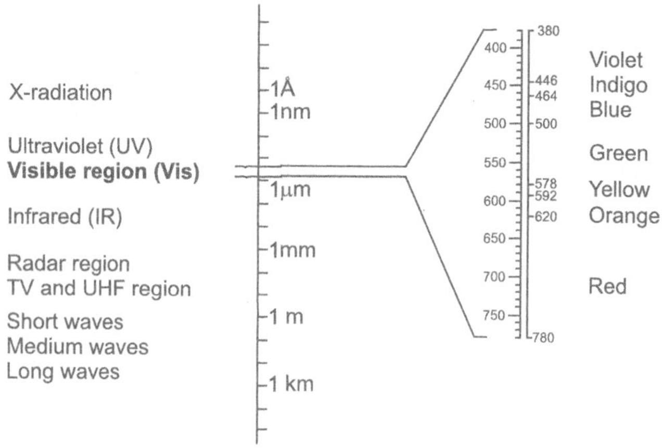

Fig. 1.13. The range of electromagnetic waves.

The resolving power (RP) of a microscope (*i.e.* its ability to reveal the fine details of a specimen) depends on wavelength. As a general rule, the shorter the wavelength of illuminating light, the greater the resolving power. Thus, with ultraviolet light as illumination, finer details can be seen than with visible light.

Refractive Index and Refraction

The nature of the medium through which light passes during the operation of the microscope affects the image seen. The rate at which light moves is not the same for all transparent media. Denser materials exert a slowing effect on light rays. This difference in velocity is expressed in the form of a *refractive index* or index of refraction. The refractive index, represented by the Greek letter η (eta), is defined by the formula

$$\eta \text{ (refractive index)} = \frac{\text{Speed of light in a vacuum}}{\text{Speed of light in the medium being tested}}$$

In making determinations of light velocity, it is important to keep the temperature constant during the testing period. The indexs of some commonly used materials are 1.00003 for air, 1.33 for water, an average of 1.6 for various glasses, and 1.55 for immersion oil.

Light rays travelling in a single medium generally move along a straight path. However, if these rays pass at an oblique angle from one material into another material having a different refractive index, the light wave changes direction. This phenomenon, called *refraction*, takes place at the boundary between the two media. The importance of refraction and refractive indexes of materials is clearly demonstrated in the use of oil immersion objective and other forms of microscopy.

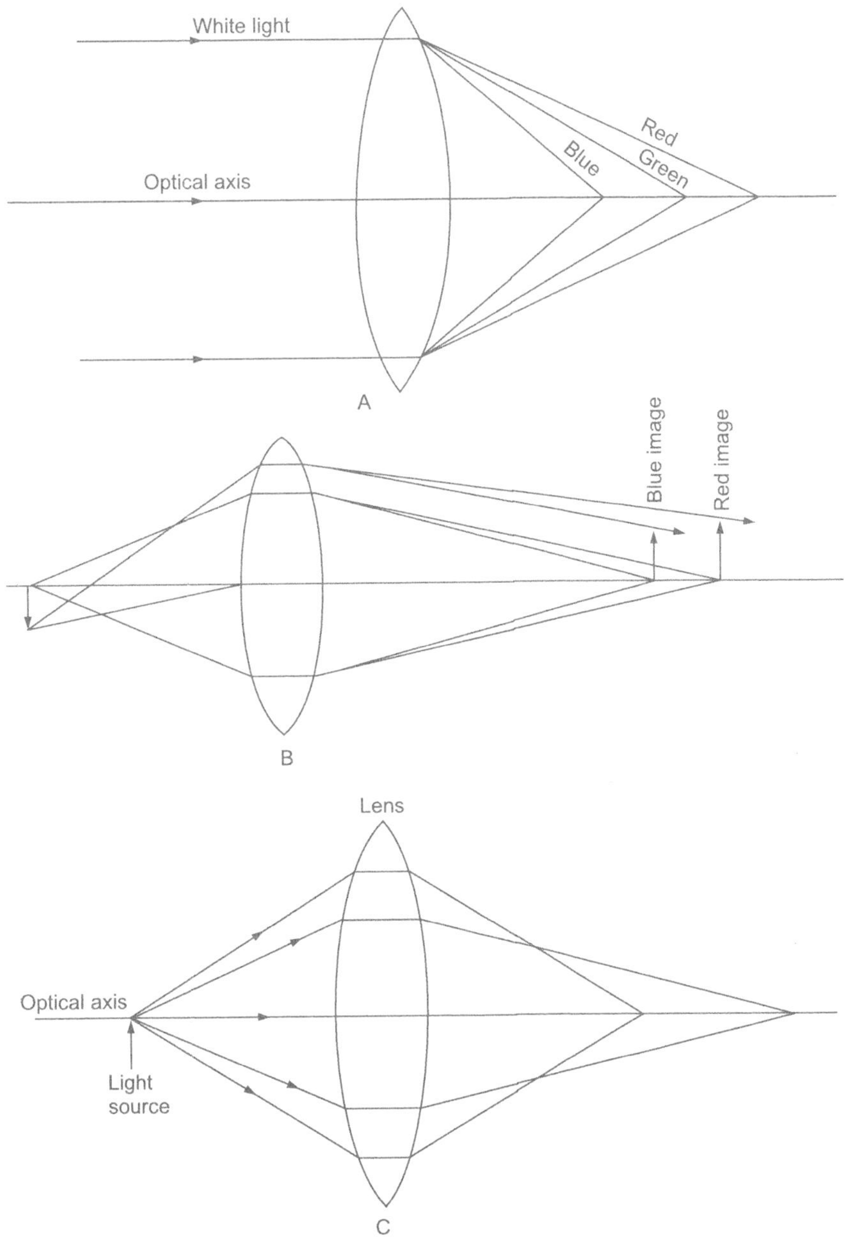

Fig. 1.14. (A) Longitudinal chromatic aberration in an uncorrected lens. Different wavelengths cause each of the three primary colours to be focussed at a different point along the optical axis.

(B) Latteral chromatic aberration. As focal length is varied, magnification changes, altering image size.

(C) Spherical aberration. Light rays passing through the outer portion of the lens are more strongly refracted than those passing through the central portion and are focussed at a different point along the optical axis. This problem can be minimized by using an aperture to restrict the light path to the central part of the objective.

Successful operation of a microscope depends upon understanding the principles involved in its operation, the function of the instruments, component, and the procedures for proper maintenance and use.

MICROSCOPES AND MICROSCOPY

Microscopes are of two categories, light (or optical) and *electron*, depending upon the principle on which magnification is based. *Light microscopy*, in which magnification is obtained by a system of optical lenses using light waves includes :

1. bright-field
2. dark-field
3. fluorescence, and
4. phase contrast microscopy

The electron microscope, as the name suggests, uses a beam of electrons of light waves to produce the image. Specimens can be examined by either *transmission* or *scanning* electron microscopy.

Each type of microscope has definite limitations. Each instrument can be used to observe cells or other objects that are at least a certain size and no smaller. For example, an ordinary light (bright-field) microscope can be used to observe objects with dimensions down to about 0.2 μm. An electron microscope can be used with specimens only one tenth that size - 15 nm (or 0.015 μm).

LIGHT MICROSCOPES : SIMPLE MICROSCOPE

A simple microscope is that which uses only one lens unit to magnify object. The most useful magnifications range from 6 to 12x. Magnifications up to 20x are available.

The magnification is accomplished according to a simple principle of physics. A ray of light is passing from one transparent medium, in which the velocity of light is high, into another of greater optical density, in which the velocity is less, is bent toward the normal to the lens surface, *i.e.* toward an imaginary line intersecting the curved surface at a right angle at the point at which the light enters. Likewise, in passing from a denser medium to a less dense medium, it is bent away from the normal. Glass is denser than air, and hence, light passing through air and striking the curved surface of a double convex lens and continuing through it into the air on the other side, is so bent on entering the lens and again on leaving it that the rays converge and thus form an image. This image may be either larger than the original object or smaller, dependent on its position with respect to the lens. With convex lenses, the greater the curvature the greater the magnifying power.

Bright-field Microscopy

Leeuwenhoek's early microscopes were simple in the sense that they used a single lens-rather like a magnifying glass. There are fundamental limitations to the magnifying power of a simple microscope, no matter how carefully it is constructed. Credit for the invention and early development of the compound microscope is still debated, but most experts acknowledge the basic contributions of Hans and Zaccharias Janssen, lens grinders of Middleburg, Hollang. Around 1590, the Janssens introduced a second lens to magnify the image formed by the primary lens. This is the principle upon which today's compound microscopes are based.

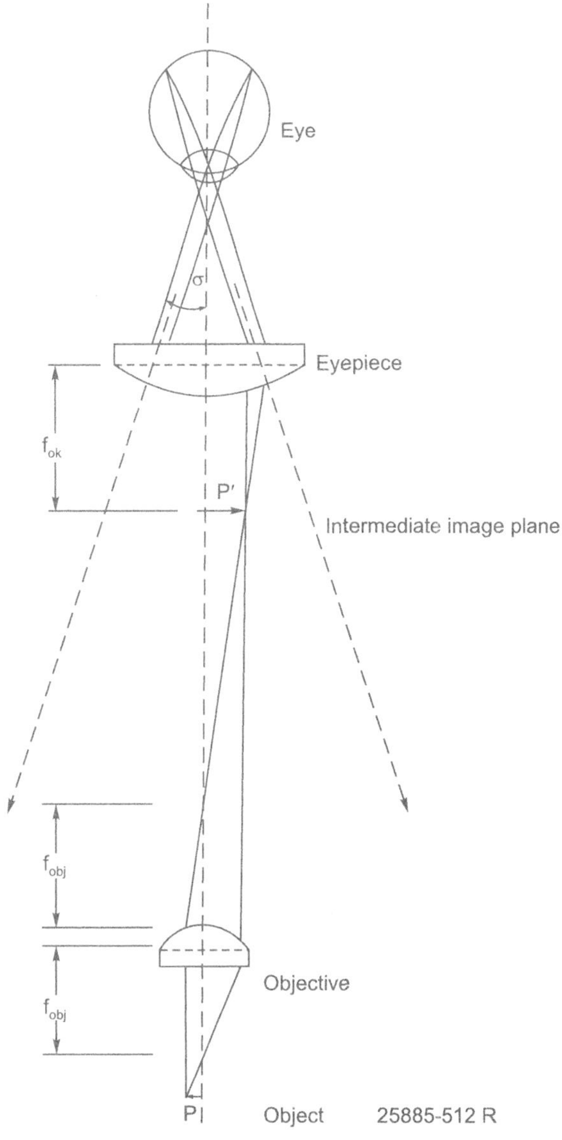

Fig. 1.15. Optical path in the microscope (diagrammatic) The objective forms a magnified, real, inverted and side-reversed image of object P at the reproduction scale 5;1. The subsequent 8x eyepiece magnifies this intermediate image another 8x. The observer thus sees the image as if he viewed the 5 × 8 = 40x magnified object from a distance of 250 mm without and instrument.

Although their microscope had both ocular and objective lenses, it was a rather crude affair that looked like a short telescope and was focused by sliding together the metal tubes of barrel. Apparently this instrument was no well received, since there are no records of any discoveries having been made with it (Walter & McBee, 1969).

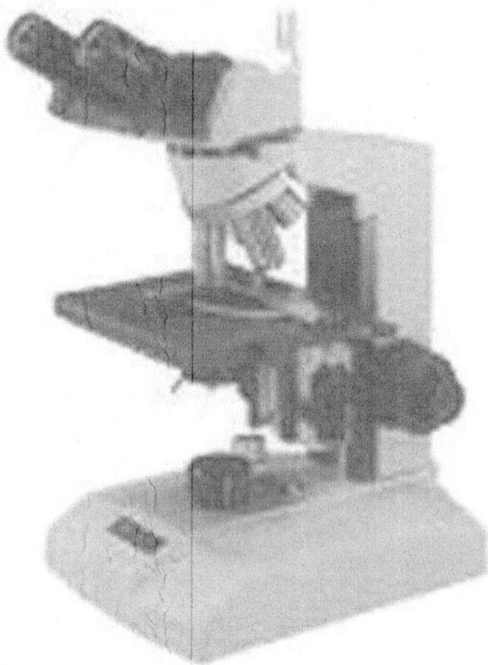

Fig. 1.16. A bright-field microscope. (Atlas, p.43)

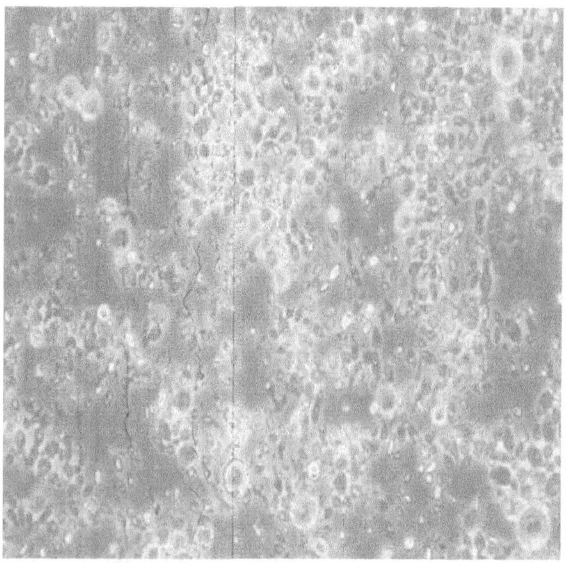

Fig. 1.17. This photomicrograph of bacteria (small rods) and yeasts (larger oval-shaped cells) was produced by using brightfield microscopy. By using bright-field microscopy, the characteristic sizes, shapes and arrangements of the microbial cells can be observe. (Atlas, p. 43).

In bright-field microscopy, the microscopic field (the area observed) is brightly lighted and the specimens appear dark because they absorb some of the light. Ordinarily, microorganisms (such as bacteria) do not absorb much light, but staining them with a dye greatly increases their light absorbing ability, resulting in greater contrast and colour differentiation. Generally microscopes of this type produce a useful magnification of about 1000x to 2000x.

A Compound Microscope

Consists of at least two lens systems, the objective, which magnifies the specimen and is close to it, and the *ocular*, or *eyepiece*, which magnifies the image produced by the objective lens. The total magnification thus obtained is equal to the product of the magnifying powers of the two sets of lenses. Under optimal conditions, maximum magnifications can range from approximately 1000 to 2000 (x) the diameter of a specimen being observed.

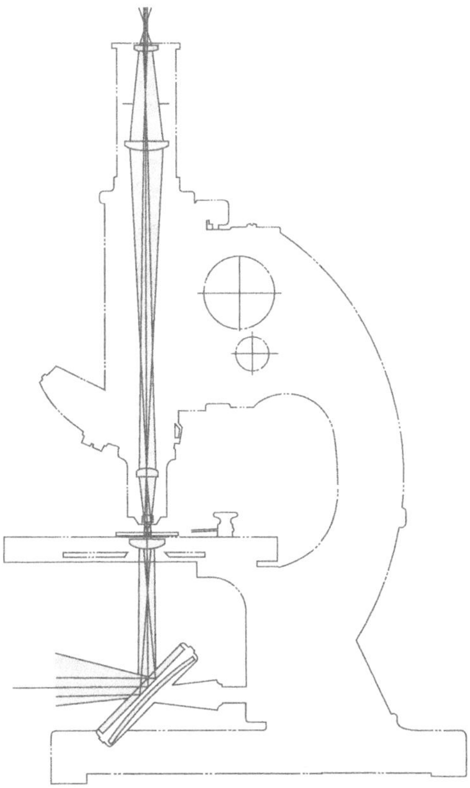

Fig. 1.18. Path of light in a compound microscope.

The compound microscope consists of a series of optical lenses (systems), mechanical adjustment parts, and supportive structures for these components. The optical lenses include the ocular, the objectives (usually three with different magnifying powers), and the substage condenser. The coarse, fine, and condenser adjustment knobs, together with the iris diaphragm level, comprise the major mechanical parts. The various components are held in position by supportive structures, such as the base, arm, pillar, body tube (barrel), and revolving nose-piece. Several of these microscope components are discussed ahead.

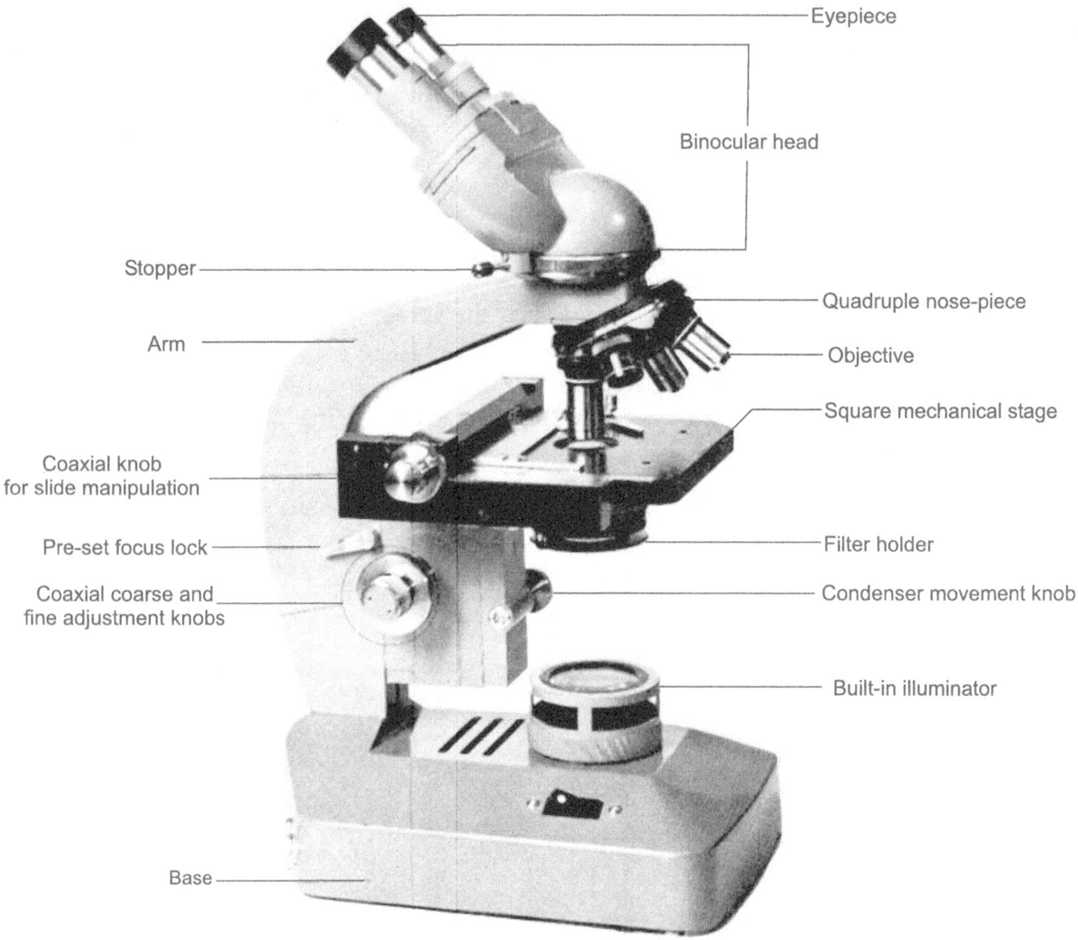

Fig. 1.19. Name of parts.

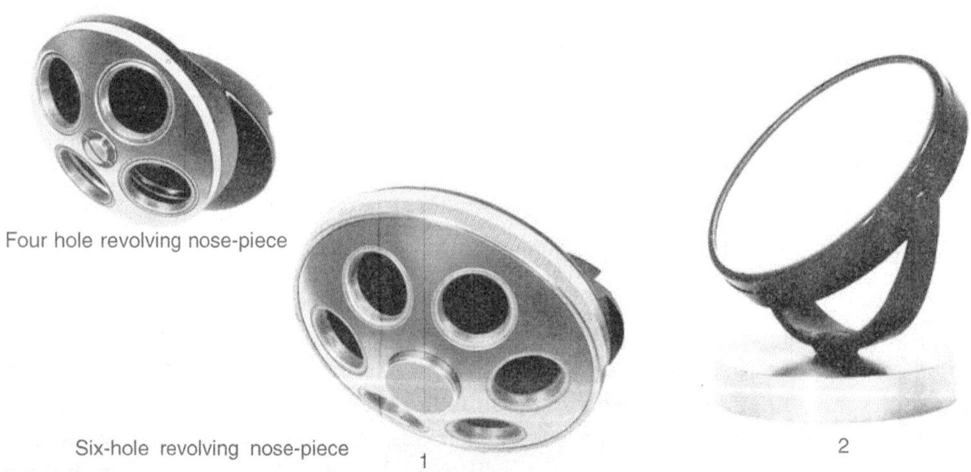

Fig. 1.20. 1. The nose-piece, 2. Plano-concave mirror.

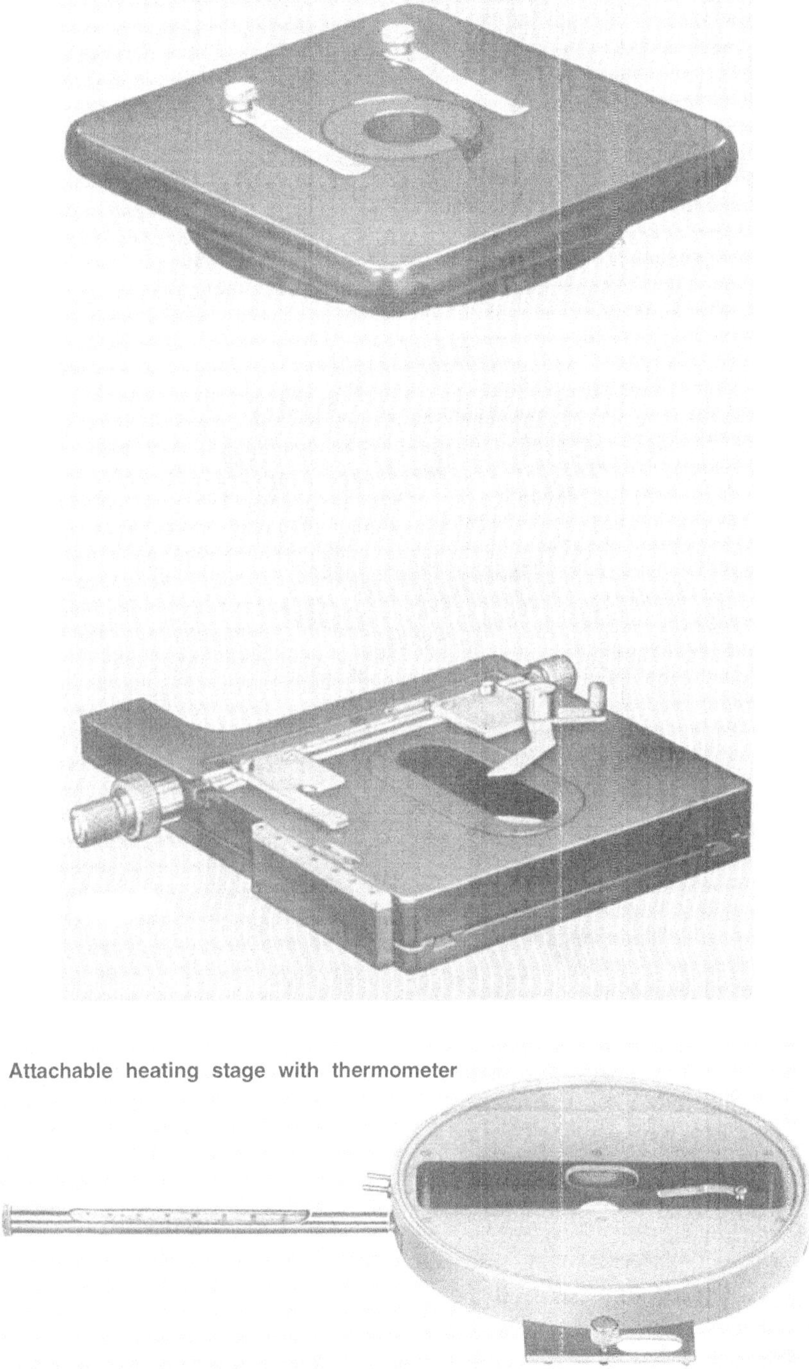

Fig. 1.21. Object stages.

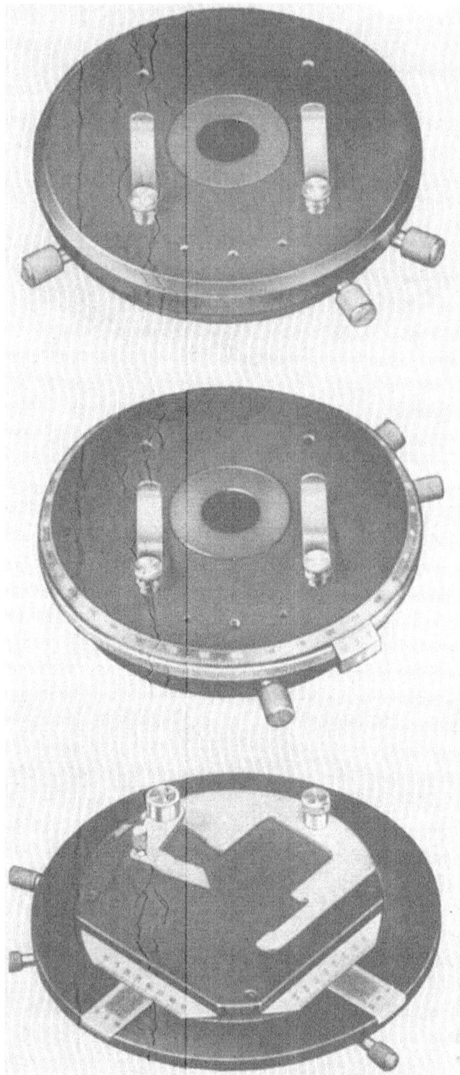

Fig. 1.22. Objective stages.

Ocular (Eyepiece)

A short tube generally containing two lenses, the ocular fits into the upper portion of the microscope body tube. Several different types of eyepieces can be used to examine specimens. The specific type used generally depends upon the objective lenses on the instrument. The magnifying power of the ocular is usually engraved on it. Common magnifications for oculars include 1x, 2x, 5x, 10x and 15x. Eyepieces are used to magnify the image of the specimen produced by the objective and to correct certain distortions produced by the objective. Commonly used oculars include compensating, Huygenian, and hyperplane eyepieces. The *compensating* type is designed to neutralize distortions introduced by some objectives because of different wavelengths of light. The *Huygenian* is used with low power objectives. *Hyperplane* eyepieces are generally used with objectives of intermediate power.

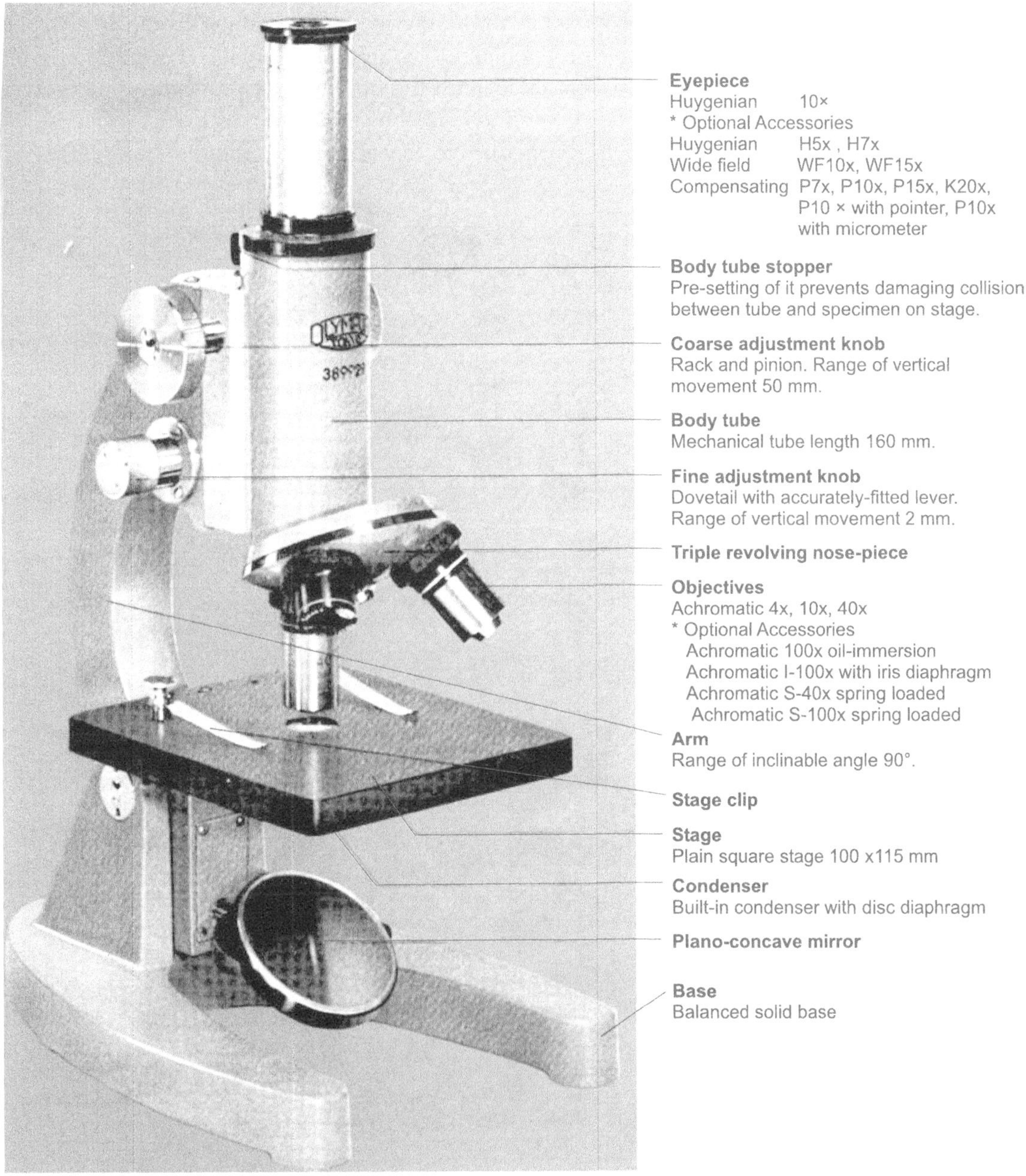

Fig. 1.23. Monocular microscope.

Objectives

These are considered the most important of the microscope's optical parts, primarily because they affect the quality of the image seen by the observer. Most microscopes are equipped with three objectives with different magnifying powers: The low-power, high-power (or high-dry), and oil-immersion lenses. The individual objectives commonly can be distinguished by their respective lengths (low-power is the shortest, oil-immersion the longest).

Objective lenses are used together, or concentrate, the light rays coming from the specimen being viewed form the image of the specimen, and magnify this image. Several important properties of a microscope are directly associated with the objectives. One of these is resolving power or resolution, which is specifically defined as the ability to distinguish clearly two points that are close together within the structure of the specimen.

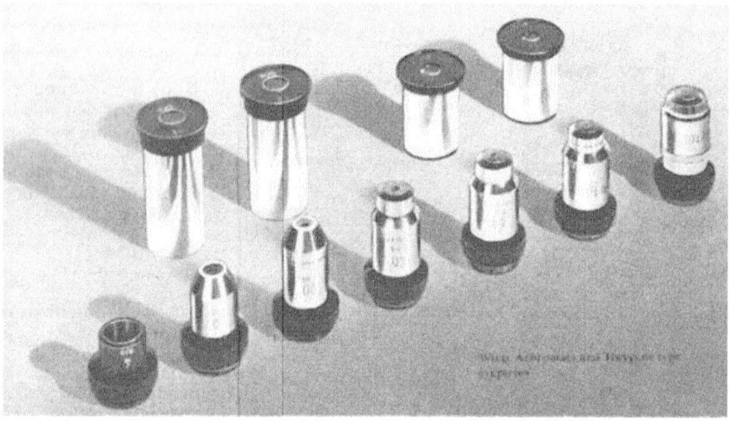

Fig. 1.24. Objectives and Eyepieces.

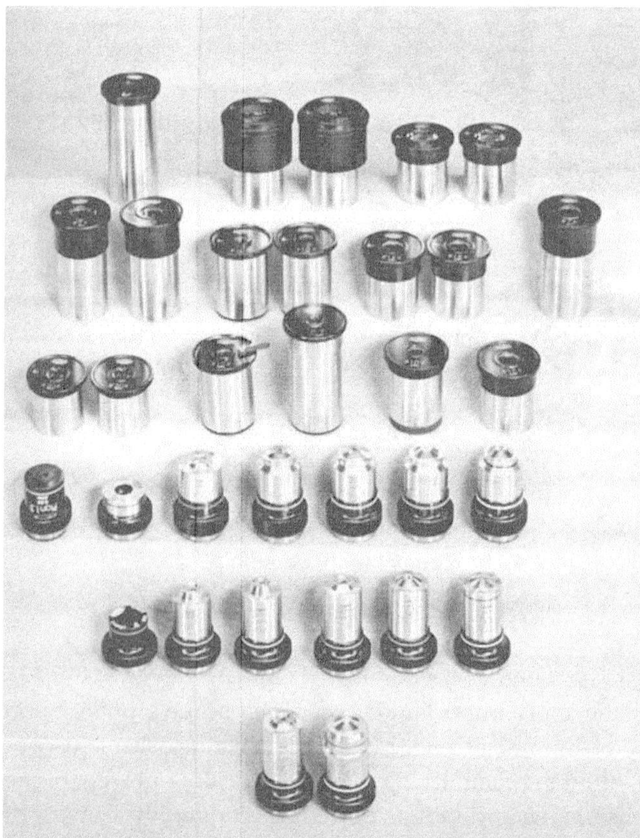

Fig. 1.25. Objectives and Eyepieces.

Another important feature is the property of *parfocal* (discussed earlier). Stated simply this means the changing of objectives without major focusing adjustments.

The Condenser and Iris Diaphragm

The illumination system of a microscope is also of considerable importance, especially when high magnifications are used. A condenser is found under the microscope stage between the source of illumination and the specimen or object to be viewed. This component frequently is called a *substage condenser*. One of the most commonly used is the Abbe condenser. It consists of two lenses that illuminate specimens with transmitted light. The condenser is important to high-resolution microscopy. By raising or lowering the condenser the plane of focus of light can be altered and a position can be chosen that leads to precise focus. Microscopic examinations of specimens with either high-power or oil-immersion objectives require adequate illumination. The condenser serves this purpose. In addition to the Abbe, variable-focus and achromatic condensers are also commonly used.

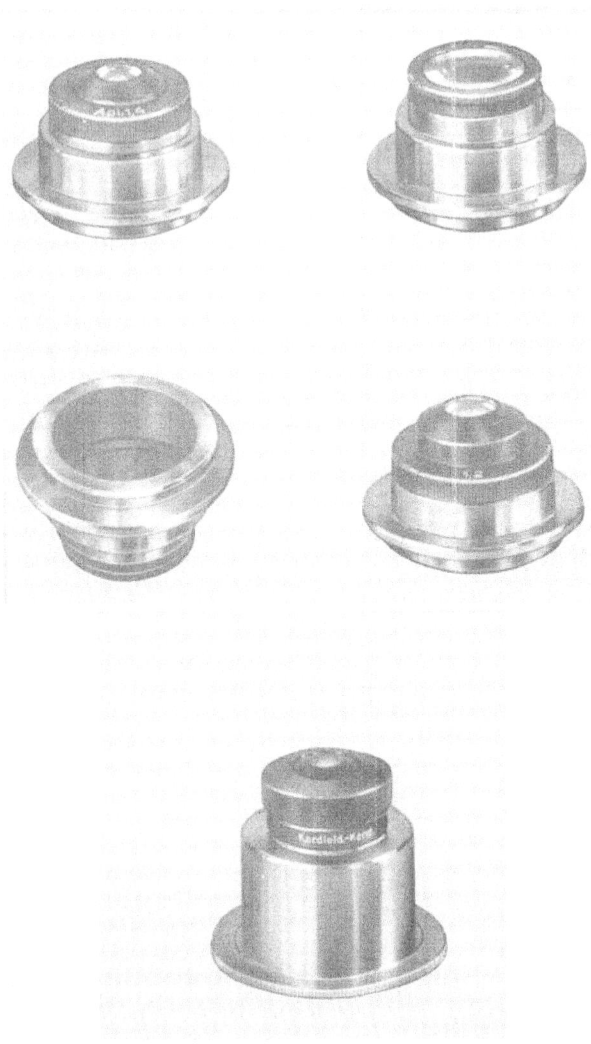

Fig. 1.26. Condensers.

Special phase condenser for long working distance

Dark field immersion condenser

Fig. 1.27. Condensers.

Occasionally, too much light may pass through the specimen and into the objective lens, significantly decreasing the contrast of the specimen. Microscope condensers are generally equipped with an *iris diaphragm* to control light intensity. This component functions to control the amount of light entering the condenser. When unstained material is to be examined the opening of the iris diaphragm generally is reduced.

Mechanical Operation

A microscope usually has a set of two to four objectives permanently installed on a revolving nose-piece. The body tube of the microscope on which the objectives and ocular are mounted is moved up and down by two mechanisms, a course adjustment which produces rapid displacement, and a fine adjustment which moves the body tube very slowly.

The objective lens is placed close to the object to be viewed and the ocular lens (eyepiece), placed close to the eye. The primary enlargement of the object is produced by objective lens, and the image so produced is transmitted to the ocular, where the final enlargement occurs.

The Real Image

Produced by the objectives is brought to a focus within the eyepiece. This is a system of 2 or 3 lenses near the top of the microscope. This real image is magnified by the topmost lens (ocular) which appears as a greatly enlarged *virtual image.*

The normal procedure is using the microscope is to locate the object with a low-power objective and then turn to the next higher power. With an objective of 10x or less in position, look into the ocular and manipulate the mirror until the field of view is uniformly illuminated. Move the body tube with the coarse adjustment until the image is visible, then bring the image into sharp focus with the fine adjustment. Search the section by moving the slide, using the fine adjustment freely to bring into sharp focus structural features at different depths in the specimen.

When it is necessary to turn to a higher magnification, centre the desired structure in the field of view and bring it into short focus with the lowest power. Without changing the focus, turn the objective of next higher magnification into position. The image should now be visible, and it should require not more than a quarter turn of the fine adjustment to bring the image into sharp focus (the high-power objective should never be used for locating the object).

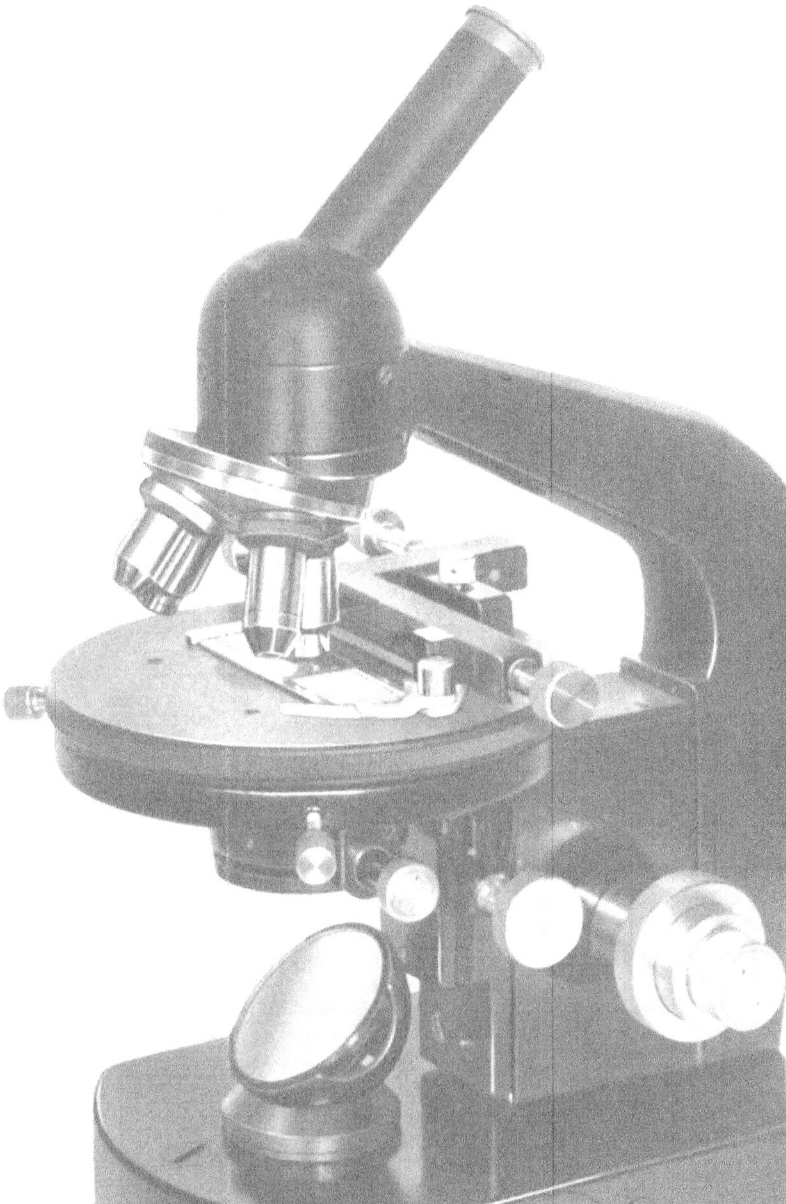

Fig. 1.28. Inclined monocular microscope.

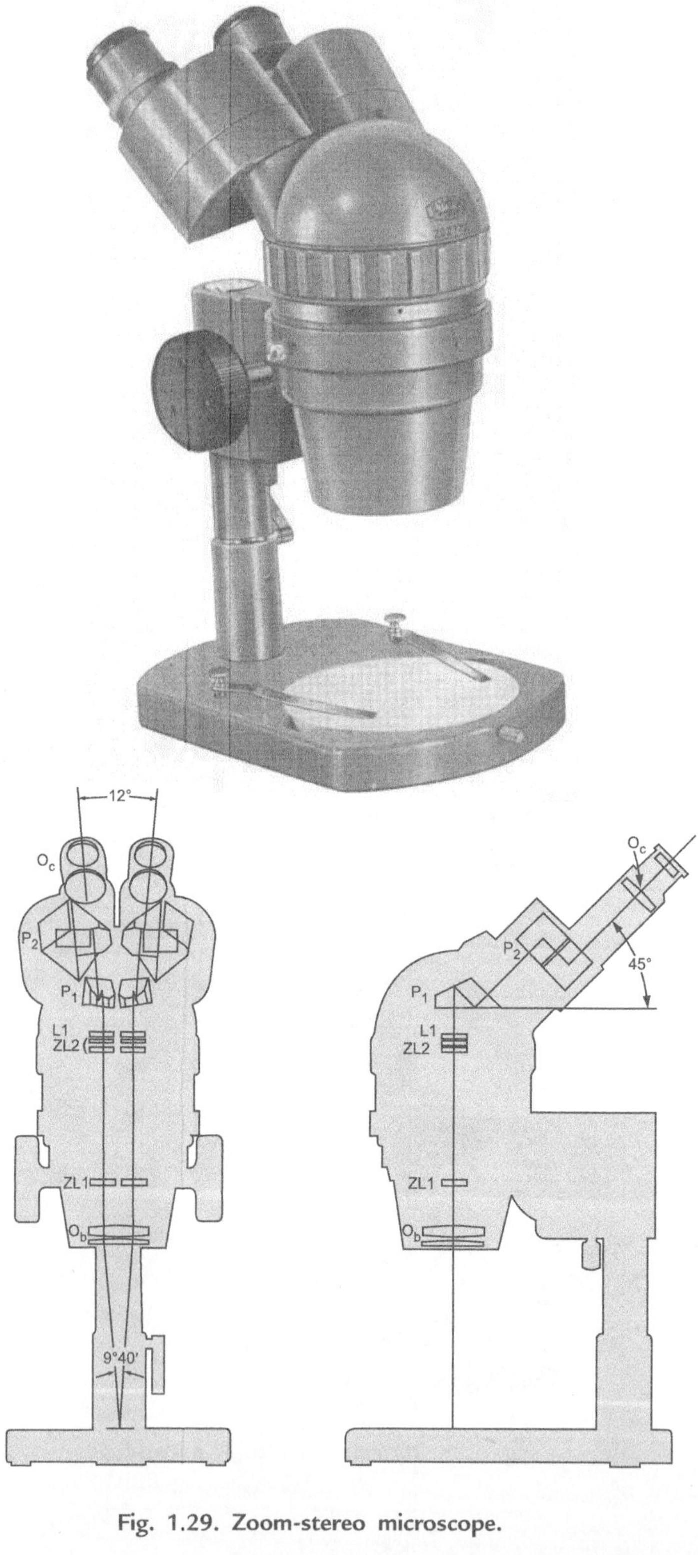

Fig. 1.29. Zoom-stereo microscope.

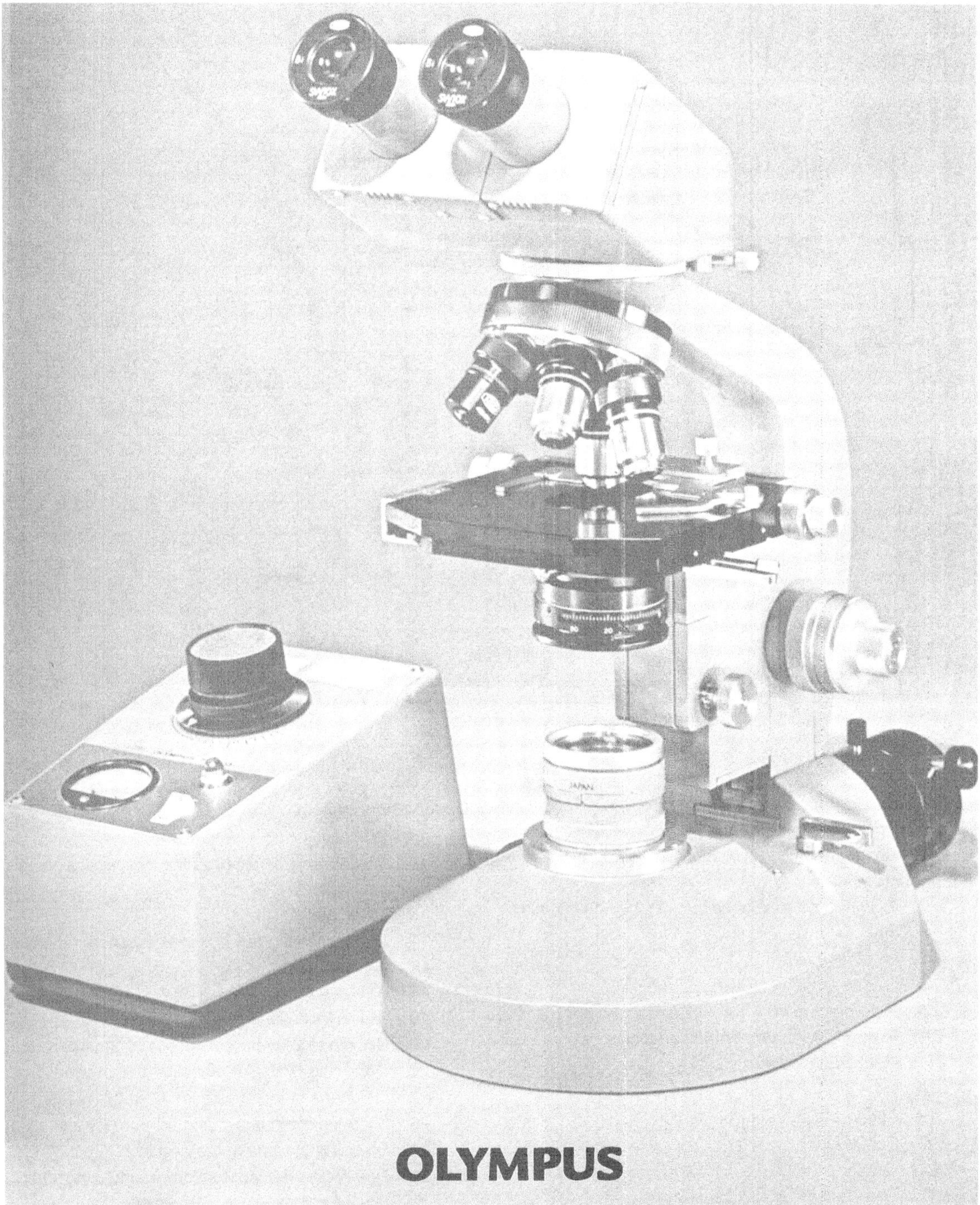

Fig. 1.30. Binocular research microscope.

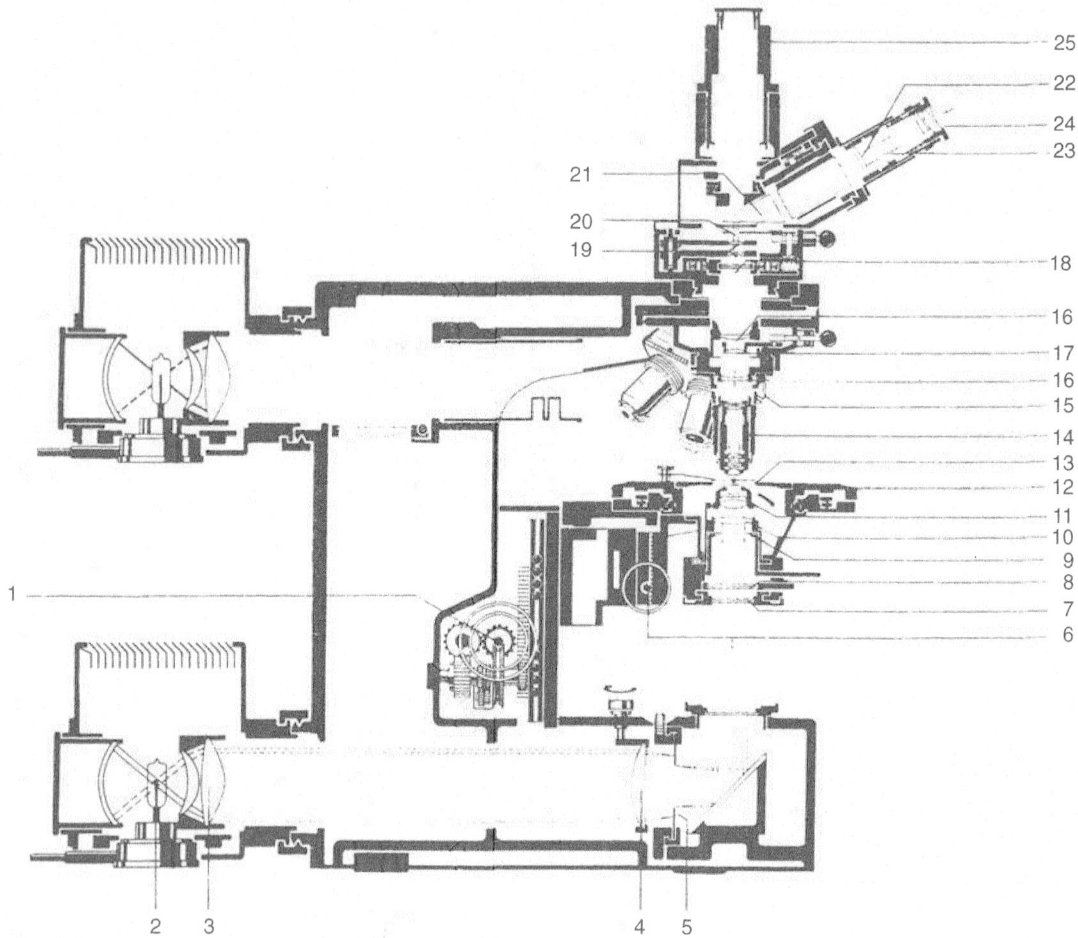

Sectional drawing of a research microscope (ORTHOPLAN POL) sectional drawing incident light

1. Coarse and fine adjustment, actuating the stage.

Illuminating system

2. Filament of the halogen lamp in the Lamp Housing 100
3. Lamp condenser
4. Hinged lens
5. Field diaphragm
6. Vertical adjustment of the condenser
7. Polarizer
8. λ/4-plate for circular polarization
9. Aperture diaphragm of the condenser
10. Condenser
11. Condenser top
12. Object stage on ball bearings
13. Specimen

Observation system

14. Objective
15. Objective centering revolving nose-piece on ball bearings
16. Tube lenses
17. Supplementary lens for the conoscopy of small particles
18. Analyzer
19. BERTRAND lens
20. Pinhole stop for the conoscopy of small particles
21. Deflecting prism for binocular observation
22. Field lens of the eyepiece
23. Intermediate-image plane with crosslines
24. Focusing eyelens of the eyepiece
25. Photo tube for photomicrography, projection, attachment eyepieces or microphotometry

Fig. 1.31

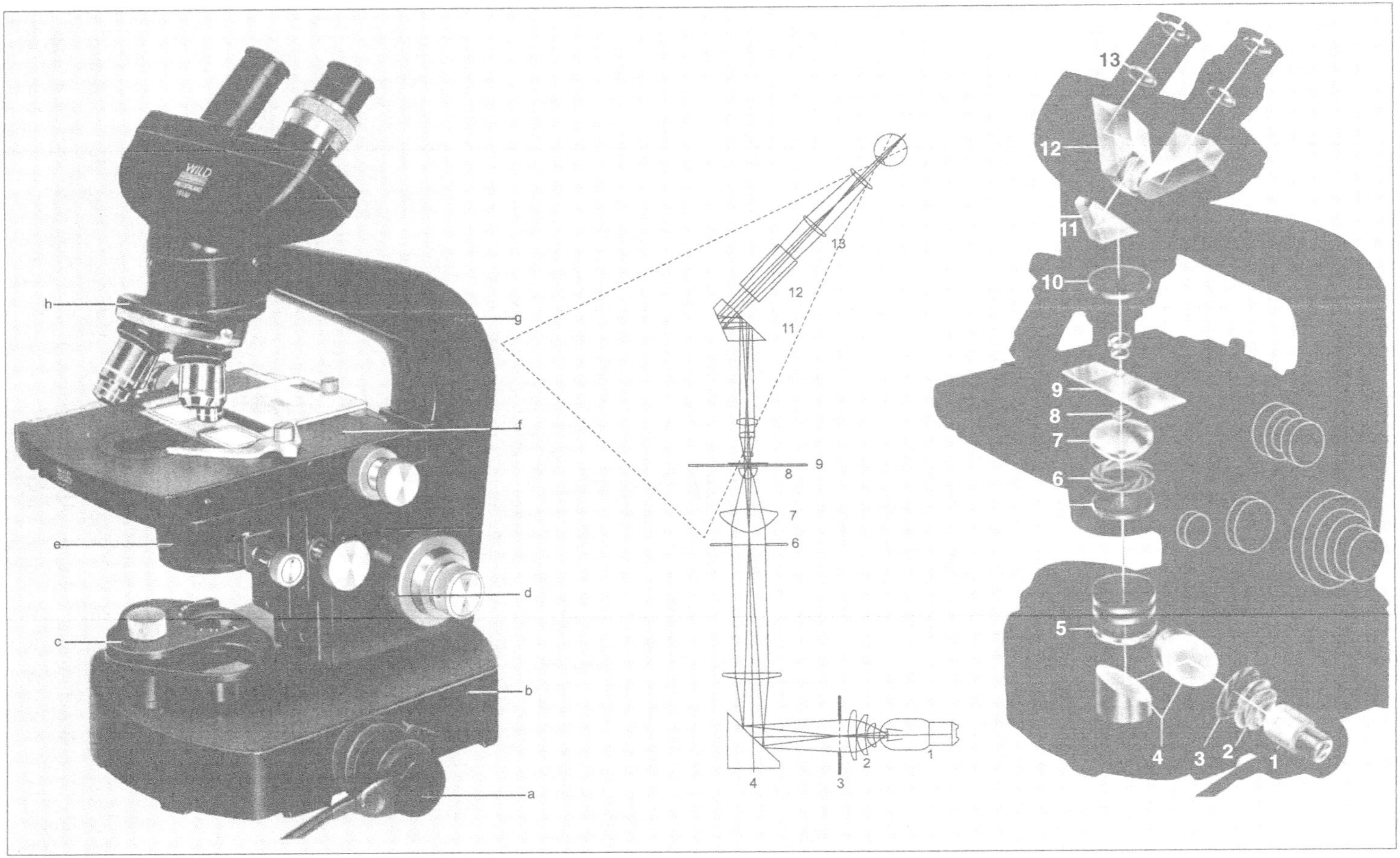

a) built-in illumination
b) base plate
c) filter exchanger
d) coarse - and fine adjustment (co-axial)
e) aplanatic swing-out condenser
f) mechanical stage
g) support
h) object nose-piece

1. Incandescent lamp
2. Lamp collector
3. Light field diaphragm
4. Deviation prism
5. Lens of centering inset
6. Aperture diaphragm (= condenser iris diaphragm)
7. A spherical condenser lens
8. Swing-out front lens
9. Object carrier and cover glass
10. Analyzer
11. Deviation prism
12. Divider for light rays
13. Collective lens

Fig. 1.32

The most common type of compound microscope consists of one objective and one ocular in working position at one time. This is known as *monocular monobjective* microscope while a *binocular monobjective* microscope uses a matched pair of ocular with a single objective. In the former type of microscope, the user employes one eye at a time, in the latter, use of both eyes diminishes eyestrain and fatigue. A special type of binocular microscope which utilizes matched pairs of objectives is the dissecting or *stereoscopic binocular.* This microscope facilitates dissection, isolation, and manipulation of the object.

Dark-field Microscopy

The dark-field microscopy was an early development permitting the visualization of transparent cells. The effect produced by the dark-field technique is that of a dark background against which objects are brilliantly illuminated. This is an effect opposite to the one obtained with bright-field microscopy, in which specimens usually appear darker than the light background.

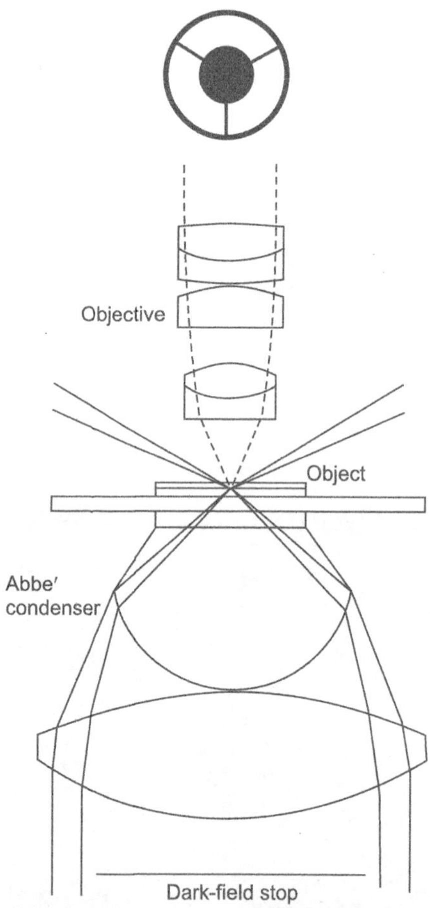

Fig. 1.33. Dark-field Microscopy. Because of the application of the dark-field stop, the specimen is illuminated only by oblique rays of light. An Abbe condenser is also shown, as well as a typical dark-field stop. (Wistreich & Lechtman p. 63).

The dark-field procedure is commonly performed by fitting an Abbe condenser with an opaque disk or "dark-field stop". This dark-field stop is placed below the condenser, thus

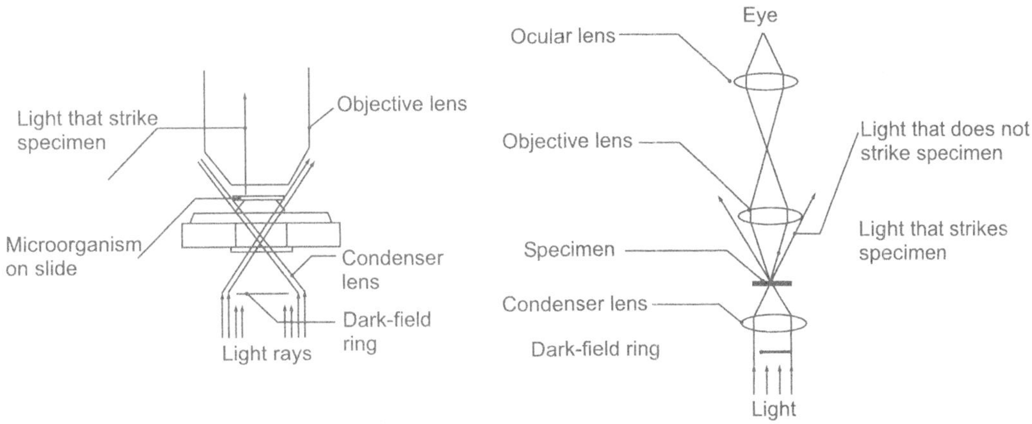

Fig. 1.34. Diagram of a darkfield microscope showing the path of light passing through the background and light striking the specimen. (Atlas, p.48)

eliminating all light from the central portion of the condenser. A thin cone of illumination reaches the specimen at an angle to the objective lens. Light is scattered by the specimen, which thus acts as a light source. It appears as a glowing object against a dark background.

The microscope slides and cover slips used in dark-field examinations must be absolutely free from dirt, dust, and scratches. Unwanted objects and marks can reflect light and could easily brighten the background.

Dark-field microscopy is used for examination of unstained microorganisms, the contents of hanging-drop preparations and colloidal solutions. In cells undergoing division, the centromeres, spindles, and chromosomes may also be observed.

Fluorescence Microscopy

An early twentieth-century advance, *fluorescence microscopy* is used for examining various cell types and cellular structures, locating chemical components in cells and tissues, and for the rapid diagnosis of otherwise difficult disease states. This types of microscopy provides a means of studying structural details and other properties of a wide variety of specimens. Instruments used in fluorescence microscopy do not differ optically or mechanically from conventional microscopes but they do require special filter systems. The different types of specimens differ in "fluorescence power" from their surroundings. This property of *fluorescence* is obvious when a substance becomes luminous upon exposure to ultraviolet light. With certain substances, including some dyes, fat, oil droplets and uranium ores, exposure to this form of radiations causes them to absorb the energy of the invisible ultraviolet light waves and remit the energy as visible light waves. Many living tissues fluoresce naturally, absorbing ultraviolet radiation and emitting green, yellow or red light. Selecting combinations of natural fluorescence with fluorescent dyes makes possible identification, location and counting of cell types and differentiation of cell parts. Fluorescence of cells can be measured and used to compute the concentration of chemical components present.

Fluorescent dyes commonly used are acridine orange R, auramine O, primulin, and thiazo-yellow G.

Polarizing Microscopy

The polarizing microscope is a light microscope with polarizing optics in the ocular and condenser, each transmitting only plane-polarized light. The condenser is then spoken of as the *polarizer* and the ocular as the *analyzer* (light can be polarized, that is, separated into two rays that vibrate at right angles to one another by passing it through a double refracting crystal such as calcite. *Nicol Prism* is one of the best known polarizers and transmits plane–polarized light).

Polarizing optics have been especially useful in study of biological membranes and of the dividing cells. They are also useful in studying fibrillar structures.

Interference Microscopy

The Nomarski interference microscopy was developed in 1955 and uses polarized light for illumination. Structures in a dividing cell such as spindle fibres and other structures in living cells could be seen more clearly by using interference microscope.

Phase-Contrast Microscopy

Phase-contrast microscopy is useful in examining the internal structures of transparent, living cells or in demonstrating their presence in certain fluid environments. It uses a conventional light microscope fitted with a phase-contrast objective and a phase-contrast condenser. The annular diaphragm used in the condenser of the phase-microscope permits light to pass through the condenser as a hollow cone, the remaining light being absorbed. This cone is focussed on the object.

The special optical system makes it possible to distinguish unstained structures within a cell which differ only slightly in their refractive indices or thickness.

The phase-contrast microscope has made it possible to observe structures previously very difficult or impossible to see. For example, in a living state, cells are difficult to see because of their transparent quality. Use of this microscope distinguishes cells and their structures from the background so that their shapes, sizes, and other features could be clearly marked out. Similarly, the behaviour of chromosomes of living cells during mitosis or meiosis can be followed with ease, and many of the other organelles may be studied.

Limitations of the Light Microscope

Ultraviolet light is not visible but when it is absorbed by certain fluorescent substances, visible light is given off. Only a small part of the radiation which comes to earth from the sun is seen by human eye. Light may be considered a wave motion, and different types of light differ in the size of the waves, or the wavelength (*i.e.*, the distance from the top of one wave to the top of the next). The pigments in the retina of the eye which are sensitive to light undergo chemical change when acted upon by certain wavelengths of light. These changes in turn induce nerve impulses in the optic nerves, which eventually, in the brain, give rise to the conscious sensation of vision. The wavelengths to which the eye is sensitive make up the *visible* portion of the sun's spectrum. Since different wavelengths within this range produce conscious sensations, they are identified as different colours; red, orange, yellow, green, blue and violet; the red having the longest wavelength, the violet the shortest. Wavelengths just

Standard type phase condenser for bright-field and phase contrast Phase condenser.

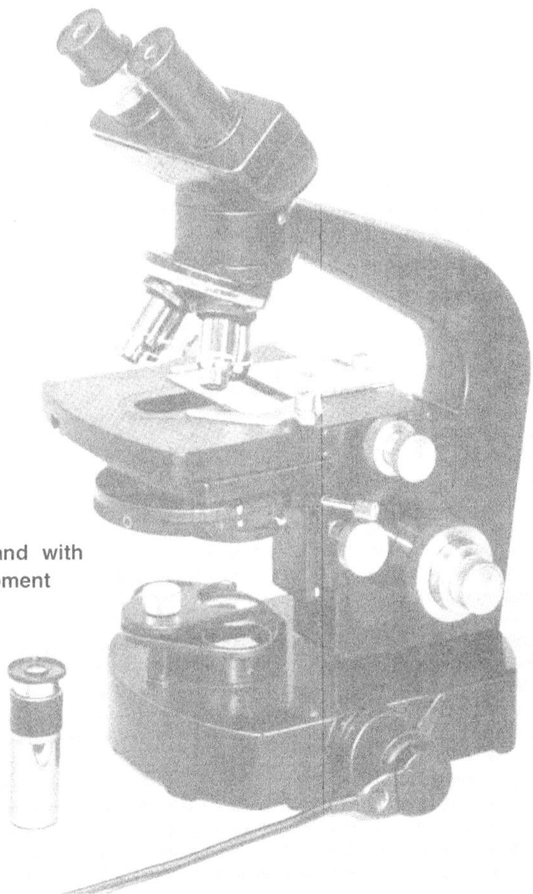

Wild M 20 KGST stand with phase contrast equipment

Fig. 1.35

longer than the red are called infrared. They produce heat, but are not visible. Those just shorter than the violet are called ultraviolet. These too are invisible, but affect man and microorganisms in a number of ways.

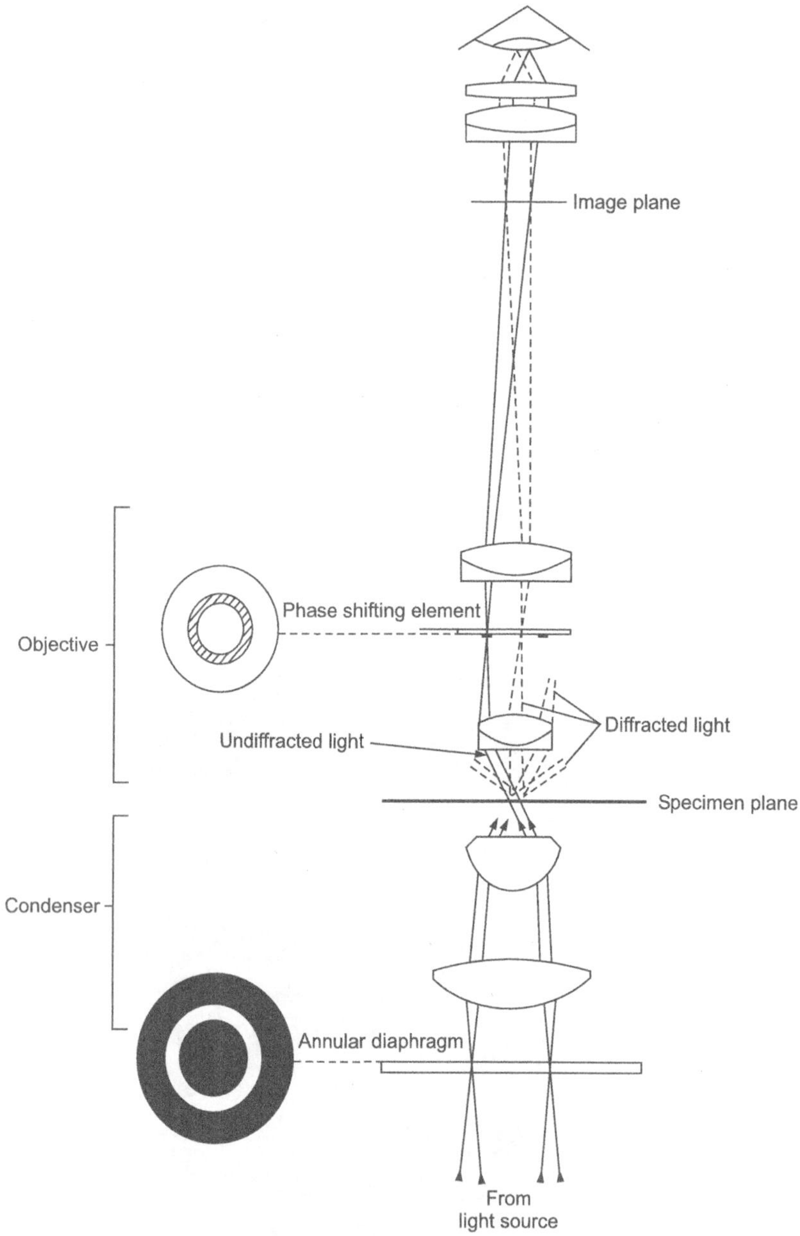

Fig. 1.36. The operation of the phase-contrast principle. (Wistreich & Lechtman, p. 66)

All this brings us to the fact that if one is to see an object it must produce a clear image with visible light. Objects of a certain size (such as bacterial cells), do this. Smaller structures, however, merely scatter the light, and thus no clear image is formed. However, such small particles may form a clear pattern with shorter wavelengths-ultraviolet, for example. The eye would not be sensitive to this, but a photographic plate coated with a chemical sensitive to ultraviolet would be. In other words, it is possible to take pictures of objects too small to be seen with visible light. The pictures may then be enlarged.

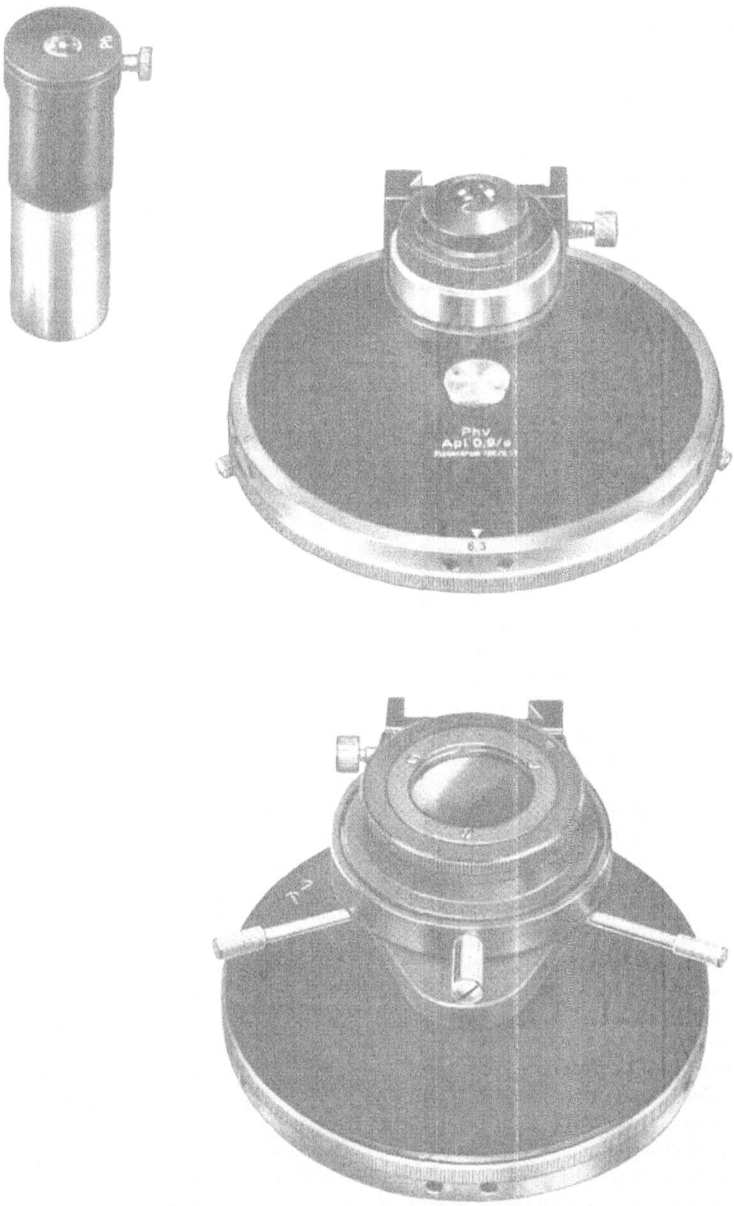

Fig. 1.37. Variable phase-contrast condensers.

The Electron Microscope

In an electron microscope, a stream of electrons passing through the vacuum chamber is scattered by objects placed in its path, as light passing up through the slide on the microscope is absorbed by the objects there. In the case of the electrons, however, the image is formed by this scattering, while the light image is formed by absorption. In the light microscope, the images are magnified by lenses which bend the light outward; in the electron microscope, electromagnets attract the electrons outward to form a large image. The image in the light microscope is finally brought to focus on the retina of the eye. In the electron microscope, it is focused on a screen coated with a material which fluoresces (gives off visible light) where electrons come in contact with it. In this way a picture is produced, which may be seen by looking in at the fluorescent screen through a window in the apparatus, or a photograph may be made and enlarged still further to make a permanent record.

The electron microscope was developed on the eve of World War II, and has proved a powerful tool for analysis of large chemical molecules, viruses, and cellular structures. Ernst Ruska of Germany was awarded the Nobel Prize in 1986 for his work going back to 1920s on constructing the first electron microscope. Gerd Binning and Heinrich Rohrer of Switzerland also shared the 1986 Nobel Prize for their work on constructing the "scanning tunneling microscope" at the IBM laboratory in Zurich.

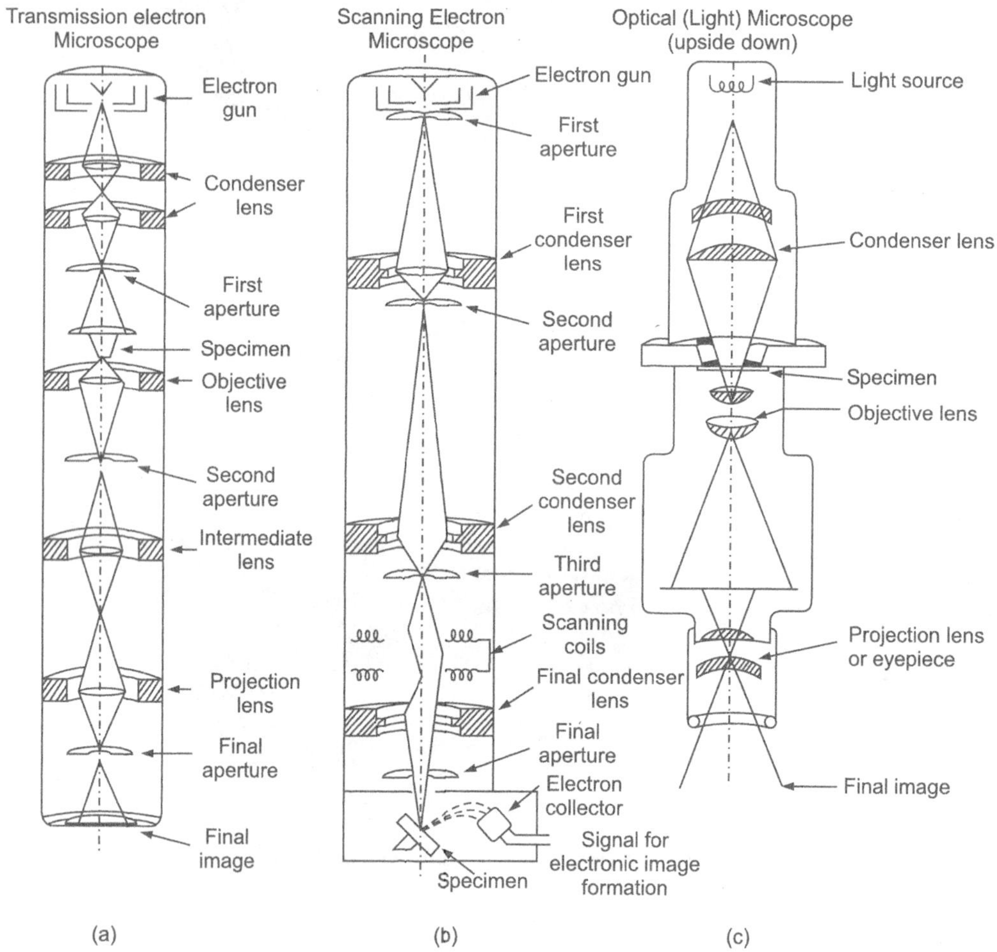

Fig. 1.38. A comparison of–

(a) the transmission electron microscope

(b) the scanning electron microscope, and

(c) the optical, or bright-field microscope. (Wistreich & Lechtman, p. 67)

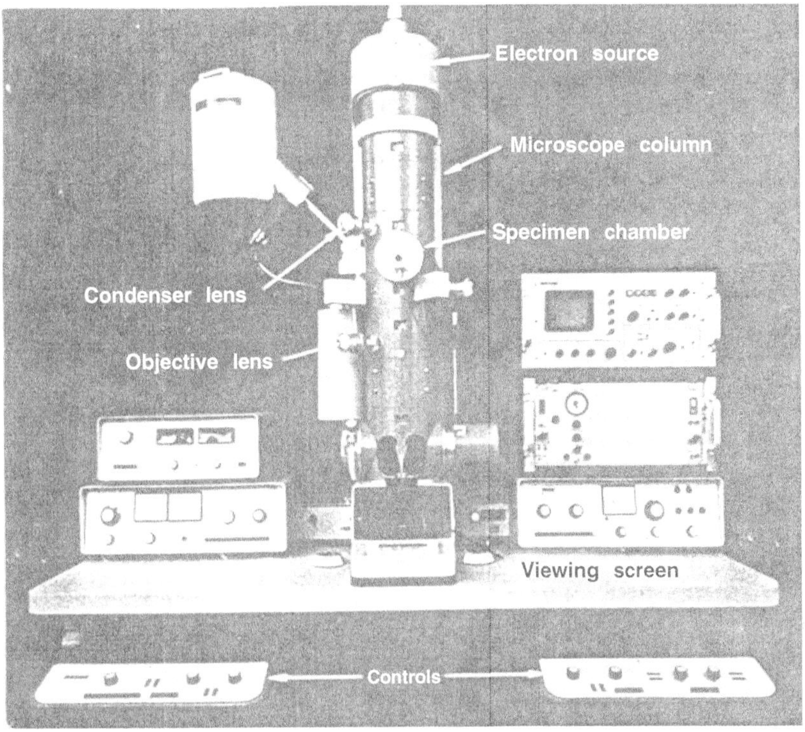

Fig. 1.39. Carl Zeiss Electron Microscope
(Wistreich & Lechtman, p. 68)

In the electron microscope, the beams of electrons are focused by magnets that serve as lenses [electron = the negatively-charged particle which forms a constituent of all atoms, its positively-charged counterpart of equal mass and charge being called the positron. However, the term electron is often used to include both negative electrons (negatrons) and positive electrons (positrons)].

The electron microscope has a resolving power 400 (theoretically 2000) times that of the light microscope, and instrumental magnifications of up to 500,000x (resolving power, as mentioned earlier, is that property of a lens which makes possible the recognition, as distinctly separated bodies, of objects that are exceedingly close together, or subtended by a small visual angle. For example, a lens of poor resolving power will show a slender chromosome as a single thread, whereas a lens of good resolving power will show the chromosome as two interwound threads).

By this device, the fine structure of bacterial cells, viruses, and other minute particles may be seen.

The image results from differential scattering of electrons from the molecular constituents of the cell. Scattering is greater than denser the material, regardless of its chemical composition. Even a gas scatters electrons; therefore, biological objects must be studied in a vacuum, since electron scattering from the gas molecules would obscure the differential scattering of the structures in the biological object.

There are two basic types of electron microscopes: The transmission electron microscope (TEM), and the scanning electron microscope (SEM).

The Transmission Electron Microscope

The realization that the light microscope had reached its limit of resolution (0.2 μm), with further improvements unlikely, prompted the development of the electron microscope.

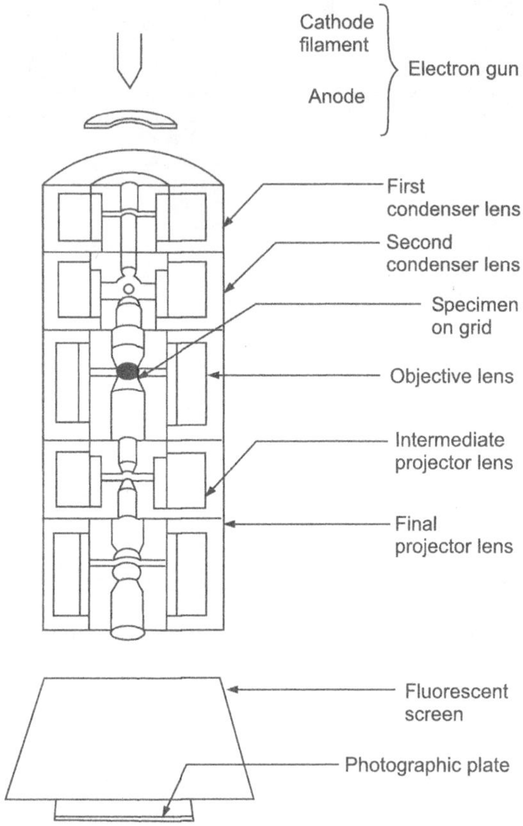

(A)

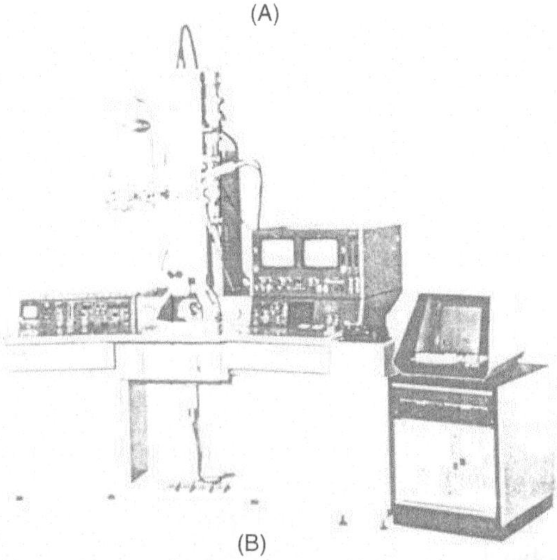

(B)

Fig. 1.40. (A) Diagram of a TEM, (B) The transmission electron microscope (TEM) allows the visualization of the fine detail of the microbial cell.

Electrons which are very small negatively charged particles behave in many respects like light waves. Accelerating electrons through 50,000 volts, for example, produces a wavelength of approximately 0.05 Å, much shorter than even ultraviolet light. In actual practice, the resolving power achieved in the electron microscope is about 4Å, or 0.0004 μm.

Most **transmission electron microscopes** are similar in basic design to the light microscope. However, certain distinct differences exist. The basic components of the transmission electron microscope include a source of illumination, optical viewing, and vacuum systems, and electronic systems designed to hold the electron voltage (wavelength) stable.

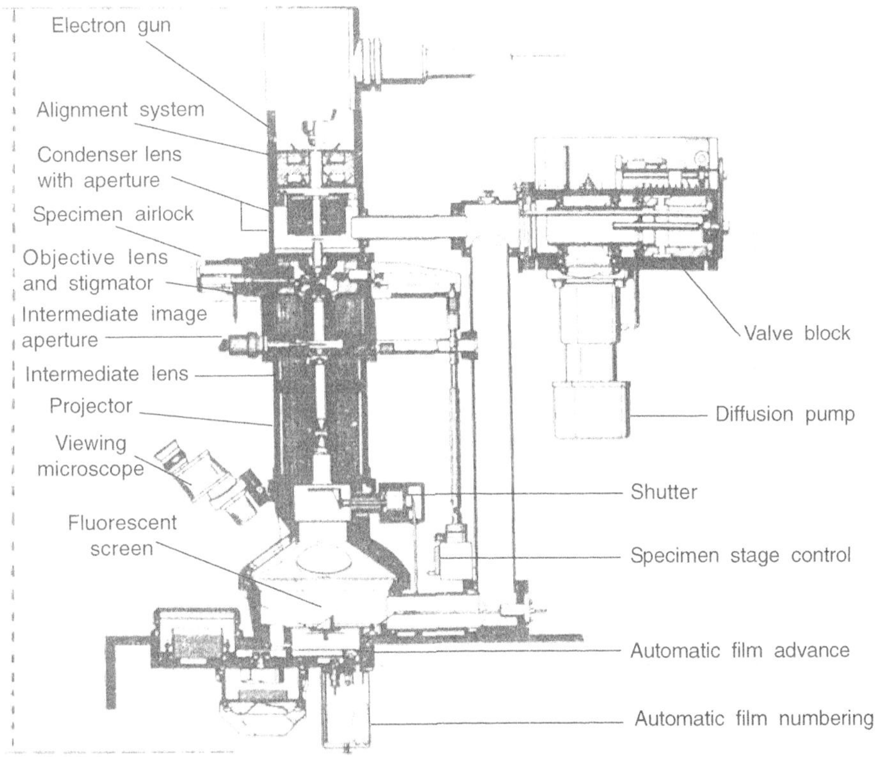

Fig. 1.41. Electron microscope.
(Carl Zeiss Inc., Oberkochen, Wurttemberg)

The source of illumination for the transmission electron microscope, the electron gun, generates an electron beam from a thin, V-shaped tungsten filament. This metal wire must be heated enough to release a sufficient quantity of electrons from the tip of the filament by thermionic emission. These electrons are concentrated and accelerated by other components of the electron gun, producing a fast-moving narrow beam of electrons. In order for the transmission electron microscope to function properly, the components of the electron gun and the lenses must be aligned with one another.

The lens system for the electron beam is magnetic rather than glass. The lens of an electron microscope consists of a lens coil formed by several thousand turns of wire encased in a soft iron casing. A magnetic field is created by passing a current through the coil. The electron beam is concentrated within the soft iron wiring and other lens components also made of soft iron, called polepieces, located in the lens coil.

Three general types of lenses are found in an electron microscope: Condenser, objective and projector lenses. The projector serves to project the final image of the specimen onto a viewing screen, in place of the ocular, or eyepiece of a conventional light microscope. The screen is coated with a phosphorus compound that fluoresces upon being irradiated by electrons. Permanent records of an image are made photographically. Specimens are generally viewed directly through clear, lead-treated thick glass with the aid of a binocular microscope.

The transmission electron microscope will function only if a vacuum is maintained in the microscope column. Otherwise molecules in the path of the electrons can deflect them and interfere with the image.

For electron microscopy, the specimen to the examined is prepared as an extremely thin dry film on small screens and is introduced into the instrument at a point between the magnetic condenser and the magnetic objective; this point is comparable to the stage of the light microscope. The magnified image may be viewed on a fluorescent screen through an air-tight "window" or recorded on a photographic plate by a camera built into the instrument.

High-Voltage Electron Microscopy

The capability of transmission electron microscopes to show minute details and to produce functional images of preparations depends on the specimen's ability to scatter electrons passed through it. This passage of electrons is influenced by two factors, the thickness of the specimen and the energy of the electrons. The latter depends on the accelerating voltage used to generate the electrons. A new instrument, the **high-voltage electron microscope**, is similar to the general transmission microscope, except that its electron voltage is considerably greater. This feature of the microscope makes it possible to penetrate thicker specimens and to observe the three-dimensional relationships of cell components.

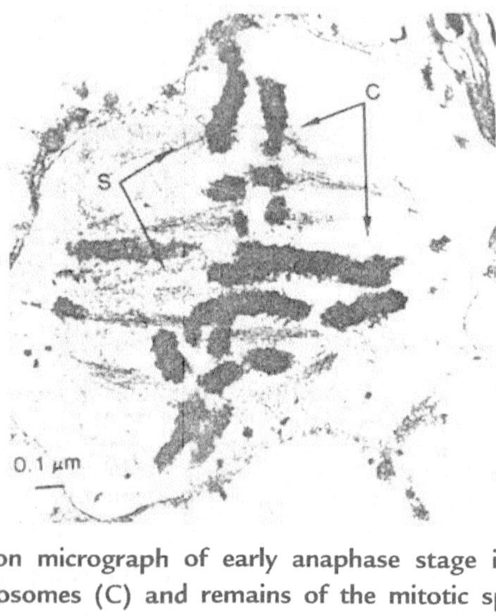

Fig. 1.42. A high-voltage electron micrograph of early anaphase stage in the green alga *Oedogonium cardiacum*. The chromosomes (C) and remains of the mitotic spindle (S) are well preserved. (Wistreich & Lechtman, p. 72)

The microscopes with the accelerating voltage higher than 150 KV were called as ultra-high voltage electron microscopes. Since these electron microscopes revealed the inner structure of thick specimens, they had rather high evaluation. Nowadays, 200 microscopes and 300 KV to 400 KV electron microscopes are going to be the popular future models.

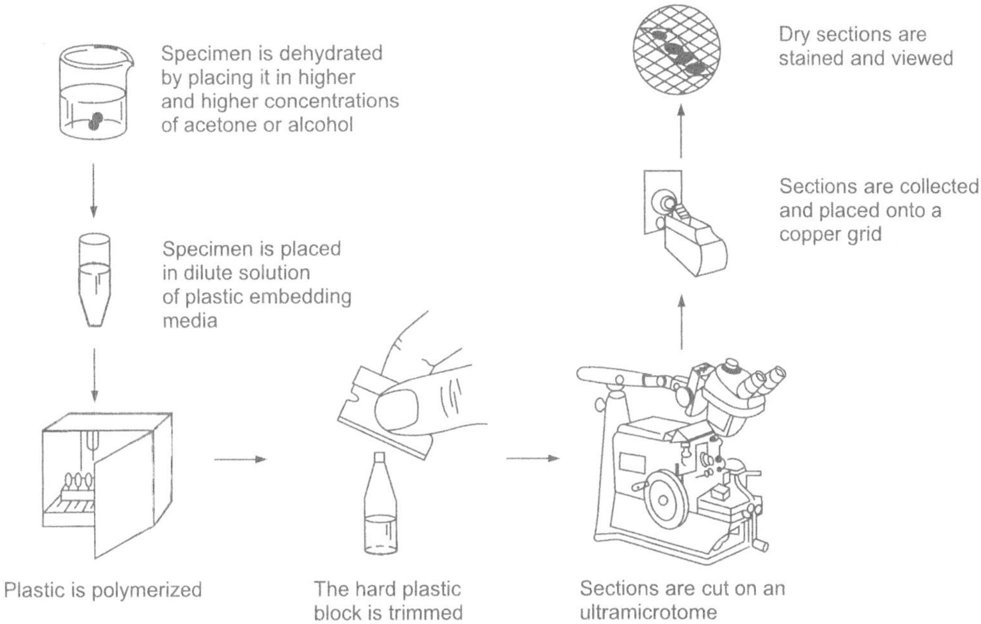

Fig. 1.43. Preparation of a specimen for viewing by transmission electron microscopy. The elaborate preparation requires far more time than the actual visualization of the specimen. (Atlas, p. 52)

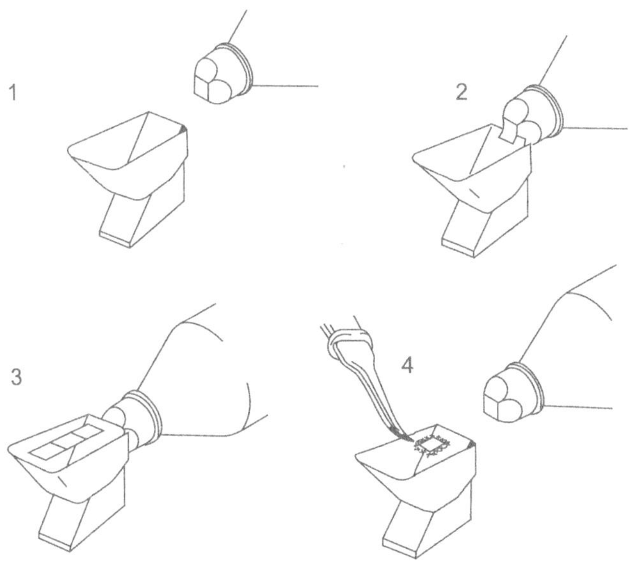

Fig. 1.44. A representation of how ultrathin sectioning is performed :

1. The specimen approaches the surface of the glass knife,
2. The beginning of a section,
3. The appearance of several sections,
4. The removal of a specimen for viewing on a coated grid.

(Wistreich & Lechtman, p. 71)

Preparation of Cells for Microscopy

Fixation : The phase contrast microscopy is a useful method for enhancing the contrast of living cells so that greater detail may be observed, and has proved to be extremely useful for some purposes. However, the examination of living cells is extremely limited in many ways. Living cells and cell organelles move and are quite thick, and this tends to obscure cellular detail. Since resolution is dependent on specimen thickness, sections of living cells and tissues are generally employed in both light and electron microscopy. Furthermore, it is often desirable to stain or process cells in various ways, although this may cause morphological damage since living cells are fragile structures. Thus, for the majority of light and electron microscope studies of cell structures, the first step is a process called **fixation**. The purpose of fixation is to kill and stabilize cells so that the structure is preserved in the subsequent treatments. Normally a chemical, known as fixative, is employed. Ideally the fixative should penetrate and kill the cell instantly, preserving the structure of the cell as it was in the living state.

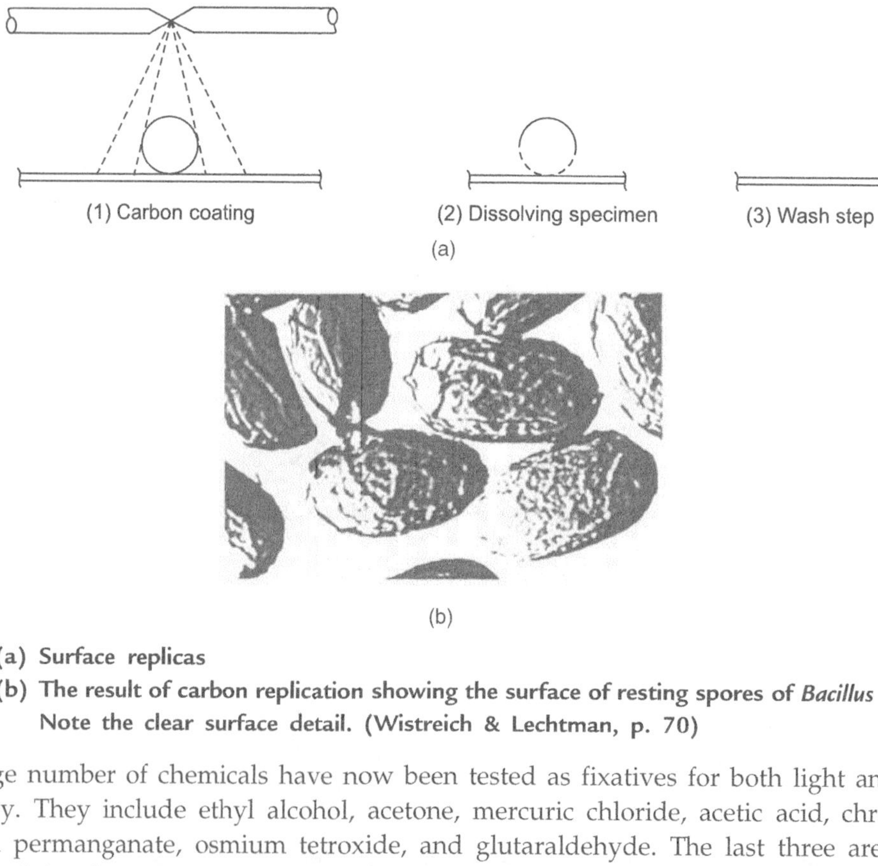

Fig. 1.45. (a) Surface replicas
(b) The result of carbon replication showing the surface of resting spores of *Bacillus megaterium*. Note the clear surface detail. (Wistreich & Lechtman, p. 70)

A large number of chemicals have now been tested as fixatives for both light and electron microscopy. They include ethyl alcohol, acetone, mercuric chloride, acetic acid, chromic acid, potassium permanganate, osmium tetroxide, and glutaraldehyde. The last three are the most widely used in electron microscope studies. There is considerable variation in the chemical nature of these fixatives and the chemistry of fixation is far from being fully understood. There is also no universally satisfactory fixative and it is now usual to use specific fixatives for the examination of specific structures. For example, potassium permanganate is a good fixative for the examination of membranes although it is destructive to nucleic acids and so does not preserve ribosomes.

Specimen Preparation for Electron Microscopy

Following are some techniques of Specimen Preparation for Electron Microscopy.

(1) Embedding : As an electron beam has a low penetrating power and so the column of microscope is kept under high vacuum. This means that water must be removed from the specimens or the vacuum will be destroyed. Therefore, specimens are dehydrated after fixation, usually by immersion in a series of ethanol-water mixtures of increasing alcohol concentration upto 100% alcohol.

(2) Sectioning : Associated with the low penetration property of the electron beam is the question of specimen thickness. The formation of the image in an electron microscope depends upon the electrons passing through the specimen. If the specimen is too thick all of the electrons will be scattered to some extent, resulting in an image with poor contrast and considerable loss of detail. Therefore it is necessary to cut thin slices or sections of the material to be examined, and these must be of the order of 50 nm, thick. To cut these ultra-thin sections the tissue must first be embedded in a hard support medium since biological material has little mechanical strength. The embedding media normally used are plastic resins (such as Epikote or Araldite). Since these are not easily miscible with ethanol, the tissue is first transferred to acetone. The unpolymerized resin is mixed with acetone and its concentration slowly increased, perhaps over a period of days, so that it fully infiltrates the specimen. The resin is then polymerized to a hard plastic by heating in an oven at about 60°C for 24 hours or so. The specimen, fully embedded and supported in plastic, can be cut into ultrathin sections by glass or diamond knives using a special apparatus called an ultramicrotome (which can make section 0.2 to 0.02 μm in thickness).

(3) Staining : When sections of biological materials are examined under the electron microscope little detail of structure is usually seen since there is little basic contrast, although potassium permanganate especially, and osmium tetroxide, do stain to so some extent during fixation. Therefore, it is necessary to stain sections to enhance contrast. But unlike the staining

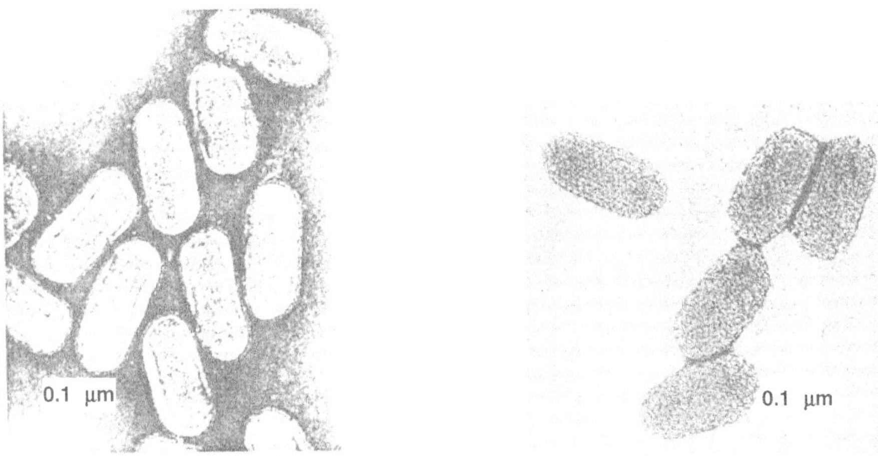

(a) (b)

Fig. 1.46. Two different negative staining procedures involving sowthistle yellow vein virus :

(a) The appearance of the virus when stained with phosphotungstic acid,

(b) A similar viral preparation stained with uranyl acetate. The surface proteins of the virus are quite evident.

(Wistreich & Lechtman, p. 70)

sections for light microscopy, the stains used here are not dyes but solutions of heavy metals which are differentially absorbed by different cell structures and thus increase the scattering of electrons from these parts. Two commonly used stains are salts of lead and uranium. The former enhance the contrast of membranes and nucleic acids, the latter of proteins and nucleic acids. This is known as **Positive Staining** and may be contrasted with the negative staining technique which is frequently used in the examination of fragments of isolated organelles and membranes and of small particles such as viruses and ribosomes. Negative staining produces an electron-dense background against which the less dense specimens are observed. Negative stains consist of the salts of heavy metals, and include potassium phosphotungstate, uranyl acetate, and ammonium molybdate. The electron-dense stain appears to dry more rapidly on the support film than on the specimen and forms a sharp boundary, the specimen appearing white on a dark background.

(4) Freeze etching : This technique has been developed for preparing material for electron microscopy as an alternative method to the standard procedure for chemical fixation, dehydration, embedding and sectioning.

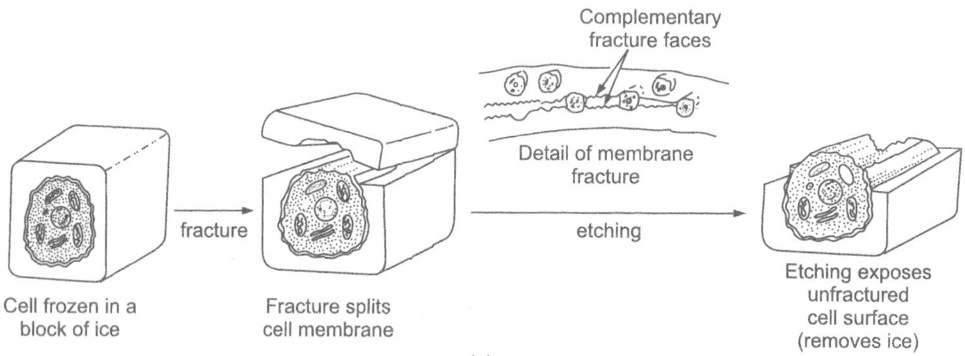

(a)

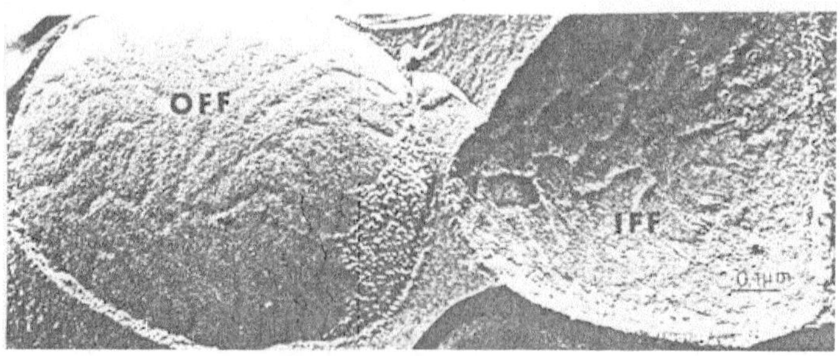

(b)

Fig. 1.47. The freeze-fracture and freeze-etching technique :

(a) The frozen cell is fractured with a sharp blow by a blade or knife. The fracture line very often runs through the interior of membranes, separating the two layers thereby revealing membrane proteins that would otherwise remain buried. The last step allows for the etching of the membrane through the sublimation of surface ice. This exposes the unfractured membrane.

(b) An example of the freeze-fracture technique. With membranes, two complementary fracture faces are produced, OFF, the outer fracture face, and IFF, the inner (concave) fracture face. Both faces of this cell membrane of *Streptococcus faecalis* show particles.

(Wistreich & Lechtman, p. 72)

Freeze-etching is based on fixation by rapid freezing. This is accomplished by immersing pieces of tissue in liquid nitrogen (about –160°C) or in certain solvents cooled by liquid nitrogen. This method results in homogenous fixation with no loss of soluble substance and little chemical change. Water must then be removed before the cells can be used for examination by electron microscopy. This is done by freeze-drying in which the ice is removed by sublimation in a vacuum at about –40°C, or by freeze substitution in which the ice is dissolved in a solvent such as ethanol or acetone at a low temperature.

The specimen is fractured to expose the inner structures. This is followed by a short period of freeze-drying or etching in a vacuum which frees the exposed surface of the specimen from the ice in which it is embedded. A replica of the exposed surface is produced by coating the surface with a film of heavy metal and carbon vapour. The specimen is thawed (ice is melted) and the replica floated off on water to be collected and examined under the electron microscope. The picture produced is however, purely morphological.

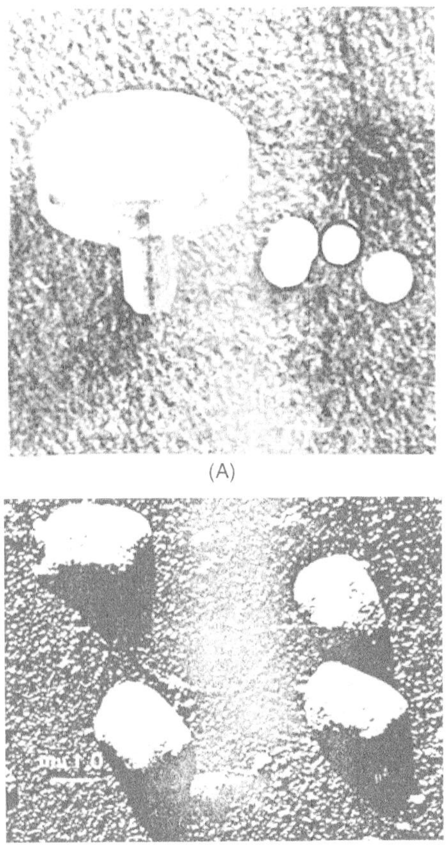

(A) (B)

Fig. 1.48. (A) A comparison of specimen holders in electron microscopy. A specimen stub used for scanning electron microscopy and a number of metal specimen grids used in transmission electron microscopy studies.

(B) An electron micrograph demonstrating the effect of shadow casting. The specimen here is vaccinia virus. The microorganism is used in vaccination against smallpox.

(Wistreich & Lechtman, p. 69)

(5) Shadow-casting : Because of the small contrast between some biological objects and the background even after staining, **shadow-casting** with heavy metals is often used, especially to bring out three dimensional topography. In this procedure, a thin film of an electron-dense material is deposited at an oblique angle on a specimen. Substances employed for this purpose are heavy metals (chromium, nickel, platinum, uranium, or alloys of gold and palladium or platinum and palladium). The metal to be used is placed on a tungsten filament or other device, which is heated to cause the electron-dense material to evaporate from the surface. The metal piles up and outlines the object.

When the object is too thick for shadow-casting, a replica can be made. In this case the object is subjected to vacuum to evaporation of platinum and carbon, and becomes coated to a depth of several nanometers. The preparation is then treated with hydroxide to remove the specimen and the replica is prepared for electron microscopic study.

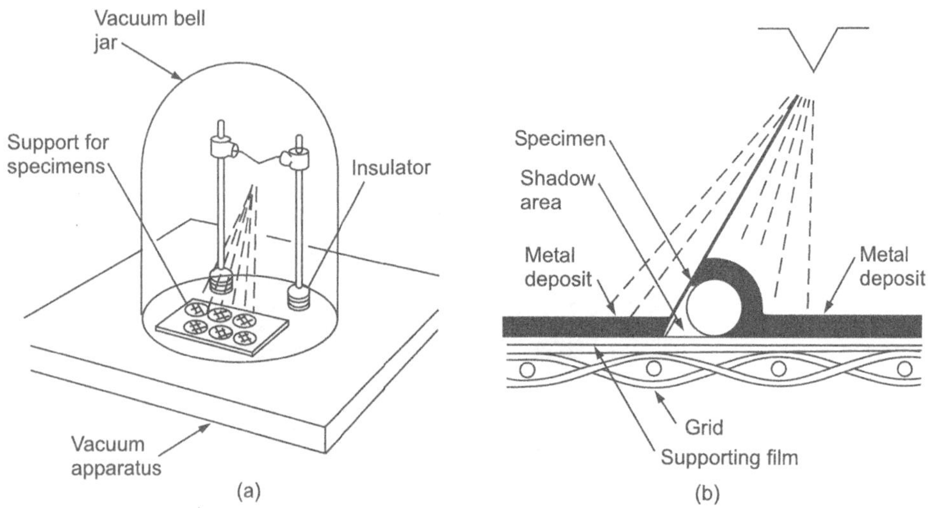

Fig. 1.49. (a) A diagrammatic representation of the shadow-casting apparatus.
(b) The distribution of metal evaporated onto a specimen. The shadow area is transparent and consequently appears light in viewing the specimen. Because most electron micrographs are printed as negatives, it registers as a dark region in electron micrographs.
(Wistreich & Lechtman, p. 69)

(6) Surface spreading : It is also a useful technique for study of cell fine structure. The spread layer is mounted on a grid (a grating—a frame of parallel bars), dried, and shadowed for electron microscopic study. In this manner, a bacterial protoplast (bacterium from which the cell wall has been removed with the enzyme lysozyme) that is ruptured osmotically on the surface of a vessel filled with distilled water is found to liberate its chromosomes as a single long fiber. The same technique has been used for a study of interphase chromosomes in cells of plants and animals and for a study of microtubules present in some cells.

Limitations of Electron Microscopy

Despite the great advantage of tremendous resolution and magnification, there are several limitations to electron microscopy. For example, the specimen being examined is in a chamber that is under a very high vacuum. Thus, cells cannot be examined in a living state. In addition, the drying process may alter some morphological characteristics. Another limitation of the technique is the low penetration power of the electron beam, necessitating the use of thin sections to reveal the internal structures of the cell.

Scanning Electron Microscope

The scanning electron microscope (SEM) is quite different in principle as well as in application from the conventional transmission electron microscope (TEM). The scanning electron microscope gives a three dimensional quality to specimen images. It has been used a great deal to study the surfaces of specimens too thick for conventional transmission electron microscopy. Specimens for the scanning electron microscope require less preparation than those for transmission instruments. Furthermore, large portions of surfaces can be seen in detail with excellent contrast.

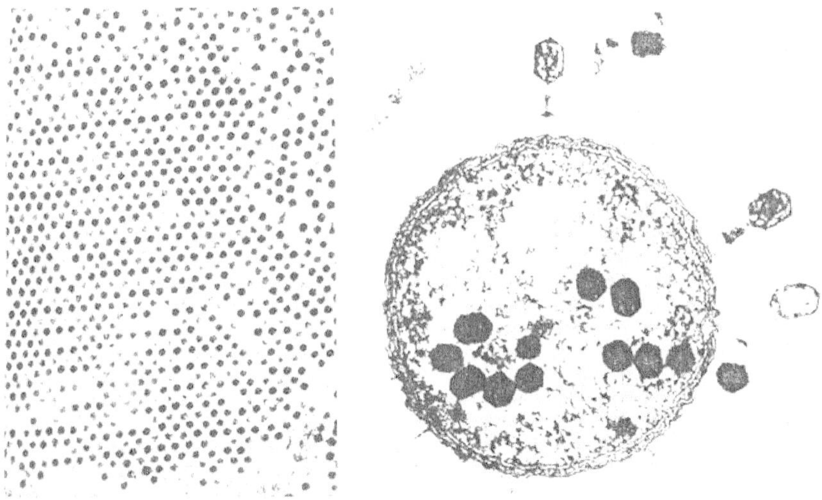

(a) (b)

Fig. 1.50. To view viruses in electron microscope is needed—

(a) Here an electron micrograph shows an intracellular crystalline array of adenoviruses viewed with a TEM (44,000 x)

(b) This electron micrograph shows phage infecting a bacterium. Some viruses are outside of the cell while new viruses are being made within the bacterium (200,000x).

(Atlas, p. 51)

In scanning electron microscopy, the specimen is subjected to a narrow electron beam which rapidly moves over (scans) the surface of the specimen. This causes the release of a shower of secondary electrons and other types of radiation from the specimen surface. The intensity of these secondary electrons depends on the shape and the chemical composition of the irradiated object. The secondary electrons are collected by a detector which generates an electronic signal. These signals are then scanned in the manner of a television system to produce an image on a cathode ray tube.

The SEM lacks the resolving power, obtainable with the TEM, but has the advantage of revealing a striking three dimensional picture. The surface topography of a specimen (such as cilia, surface pits, and contours, etc.) can be revealed with a clarity and a depth of field not possible by any other method.

A great advantage of scanning electron microscope is that the cells need not be fixed, and therefore they are observed in a more natural condition. Most scanning electron microscopes are capable of achieving resolution in the 1–10 nm, range, and thus useful magnification of the SEM is of the range 10,000—100,000x.

More recent types of microscopes such as ion, X-rays and ultrasound microscopes are being developed and used for a variety of specialised studies (Giese, 1979).

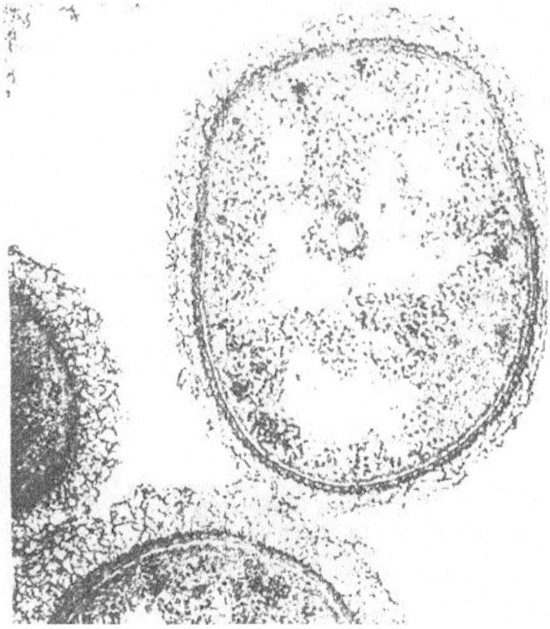

Fig. 1.51. The detailed structure of a bacterium is seen in this electron micrograph of a thin section of *Pseudomonas aeruginosa* (120,000) viewed with a transmission electron microscope. (Atlas, p. 52)

TABLE 1.2. Differences between the Ordinary Light Microscope and a Typical Transmission Electron Microscope

S. No.	Property or procedure	Light microscope	Transmission electron Microscope. [Most of these properties are also true for scanning instruments. Specimens, however are supported on the round, solid surfaces of stubs. (Wistreich & Lechtman, p. 67]
1.	Source of radiation for image formation	Visible light	Electron
2.	Medium through which radiation travels	Air	Vacuum (approx. 10^{-4} mm mercury)
3.	Specimen Mounting	Glass slides	Thin films of collodion or other supporting material or copper grids.

4. Nature of Lenses	Glass	Magnetic fields or electrostatic lenses
5. Focussing	Mechanical (*i.e.*, raising or lowering objectives)	Electrical, *i.e.*, the current of the objective lens coil is changed.
6. Magnification adjustments	Changing objectives	Electricals, *i.e.*, changing current of the projector lens coil
7. Major means of providing specimen contrast	Light absorption	Electron Scattering

TABLE 1.3. Comparison of Various Types of Microscopes

Types of Microscope	Maximum useful magnification	Resolution	Comments
Bright-field	1500x	200 nm	Extensively used for the visualization of microorganisms; usually necessary to stain specimens for viewing.
Dark-field	1500 x	200 nm	Used for viewing microorganism, particularly those with characteristic morphology; staining not required; specimen appears bright on dark background.
Ultraviolet	2500 x	100 nm	Improved resolution over normal light microscope; largely replaced by electron microscopes.
Fluorescence	1500 x	200 nm	Use fluorescent staining; useful in many diagnostic procedures for identifying microorganisms.
Phase-contrast	1500 x	200 nm	Used to examine structures of living microorganisms; does not require staining.
Interference	1500 x	200 nm	Used to examine structure of microorganisms particularly those near the edges; produces sharp multicoloured image with three-dimensional appearance.
Transmission electron microscope (TEM)	500,000 – 1,000,000 x	1 nm	Used to view ultra-structure of micro-organisms, including viruses; much greater resolution power and useful magnification than can be achieved with light microscopy.
Scanning electron microscope (SEM)	10,000 – 100, 000 x	10 nm	Used for showing detailed surface structures; produces three-dimensional image

(Atlas, p. 56)

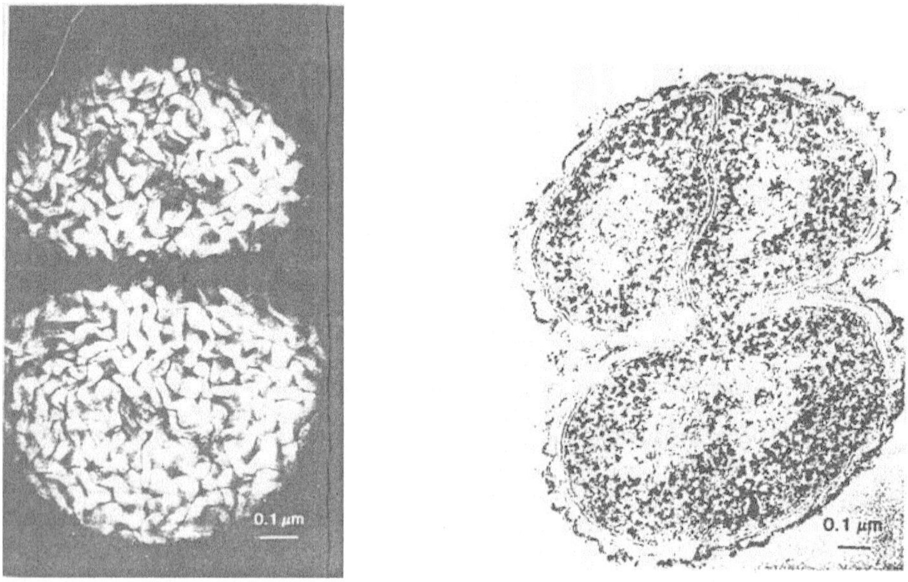

(a) (b)

Fig. 1.52. Electron micrographs of *Veilonella* sp. comparing two techniques used in electron microscopy—

(a) A negatively stained whole cell preparation. Note the convoluted surfaces of the *diplococci*.

(b) A thin section. Note the internal appearance of the organisms as well as their convoluted outer regions.

The cell on the right has developed a cross wall that will eventually result in division. (Wistreich & Lechtman, p. 71)

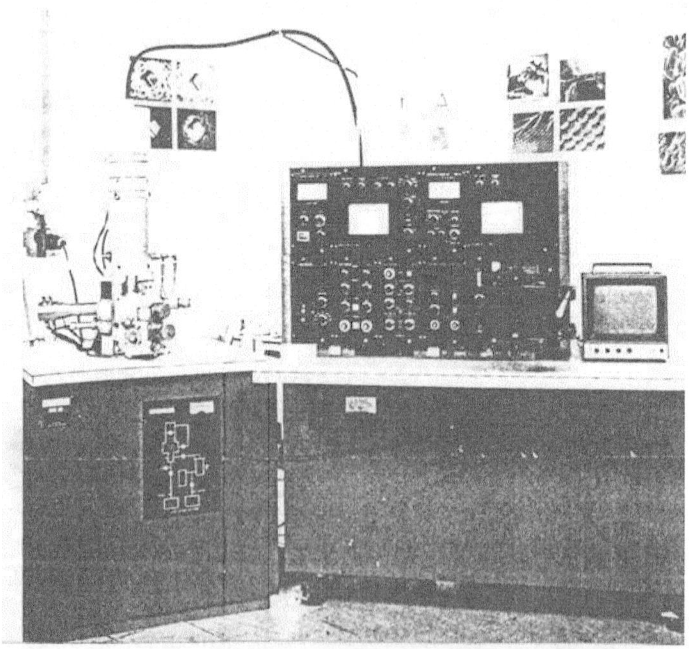

Fig. 1.53. Scanning Electron Microscope
(Wistreich & Lechtman, p. 73)

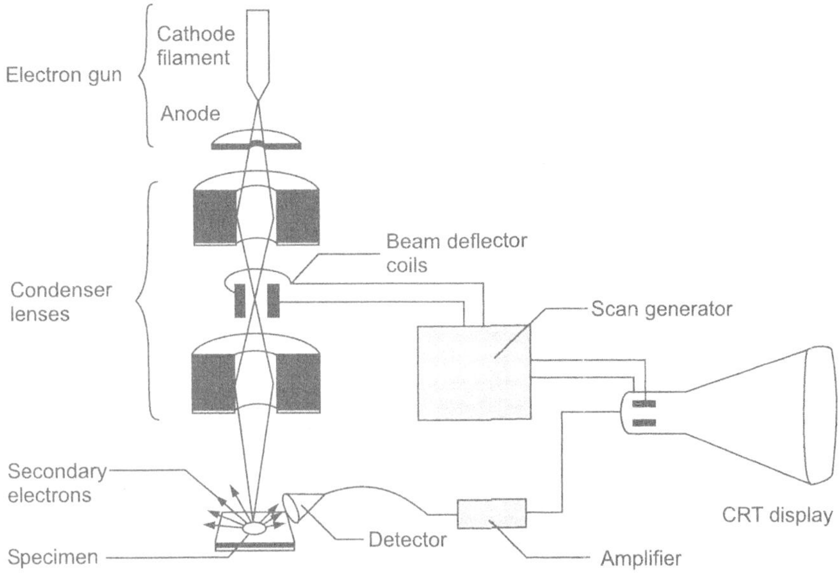

Fig. 1.54. Diagram of SEM. The scanning electron microscope (SEM) is used for viewing surface structures and their three-dimensional spatial relationships.

(Atlas, p. 54)

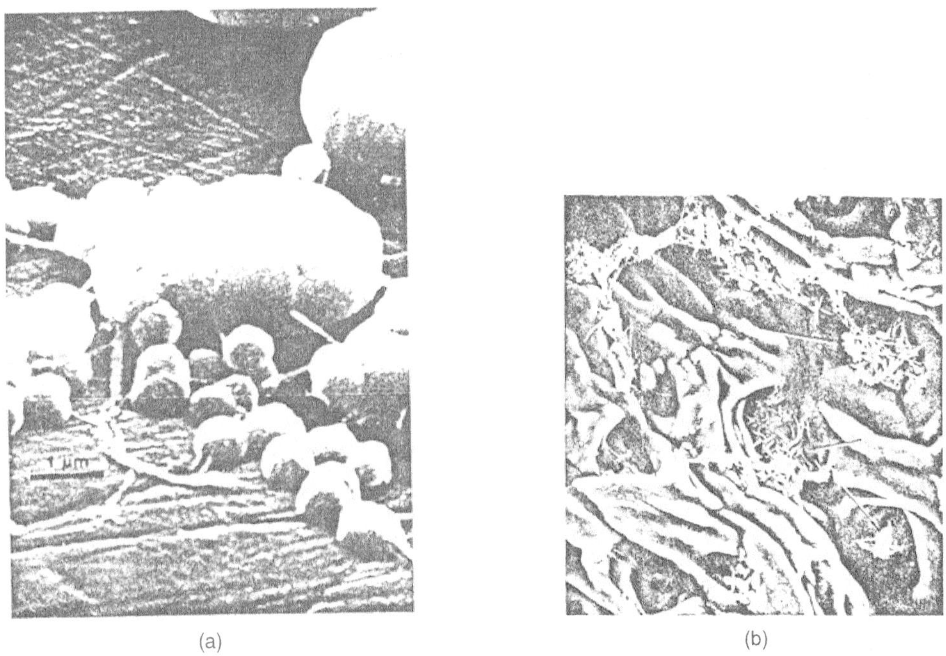

Fig. 1.55. (a) A scanning micrograph showing the clumping of red blood cells by cells of the bacterium *Neisseria catarrhalis.* Note the difference in size between the red blood cells and the bacteria.

(b) A scanning electron microscopic view of a crypt of Lieberkuhn in the colon of an adult mouse. Note that the opening of the crypt is filled with masses of spiral- and spindle-shaped microorganisms. (Wistreich & Lechtman, p. 73)

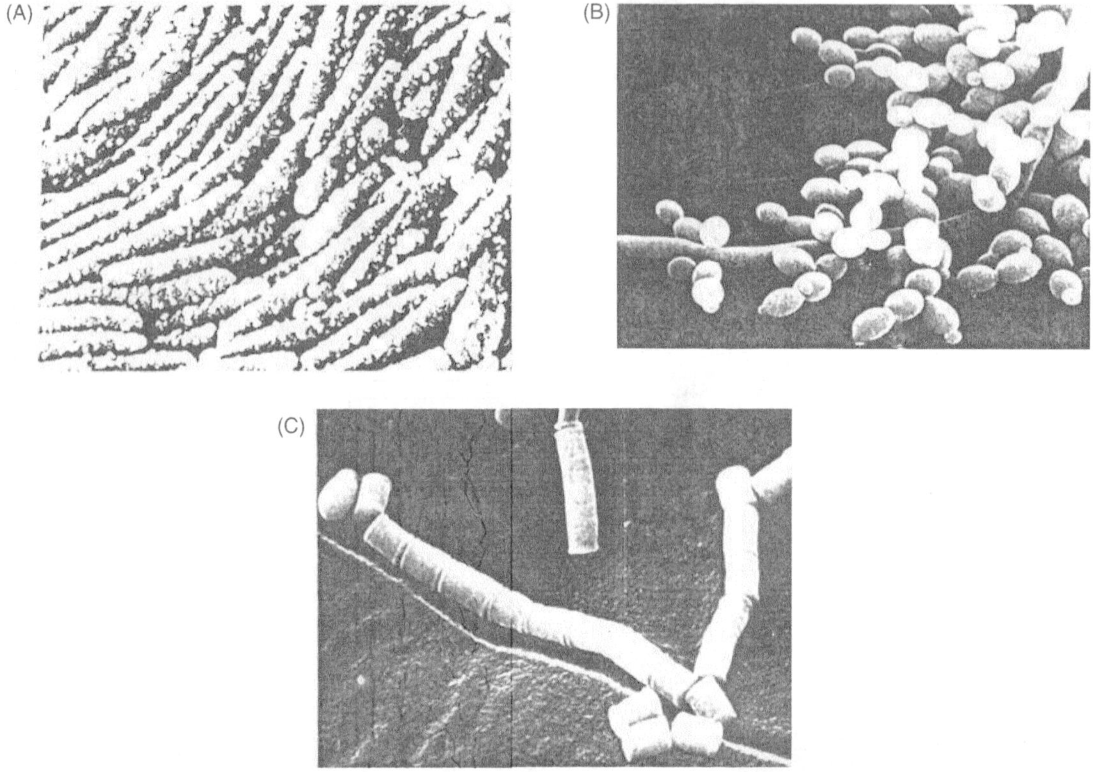

Fig. 1.56. Scanning electron micrographs of microorganisms showing their three dimensional appearance.

(A) Note the bumps on the outer membranes of these rod-shaped bacteria;

(B) *Candida albicans* yeast and hyphal phases (3400 X); and

(C) *Geotrichum candidum,* a fungus associated with food spoilage, forming arthrospores (3250 X). (A from BPS: Z. Skobe Forsythe Dental Centre; B and C from BPS: Garry T. Cole, University of Texas, Austin.)

(Atlas, p. 54)

UNITS OF MEASUREMENTS

In the biological sciences, the two major systems of units used to express measurements are the **English system** (pounds, feet, and inches) and the **metric system.** The latter is more widely used internationally and for scientific work. Since around 1960, an International system of units based upon the metric system has been accepted. It is referred to as **SI system**, from the French *Systeme International d' Unites.*

The metric system has the great advantage of being simple, since using it requires remembering just a few basic terms. The basic unit of length is the **meter** (m); the basic unit of volume is the **liter** (1); the basic unit of mass (weight) is the **gram** (g) and the basic unit of temperature is the **celsius** degree (C°).

The meter, the basic unit, is equivalent to approximately 39 inches. It is divided into 10 decimeters or 100 centimeters or 1000 millimeters or 1 million micrometers or 1 billion nanometers, and so forth. The relationship among these units may be illustrated as next page.

1 meter (m)
10 decimeters (dm)
100 centimeters (cm)
1000 millimeters (mm)
1,000,000 micrometers (µm)
1,000,000,000 nanometers (nm)

The old term **micron (µ)** has been replaced by the term *micrometer* (µm); the term **millimicron** (mµ) has been replaced by the term *nanometer* (nm); and the angstrom (Å is 0.1 nanometer (0.1 nm), according to the International system adopted by scientists worldwide (Anders Jons Ångstrom was a famous Swedish physicist, 1814–1874).

An additional feature of the metric system is that it involves decimal counting. Units are defined so that every unit is some power of 10 (10; 100; 0; 01 and so on) times the basic unit. A meter, for example, consists of 100 centimeters (cm), or 1000 millimeters (mm); and one centimeter can then be seen to consist of 10 millimeters. Converting from one unit to another in the metric system merely involves shifting of the decimal point. With suitable prefixes the meter can express a useful unit for every very small and very large distances. The same prefixes are applied to the measurement units of mass and volumes as well.

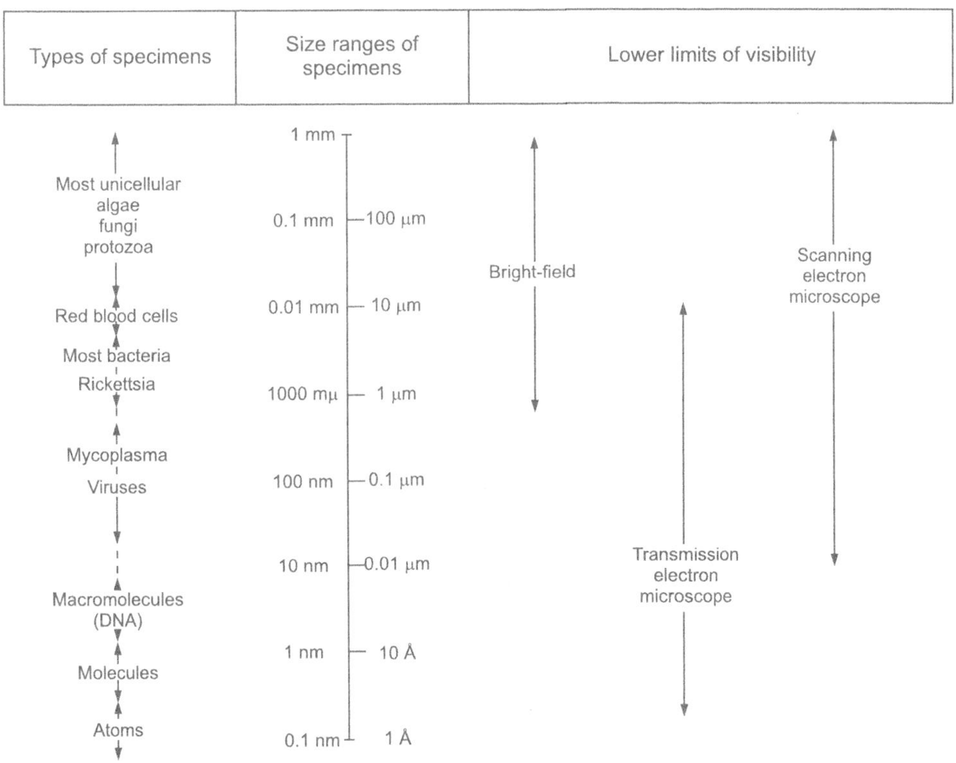

Fig. 1.57. A diagrammatic representation of the types of specimens and their size ranges that can be viewed with different types of microscopes. (Atlas, p.55)

TABLE 1.4. Some Metric Units

Unit	Symbol	Equivalents
Meter	m	1 m = 39.37 inches
Centimeter	cm	1 cm = 10^{-2} m
Micrometer	μm	1 μm = 10^{-6} m
Nanometer	nm	1 nm = 10^{-9} m
Angstrom	Å	1 Å = 10^{-10} m
Liter	1 or L	1 1 = 1 dm^3 = 0.001 m^3
Milliliter	ml or mL	1 ml = 0.0011
Degree Celsius	°C	1 °C = 1 K = 1.8 °F
		0 °C = 27.3.15 K = 32 °F

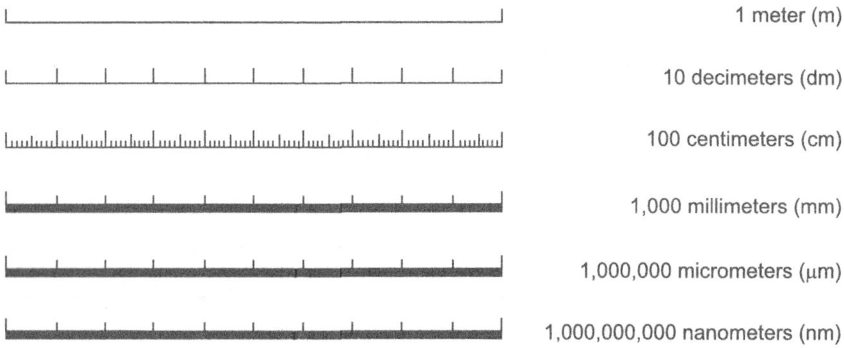

Fig. 1.58. Representation of some metric units. (Burton, p. 21)

TABLE 1.5. SI Units and Some Common Prefixes

Quantity	SI base unit	Symbol
Length	meter	m
Mass	kilogram	kg
Time	Second	s
Temperature	Kelvin	k
Prefix	**Means multiply by**	**Symbol**
Giga	10^9 = 1,000,000,000	G
Mega	10^6 = 1,000,000	M
Kilo	10^3 = 1,000	K
Hecto	10^2 = 100	h
Deca	10^1 = 10	da
Deci	10^{-1} = 0.1	d
Centi	10^{-2} = 0.01	c

Milli	10^{-3} = 0.001	m
Micro	10^{-6} = 0.000.001	μ
Nano	10^{-9} = 0.000,000,001	n
Pico	10^{-12} = 0.000,000,000,001	p

For Example :

1 Millimeter = 10^{-3} meter, or 0.001 meter

1 Nanometer = 10^{-9} meter, or 0.000,000,001 meter

1 Kilogram = 10^{-3} gram, or 1000 grams

In symbols 1 mm = 10^{-3} m,

1 nm = 10^{-9} m,

1 kg = 10^{3} g

In the study of biological materials, the most commonly used measurements of length are the **centimeter** (1 cm = 10^{-2} m), the **millimeter** (1 mm = 10^{-3} m) the **micrometer** (1 μm = 10^{-6} m), the **nanometer** (1 nm = 10^{-9} m), and the **Angstrom unit** (1 Å = 10^{-10} m). Most animal, plant, and microbial cells are measured in micrometers, cellular parts and viruses in nanometers.

Techniques for Making Microscopic Measurement

Following are some techniques for making microscopic measurement :

(1) The Microscopic Field : The field (the circle of visible area) seen while looking through a compound microscope at low power (with a 10x eyepiece and a 10x objective) has a diameter of approximately 1.5 mm. If objects on the field are relatively large, their size may be estimated by comparison with the field diameter. If they are small, however, such as the individual cells of roots, stems, or leaves, then it is necessary to use more precise techniques. The most accurate technique involves the use of an eyepiece micrometer.

(2) Calibration of Eyepiece Micrometer : Very small objects can be measured under the microscope by using an eyepiece micrometer, a small glass disc with very finely etched lines, uniformly spaced, that can be placed in the eyepiece of a microscope.

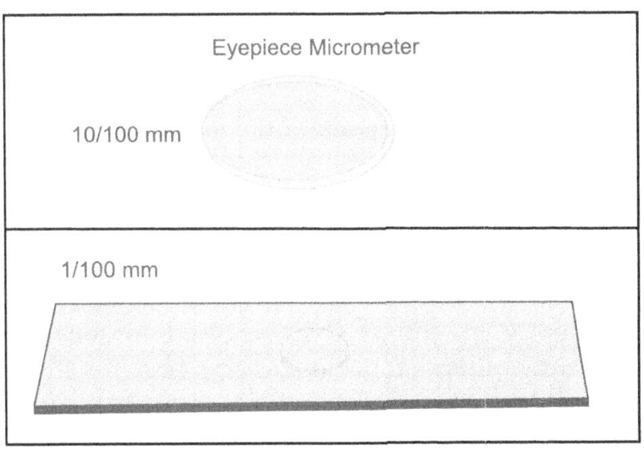

Fig. 1.59. Objective micrometer.

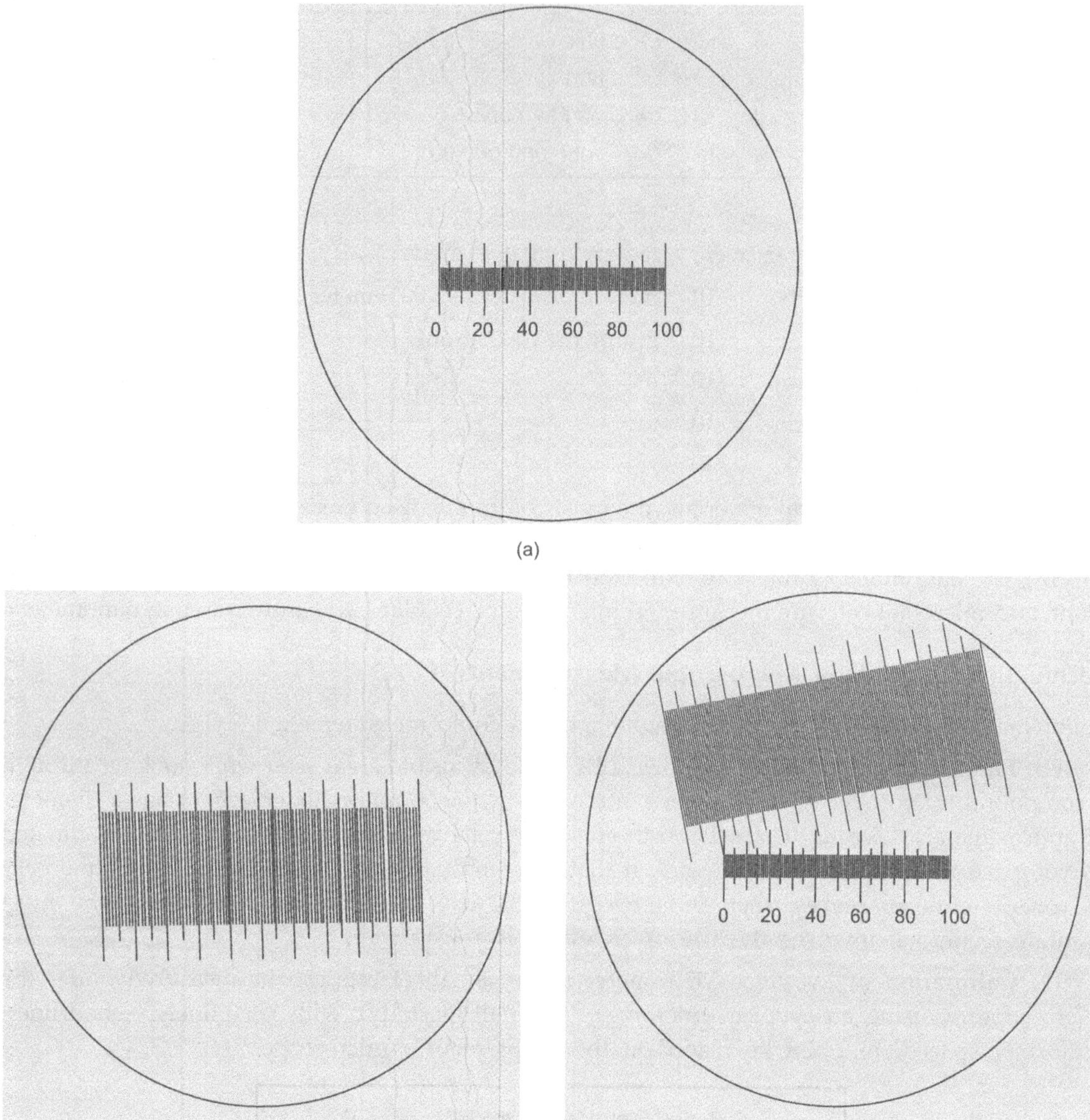

(a) (b) (c)

Fig. 1.60. (a) An eyepiece micrometer as it appears in the microscopic field.

(b) A stage micrometer as it appears in the microscopic field under low power.

(c) An eyepiece micrometer (below) and a stage micrometer (above) as they appear in the microscopic field under low power.

(Lee, p. 78)

The eyepiece micrometer is the ruler which is used to measure microscopic objects. Microscopes may have slightly different magnifications because the magnifying powers of oculars and objectives are not always consistent. It is desirable, therefore, to calibrate each micrometer eyepiece with the specific microscope and the oculars and objectives with which it will be used. On the other hand, if the microscopes used are of a standard make and

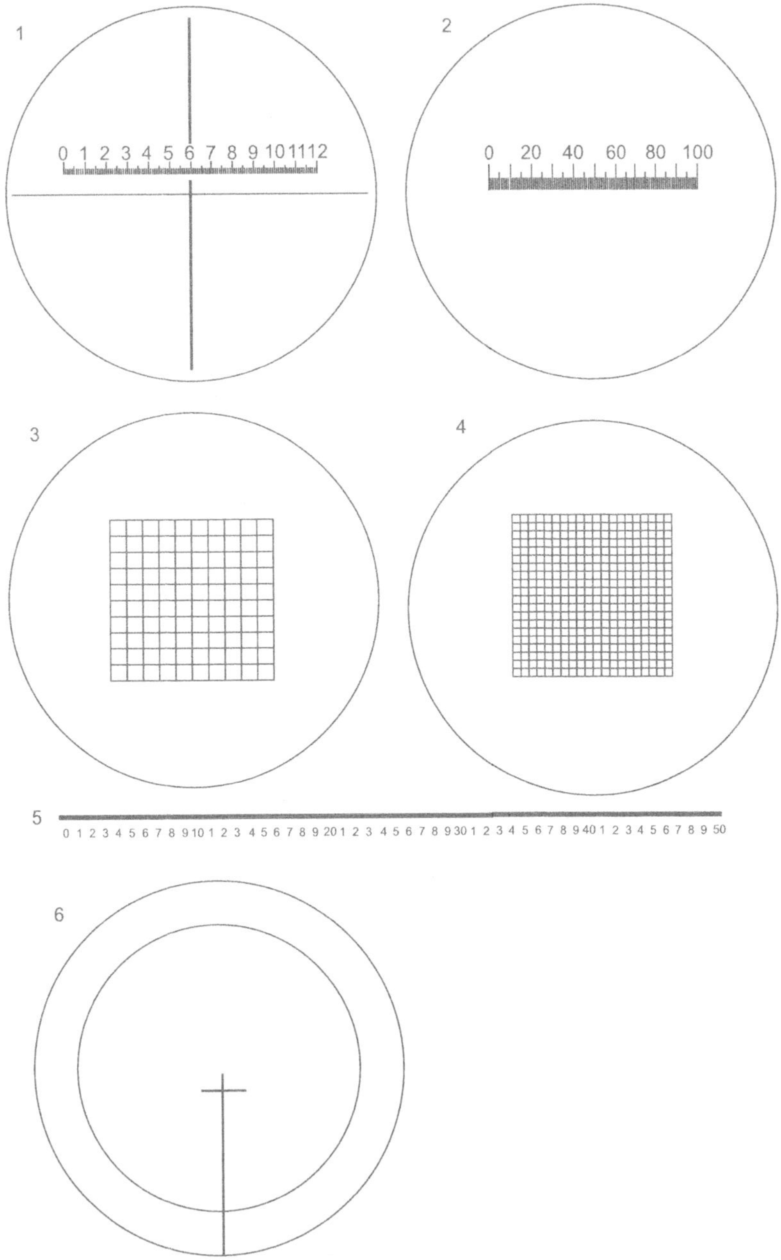

Fig. 1.61. Length determination is one of the commonest procedures in quantitative microscopy. Two graticules (reticles) commonly used are:

- Graticule with scale 12 mm:120 and crosshair (Fig. I)
- Graticule with scale 5 mm:100 (Fig. 2)

Graticules with grids are used to count number of cells, nuclei etc. within a given area :

- Graticule with grid 100 × I mm^2 (Fig. 3)
- Graticule with grid 400 × 0.25 mm (Fig. 4)

To calibrate the graticules, a stage micrometer (Fig. 5) with 50 mm scale bearing 0.1 mm and 0.01 mm graduations is necessary.

model, the eyepiece micrometer calibrations may be consistent enough that it may be needed to calibrate only one or two of them and use the same measurements for a given set of microscopes with a given set of objectives. For this calibration (determining the exact distance between each of the lines in focus on an object with a given objective), a stage micrometer is used. These are special slides with uniform lines precisely etched at standard distances. These distances are labeled. Most stage micrometers have at least a total distance of 1 mm marked in 0.01 mm units.

For calibration, turn the ocular until the lines of the eyepiece micrometer are lined up (parallel) with those on the stage micrometer. Match the end lines of the ocular and stage micrometer by moving the latter. Based on the known distance between lines on the stage micrometer, calculate the distance in microns between the lines of the eyepiece micrometer when superimposed on an object in focus with a given objective. Use the form below :

(1.0 mm = 1,000 microns)

10 units of an eyepiece micrometer = (x) units on a stage micrometer known to be 0.01 mm apart.

$$\text{1 eyepiece micrometer unit} = \frac{(x)\times 0.01\ \text{mm}}{10}$$

= ______ mm, or

= ______ microns

Check the accuracy of your calculations by recording and calculating as follows :

100 eyepiece micrometer units + (x) units on stage micrometer known to be 0.01 mm apart.

$$\text{1 eyepiece micrometer unit} = \frac{(x)\times 0.01\ \text{mm}}{100}$$

= ______ mm or

= ______ microns

If the answer is the same or quite close, the chances are that these calibrations are satisfactory. The stage micrometer may now be put away and the eyepiece micrometer used for the actual measurements. For example, if the above mentioned calibration indicated that each of the smallest units, on the eyepiece micrometer was equivalent to 7 microns on the microscope that is being used, and that a cell was measured that was 20 of these units long, then the actual length of the cell would be 140 micron long.

STUDY OUTLINE

(A) Light Microscopy

1. The microscope is the basic tool employed for the visualization of histological structures and the microbial world.
2. Magnification is the product of magnifying power of ocular lens times the magnifying power of the objective lens.
3. The typical light microscope magnifies 1000 times.
4. Modern microscope lenses correct for optical defects.

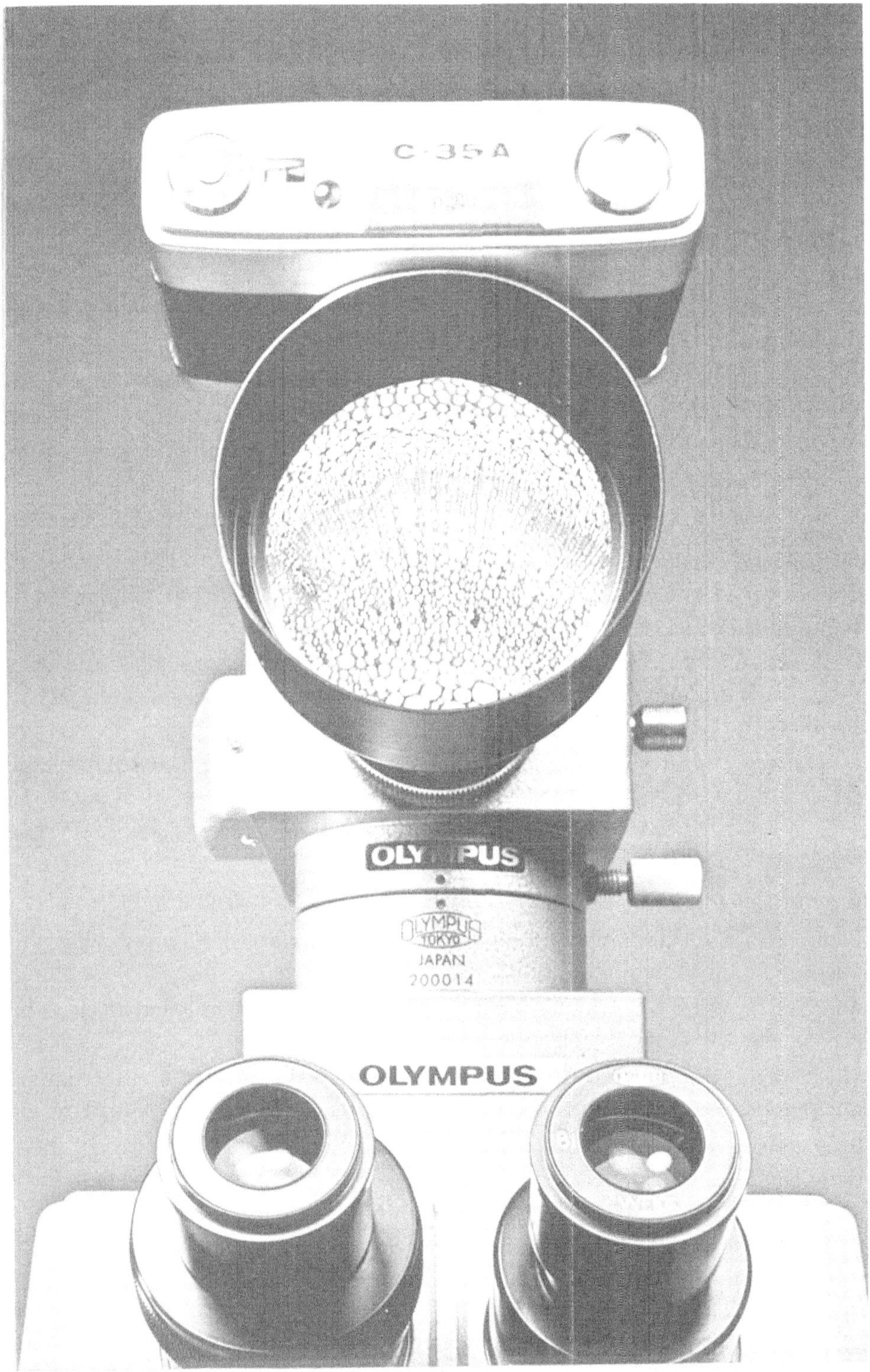

Fig. 1.62. Photomicrographic system camera.

5. Resolving power is the degree to which the detail present in the specimen is retained in the magnified image.
6. Resolution is dependent upon the wavelengths of light and the numerical aperature of the lens.
7. The best resolution for a light microscope is 200 nm.
8. The shorter the wavelength of the illuminating source, the better the resolving power.
9. The greater the numerical aperture of the lens, the better the resolving power.
10. Contrast is necessary to differentiate a structure from the surrounding background.
11. Staining is used to increase contrast.
12. In a positive simple stain procedure the dye is attracted to the tissue (or the microorganism), leaving a clear background.
13. In a negative simple stain procedure the dye is repelled by the sample, leaving the tissue clear against a coloured background.
14. A differential stain procedure distinguishes (differentiates) one structure from another, or one microbial structure from the rest.
15. In brightfield microscopy, the object is dark and the background is bright.
16. In darkfield microscopy, the object appears bright on a dark background.
17. The darkfield condenser focuses light on the specimen at an oblique angle so that only light that is reflected off the specimen will be seen.
18. In fluorescence microscopy objects are stained with dyes that fluoresce, that is, dyes that, when illuminated by light of one wavelength, give off light at a different wavelength.
19. In phase contrast microscopy contrast is enhanced by shifts in the phase of light passing through the specimen.

(B) Electron Microscopy

1. Electron microscopes use an electron beam instead of visible light.
2. Better resolution and higher useful magnification can be achieved with the electron microscopes than with the light microscope.
3. There are two basic types of electron microscopes, the transmission electron microscope (TEM) and the scanning electron microscope (SEM).
4. The design of a transmission electron microscope is similar to the compound light microscope. But electromagnets are used for the lens systems instead of glass.
5. In a transmission electron microscope the electron beam passes through the specimen; the beam is enlarged by electromagnetic objective and projector lenses and projected onto a fluorescent screen or photographic film.
6. The theoretical resolution of the transmission electron microscope is about 0.2 nm; a thousand times better than a light microscope.
7. The useful magnification of most transmission electron microscopes is over 400,000x so that even the smallest microbes can be visualized.
8. Extensive sample preparation is necessary for electron microscopy in order to maximize observation of detail.

Microscope M 20 with photomicrographic camera I, focusing eyepiece and roll film frame.

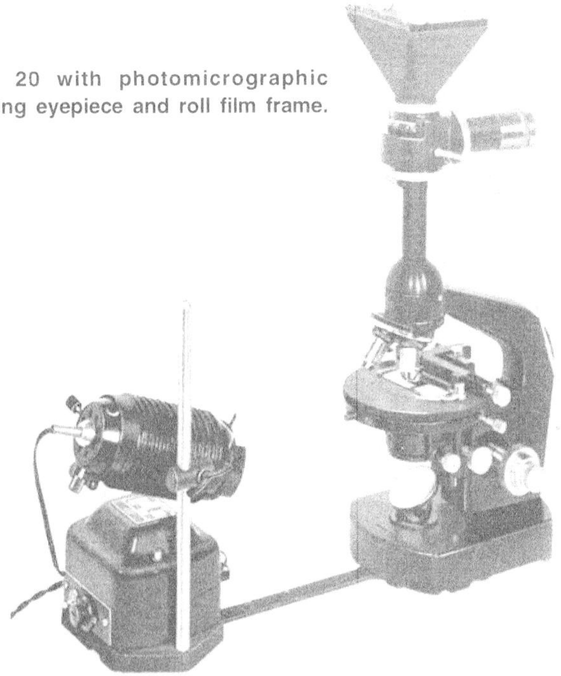

Model M 20 RdCFL with low voltage microscope lamp and projecting mirror.

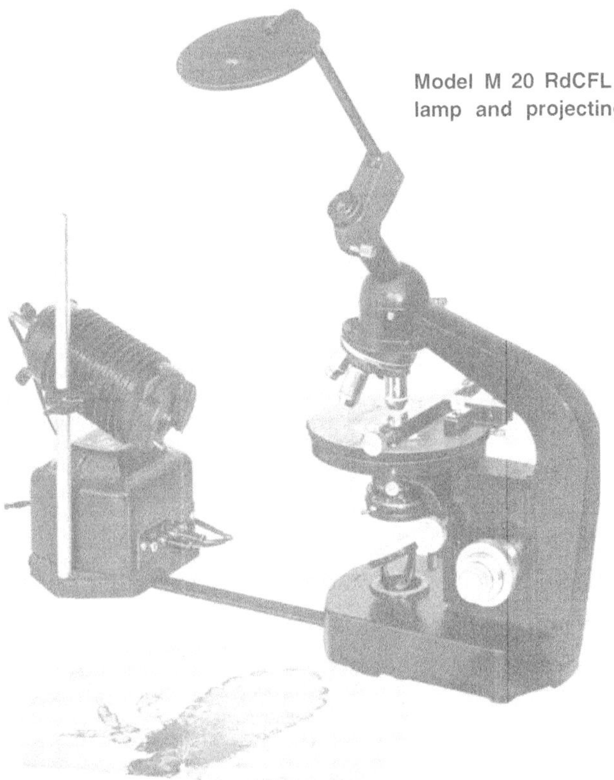

Fig. 1.63

9. In a scanning electron microscope the electron beam is scanned across the surface of the specimen. The electrons emitted from the surface of the specimen are used to generate an image on a cathode ray tube (CRT).
10. The intensity of the image seen with scanning electron microscopy reflects the composition and topography of the specimen surface.
11. The useful magnification of a scanning electron microscope generally is 10,000–100,000x.
12. The resolution obtained with an SEM is 1–10 nm.

(Atlas, p. 57–58)

SUMMARY

Microscopes and Microscopy

1. Depending upon the magnification principle involved, microscopes fall in one of two categories, *light* (optical) or *electron.*
2. Light microscopes include **bright-field, dark-field, fluorescence,** and **phase-contrast** instruments.
3. Electron microscopes include the **transmission** and **scanning** instruments.

Units of Measurements

1. **Le Systeme International d' Unities,** or **SI system** is a widely accepted system for measurements.
2. The units used for length, volume, mass, and temperature are the meter, liter, gram and Celsius degree, respectively.
3. Commonly used metric units for the measurements of length include the meter (m), centimeter (cm), millimeter (mm), micrometer (μm), nanometer (nm), and Angstrom (Å).
4. The **bar marker** on photographs serves as a size reference and represents that length of a metric unit at a particular magnification.

Properties of Light

1. **Light waves** are described in terms of their amplitude, frequency, and wavelengths.
2. **Amplitude** refers to the maximum displacement of a light wave from its equilibrium position.
3. **Frequency** is the number of vibrations that occur in one second.
4. **Wavelength** is defined as the distance between two corresponding points on a wave.
5. The **resolving power** of a microscope refers to its ability to reveal detail. It depends on the wavelengths of light used.
6. The type of material or medium through which light passes during the operation of a microscope plays a major role in the image seen.

Light Microscopes

Bright-field Microscopy

1. Two types of bright-field microscopes are known, the **simple** (one-lens) microscope and the **compound** microscope.
2. A compound microscope consists of at least two lens systems the **objective** which forms and magnifies the image of the specimen, and the **eyepiece, or ocular,** which magnifies the image produced by the objective lens.
3. **Resolving power** (the ability to show details of specimens) and **parfocal** (changing objectives without major focusing adjustments) are two important microscope features.
4. The highest—magnification objective used in general courses is the oil-immersion objective, which for best results should be immersed in a medium having the same **index of refraction** as glass.
5. Condensers serve to concentrate light for specimen viewing. Most condensers are equipped with a shutter-like device, the **iris diaphragm,** to control light intensity.

Dark-field Microscopy

1. Specimens examined by **dark-field microscopy** usually appear as bright objects against a dark background.
2. Dark-field microscopy is useful with unstained preparations and may have diagnostic importance.

Fluorescence Microscopy

1. **Fluorescence microscopy** is used for the examination of various cell types and cellular structures, the location of chemical components in cells, and the rapid diagnosis of certain diseases.
2. Ultraviolet light, fluorescent chemicals, and special microscope filter systems are required for this type of microscopy.

Phase Microscopy

1. **Phase microscopy** is of value in the examination or detection of living cells in certain fluid environments.
2. The **phase-contrast** principle makes cells and their parts stand out from the background.

Micrometry

Micrometry refers to procedures used to measure the dimensions of cells or, if possible their parts.

Electron Microscopy

The Transmission Electron Microscope

1. Most **transmission electron microscopes** are similar in basic design to the light microscope.
2. The fundamental difference is that the electron microscopes use electrons as the source of illumination. Therefore, they have different lens construction and specimen preparation procedures. They can produce higher magnification with better resolving power.

Specimen Preparation for Electron Microscopy

1. Because of their chemical composition, untreated biological specimens are difficult to see in the transmission electron microscope.
2. Special treatment procedures are used to increase the contrast of such specimens; these include shadow casting, replicas and electron staining.
3. **Shadow casting** involves depositing at an angle, a film of heavy metal on a specimen.
4. **Surface replicas** are surface impressions of specimens.
5. **Electron stains** are solutions that contain heavy metal elements that may combine with certain components of cells.
6. **Thin sectioning, or slicing** is used with specimens that are too thick for direct examination in the electron microscope.
7. Thin sectioning can be used to study the internal organization, structural arrangement, and related properties of cells.
8. **Freeze-fracture** can be used to study biological specimens without the preparation associated with other techniques.

High-Voltage Electron Microscopy

This form of microscopy makes it possible to penetrate thicker specimens and to observe three-dimensional relationships among cell components.

The Scanning Electron Microscope

1. This instrument provides a three-dimensional quality to the images of the specimens viewed.
2. Large portions of specimen surfaces can be seen in detail with excellent contrast.
3. Specimens are metal—coated for examination.

REFERENCES

1. Atlas, R.M. 1986. *Basic and Practical Microbiology,* Macmillan Publishing Company, New York.
2. Birge, E. 1992. *Modern Microbiology,* Wm, C. Brown Publishers, Dubuque, U.S.A.
3. Burton, G.R.W. 1992. *Microbiology for the Health Sciences,* Lippincott Company, Philadelphia.
4. Dutta, A.C. 1964. *Botany for Degree Students,* Oxford University Press, Bombay.
5. Frobischer, M. 1967. *Fundamentals of Microbiology,* Charles E. Tuttle Company, Japan.
6. Fuerst, R. 1983. *Microbiology in Health and Disease,* W.B. Saunders Company, Philadelphia.
7. Giese, A.C. 1979. *Cell Physiology,* W.B. Saunders Company, Philadelphia.
8. Hall, J.C., T.J. Flowers, and R.M. Roberts., 1984. *Plant Cell Structure and Metabolism,* Longman Group Ltd., Essex, England.

9. Lee, A.E. 1963. *A Laboratory Block, Plant Growth and Development,* D.C. Health and Company, Boston.
10. Pelczar M.J., E.C.S. Chan. and N.R. Krieg, 1986. *Microbiology*, McGraw-Hill Book Co., Singapore.
11. Sass, J.E. 1964, *Botanical Microtechnique.* Oxford & IBH Publishing Co. Calcutta.
12. Walter, W.G. and Richard H. Mc. Bee, 1969. *General Microbiology*, Affiliated East-West Press Pvt. Ltd. New Delhi.
13. Wistreich, G.A. and Max. D. Lechtman. 1980. *Microbiology*, Macmillan Publishing Co. Inc. New York.
14. Young, G.G. 1961. Witton's *Microbiology*, McGraw-Hill Book Co. Inc. (Kogakusha Company Ltd. Tokyo).

Chapter 2

CHEMISTRY FOR BIOLOGISTS

INTRODUCTION

Chemistry is the science that deals with the composition and structure of matter and with the transformations that matter undergoes. Chemistry is a rather broad field; at one extreme, in theoretical chemistry and spectroscopy, it borders on physics, and at the other extreme, in organic chemistry, it borders on biochemistry and biology.

ATOMS, MOLECULES, ELEMENTS, COMPOUNDS

All substances, whether gas, liquid, or solid, have certain fundamental characteristics in common. If one could break down any substance into its smallest elemental units, it would be composed of atoms. Although it is difficult to observe atomic structure, atoms are known to consist of a mass of positively charged protons, noncharged (or neutral) neutrons, tiny negatively charged electrons, and other even smaller subparticles. An atom has equal numbers of electrons and protons; thus it has a total charge that is zero. When an atom loses or gains electrons it becomes a negatively or positively charged *ion* (*e.g.*, Na^+, Cl^-). The protons and neutrons are found in a central nucleus, and the electons float around the nucleus, like negatively charged satellites attracted to a positively charged planet.

TABLE 2.1. The First 20 Elements

Element	Symbol	Atomic Number	Atomic Weight
Hydrogen	H	1	1
Helium	He	2	4
Lithium	Li	3	6.9
Beryllium	Be	4	9

Boron	B	5	10.8
Carbon	C	6	12
Nitrogen	N	7	14
Oxygen	O	8	16
Fluorine	F	9	19
Neon	Ne	10	20.1
Sodium	Na	11	23
Magnesium	Mg	12	24.3
Aluminium	Al	13	27
Silicon	Si	14	28.1
Phosphorus	P	15	31
Sulphur	S	16	32.1
Chlorine	Cl	17	35.5
Argon	Ar	18	40
Potassium	K	19	39.1
Calcium	Ca	20	40.1

(Burton, p. 79)

When a multitude of atoms with identical numbers of electrons and protons and the same chemical properties exist together, the substance is known as an *element*, for example, carbon (C), hydrogen (H), oxygen (O), nitrogen (N), sulphur (S).

The atoms of each element have their own special weight (or mass) called *atomic weight,* which is the total number of protons and neutrons each of which is assigned a value of 1. Electrons are negligible in weight. The atomic weight of a specific element is an average of the weights of all isotopes of that element as they occur in nature. *Isotopes* are atoms of the same element, with identical chemical characteristics but with different atomic weights. They differ in weight because they have various numbers of neutrons within the atoms. For example, carbon – 12 has six protons and six neutrons for an atomic weight of 12, whereas carbon – 14 has six protons and eight neutrons for an atomic weight of 14; both are isotopes of the element carbon. Elements are also identified by a number that indicates the number of positive charges, or protons, in the nucleus. This is called the *atomic number.* Thus the atomic number of carbon is 6. Atomic weights and atomic numbers can be found in a periodic table of the elements in any chemistry book.

Two or more atoms of the same or different elements that combine into a single stable unit form a *molecule;* examples include carbon dioxide (CO_2), water (H_2O), or sulphur dioxide (SO_2). When many molecules exist together in a substance that can be seen or weighed, the substance is called a compound, *e.g.,* carbon dioxide gas (CO_2).

Although there are only approximately 106 known elements, the number of compounds that are known is already in the millions, and new ones are being created and discovered each year. The component parts of living cells and systems are composed of many macromolecules that interact to produce and maintain life.

Periodic table of the elements
Simplified atomic weights for practice in chemical calculations (*Not sufficiently accurate for quantitative work.*)
Values in parentheses indicate mass numbers of the most stable or best known isotope.

IIA	IIIB	IVB	VB	VIB	VIIB	VIII	VIII	VIII	IB	IIB	IIIA	IVA	VA	VIA	VIIA
4 9.0 Be Beryllium											5 10.8 B Boron	6 12.0 C Carbon	7 14.0 N Nitrogen	8 16.0 O Oxygen	9 19.0 F Fluorine
12 24.3 Mg Magnesium	TRANSITION ELEMENTS										13 27.0 Al Aluminium	14 28.0 Si Silicon	15 31.0 P Phosphorus	16 32.0 S Sulfur	17 35.5 Cl Chlorine
20 40.0 Ca Calcium	21 45.0 Sc Scndium	22 48.0 Ti Titanium	23 51.0 V Vanadium	24 52.0 Cr Chromium	25 55.0 Mn Manganese	26 56.0 Fe Iron	27 59.0 Co Cobalt	28 58.7 Ni Nickel	29 63.5 Cu Copper	30 63.4 Zn Zinc	31 69.7 Ga Gallium	32 72.6 Ge Germanium	33 75.0 As Arsenic	34 79.0 Se Selenium	35 80.0 Br Bromine
38 87.6 Sr Strontium	39 89.0 Y Yttrium	40 91.0 Zr Zirconium	41 93.0 Nb niobium	42 96.0 Mo Molybdenum	43 (99) Tc Tehnetium	44 101 Ru Ruthenium	45 103 Rh rohdium	46 106 Pd Palladium	47 108 Ag Silver	48 112 Cd Cadmium	49 115 In Indium	50 119 Sn Tin	51 122 Sb Antimony	52 128 Te Tellurium	53 127 I Iodine
56 137 Ba Barium	57 139 La Lanthanum	72 178 Hf Hafnium	73 181 Ta Tantalum	74 184 W Tngsten	75 186 Re Rhenium	76 190 Os Osmium	77 192 Ir Iridium	78 195 Pt Platinum	79 197 Au Gold	80 201 Hg Mercury	81 204 Ti Thallium	82 207 Pb Lead	83 209 Bi Bismuth	84 (210) Po Polonium	85 (210) At Astatine
88 (226) Ra Radium	89 (227) Ac Actinium														

6	*Lanthanides	58 140 Ce Cerium	59 141 Pe Praseodymium	60 144 Nd Neodymium	61(147) Pm Promethium	62 150 Sm Samarium	63 152 Eu Europium	64 157 Gd Gadolinium	65 159 Tb Terbium	66 162 Dy Dysprosium	67 165 Ho Holmium	68 167 Er Erbium	69 169 Tm Thulium	70 173 Bb Ytterbium	71 175 Lu Lutetium
7	†Actinides	90 232 Th Thorium	91 (231) Pa Protactinium	92 238 U Uranium	93 (237) Np Neptunium	94 (242) Pu Plutonium	95 (243) Am Americium	96 (247) Cm Curium	97 (249) Bk Berkelium	98 (251) Cf Californium	99 (254) Es Einsteinium	100 (253) Fm Fermium	101 (256) Md Mendelevium	102(253) No Nobelium	103(257) Lw Lawrencium

Fig. 2.1

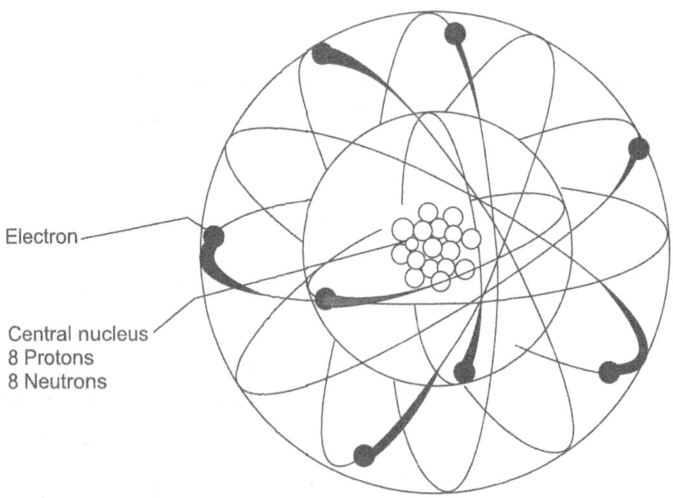

Fig. 2.2. Representation of an oxygen atom. Eight protons and eight neutrons are tightly bound in the central nucleus, around which the eight electrons revolve.

(Burton, p. 78)

TABLE 2.2. Chemical Elements Found in the Human Body

Element	%	Element	%
Oxygen	65%	Potassium	0.35%
Carbon	18.5	Sulphur	0.25
Hydrogen	9.5	Chlorine	0.20
Nitrogen	3.3	Sodium	0.15
Calcium	1.5	Magnesium	0.05
Phosphorus	1.0		

TABLE 2.3. The Mineral Elements in Nutrition

1.	Elements needed daily in large quantities.	Sodium, calcium, magnesium, potassium (phosphorus)
2.	Elements needed in smaller quantities.	Iron, copper, chlorine, iodine, cobalt, manganese, zinc, molybdenum.
3.	Mineral elements whose need is questionable.	Bromine, fluorine, selenium, tin, silicon, arsenic, boron, vanadium.
4.	Mineral elements which are toxic in relatively small quantities.	Tin, arsenic, barium, beryllium, bismuth, cadmium, lead, mercury, selenium, silver, tellurium, thorium.

(De Bey, p. 161)

WEIGHTS AND MEASURES

The quantity of matter in any given system is the *mass* of that system, where mass is measured in terms of a standard unit, the *kilogram* (kg). Two other important standard units of measurements in the sciences are that of length, the *meter*, and that of time, the *second*. These units have been variously defined in the past. At the present time, a kilogram is defined as a mass equal to a block of platinum – iridium alloy kept at Sevres, France; the meter is defined as a length equal to 1,650, 763.73 times the wavelength of the orange-red light emitted by the element Krypton (Kr^{86}), and the second is defined as a time interval equal to 1/86,400 of the mean solar day (24 hours = 1440 minutes = 86,400 seconds). Two important subunits of this system are the *gram* (g), which is 1/1000 of kilogram, and the *centimeter* (cm), which is 1/100 of a meter.

The kilogram, meter, and second are the fundamental units of the *metric system* of measurement, a system used by scientists throughout the world. In England, the *British system* of measurement is used; the fundamental units of this system are the pound, the foot, and the second.

CHEMICAL BONDING

In general, atoms form molecules by gaining or losing electrons (forming ionic bonds) or by sharing electrons (forming covalent bonds). An *ionic bond* holds together molecules, such as sodium (NaCl, common table salt) and many others, that ionize in water. When sodium chloride dissolves in water, the sodium atom (Na) loses a negative electron and assumes a positive charge, thus becoming a positive ion (Na+) in the water. Likewise, the chlorine atom (Cl) accepts one electron from the donor sodium atom (Na) to become a chloride ion (Cl^-). The ionic bonds between sodium and chlorine atoms are easily broken; however, they re-form in the water because the unlike charges are attracted to each other. They may temporarily form NaCl molecules, and then dissociate into ions again. Positively charged ions are called *cations;* and negatively charged ions are *anions.*

Other molecules are held together by covalent bonds. A *covalent bond* is formed by atoms sharing one or more pairs of electrons. The methane (CH_4) molecule has hydrogen atoms that each share two electrons with the carbon atom. This is a much stronger bond than an ionic bond; thus the molecule is much more stable. Carbon compounds are especially good examples

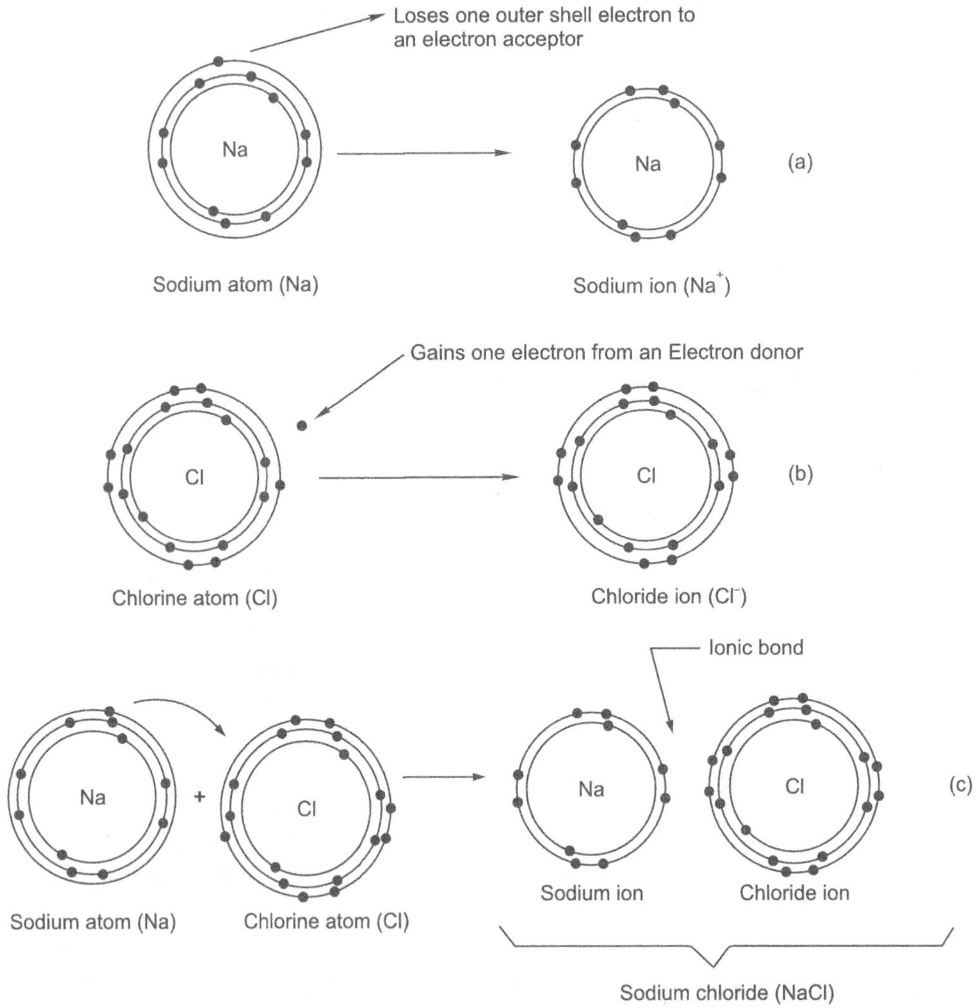

Fig. 2.3. Ionic bond formation

(a) A sodium atom loses an electron to become a positively charged sodium ion (Na^+).

(b) A chlorine atom gains an electron to become a negatively charged chloride ion (Cl^-).

(c) The positive sodium ion is attracted to the negative chloride ion to form the sodium chloride molecule which is held together by a weak ionic bond.

of covalent bonding. All of organic chemistry is based on the stable covalently bonded carbon molecules and how they react with other molecules. Biochemistry, the chemistry of living cells and systems, is a study of the large macromolecules composed of covalently bonded carbon, hydrogen, oxygen, nitrogen, phosphorus, and sulphur atoms. The loosely bound electrons on the surface of the macromolecules enable them to react and bind with other molecules in living organisms. Peptide bonds, which hold amino acids together in a protein, and glycosidic bonds, which bind the sugar groups of polysaccharides together, are also covalent bonds.

Hydrogen bonding occurs when a hydrogen atom that is covalently bonded to an oxygen or nitrogen atom is attracted to another oxygen or nitrogen atom on a different molecule or group, as seen in the α-helix of DNA and protein structure.

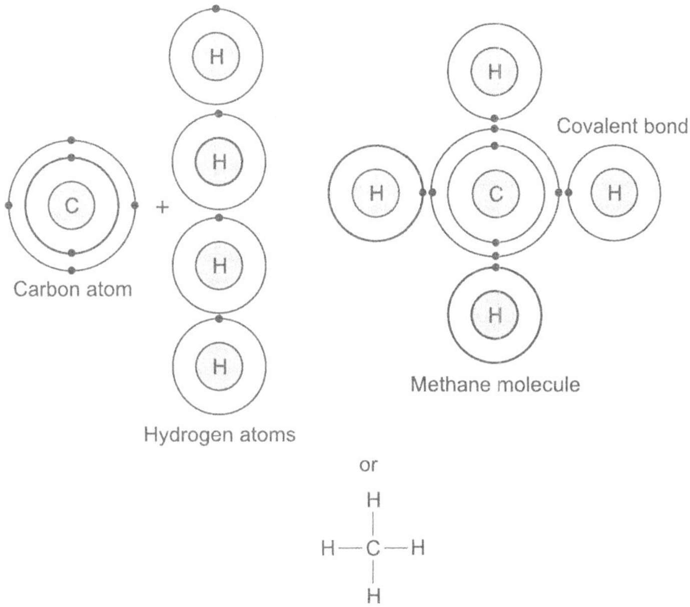

Fig. 2.4. Covalent bond formation. The methane molecule with four hydrogen atoms sharing electrons with a carbon atom, forming four covalent bonds.
(Burton, p. 82)

ORGANIC COMPOUNDS

All carbon compounds (other than those, such as carbonates, which are of mineral origin), are termed as organic compounds.

Classification of Carbon Compounds

(1) **Aliphatic compounds :** Which contain "open" (non-cyclic) carbon chains, *e.g.*, pentane.

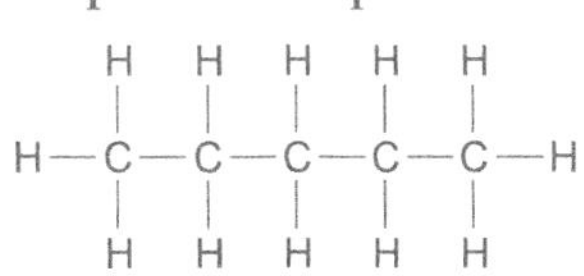

or

$CH_3CH_2CH_2CH_2CH_3$

(2) **Alicyclic compounds :** Which are the cyclic analogues of aliphatic compounds, *e.g.*, cyclopentane.

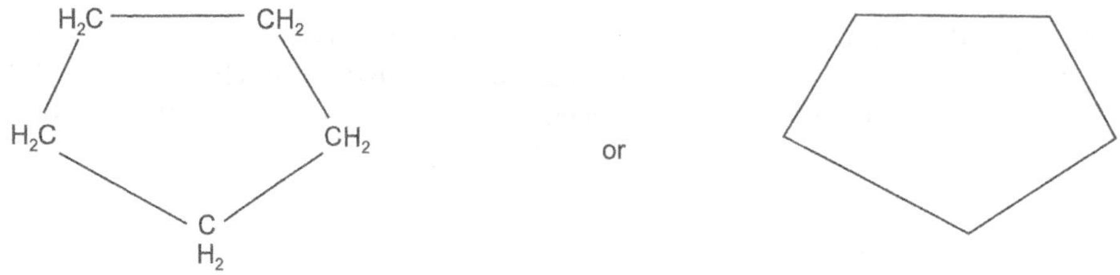

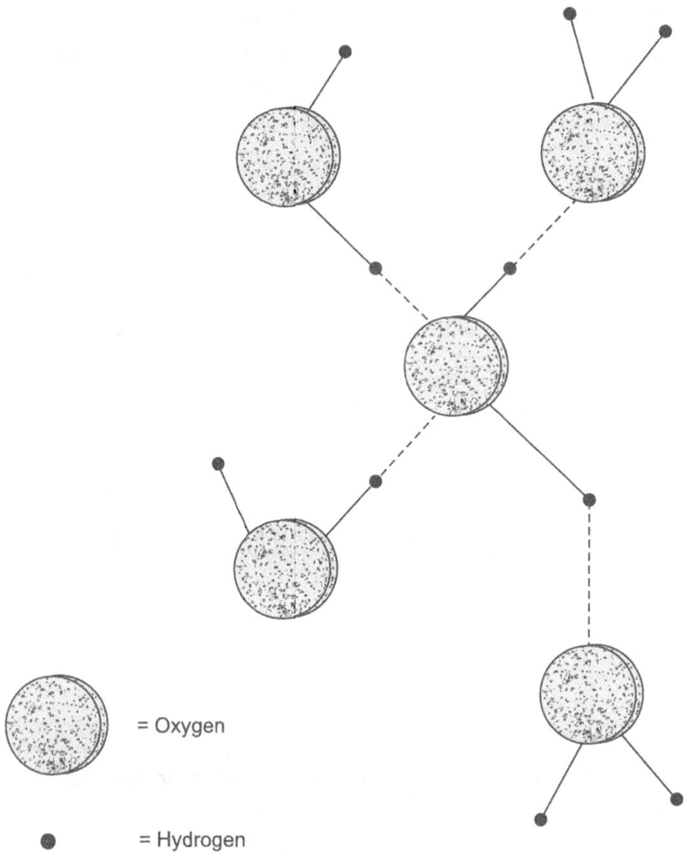

Fig. 2.5. Hydrogen bonded water molecules.

(3) **Aromatic compounds :** Which contain an "aromatic" or "benzoid" ring. The simplest compound of this type is benzene.

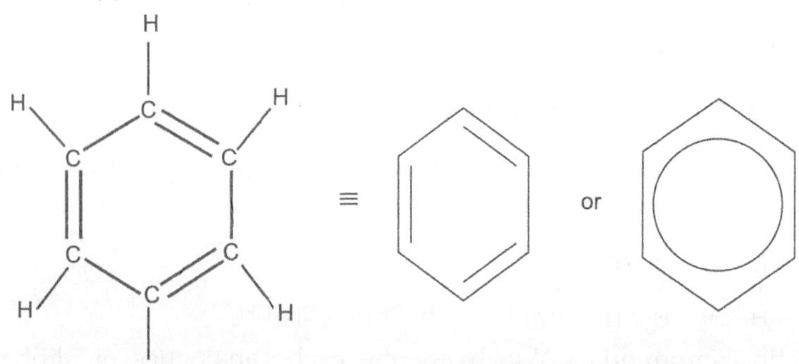

(4) **Heterocyclic compounds :** Which are alicyclic or aromatic compounds in which one or more carbon atoms has been replaced by an oxygen, nitrogen or sulphur atom *i.e.*, by a "hetero" atoms.

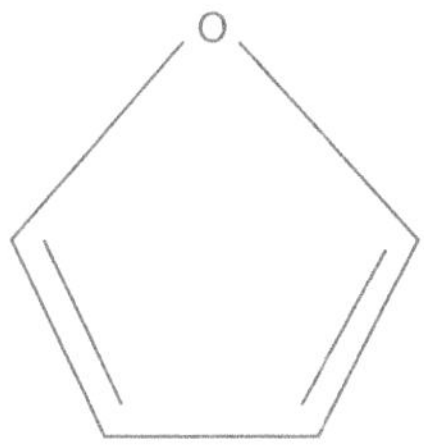

Furan

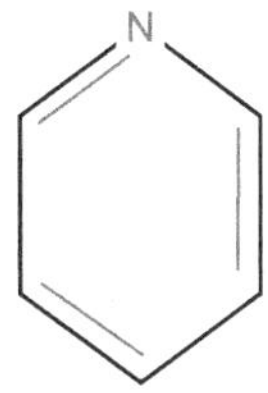

Pyridine

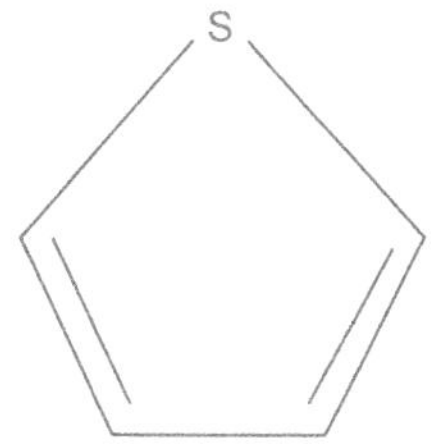

Thiophene

FUNCTIONAL GROUPS AND THEIR NOMENCLATURE

The four main types of organic compound may contain functional groups which give characteristic properties to a compound.

TABLE 2.4

Structure of Functional group	Name of Functional group	Compound containing Functional group
$>C=C<$	double bond	alkenes
$—C\equiv C—$	triple bond	alkynes
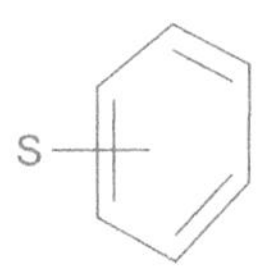	Benzene ring (may be substituted - S, indicates a substituent in an undefined position)	arenes
$—\overset{\vert}{\underset{\vert}{C}}—X$ X = halogen	Halogeno	halogeno-alkanes halogeno-arenes
- OH	Hydroxy	alcohols (aliphatic) phenols (aromatic)
- OR	Alkoxy or aryloxy	ethers
- SH	Sulphydryl	thiols or mercaptans
- SR	Alkythio or arylthio	Thioethers
- NH_2	Amino	Primary amines
- NHR	Alkylamino or arylamino	Secondary amines
- NR^1R^2	Dialkylamino, Diarylamino or Alkarylamino	Tertiary amines
$\underset{Y}{>}C=X$ X = O, Y =unspecified	Carbonyl	See ahead

H -	Aldehydo or Formyl	Aldehydes
alkyl or aryl -	Oxo or keto	Ketones
HO -	Carboxyl	Carboxylic acids
RCOO -	A cyloxycarbonyl	Acid anhydrides
$(HO)_2P(=O)—O—$	Phosphoryl - oxycarbonyl	Acid anhydrides or Acylphosphates
RO -	Alkyloxy - carbonyl or aryloxycarbonyl	Esters
RS -	(alkylthio) carbonyl or (arylthio) carbonyl	Thiol esters
X = NH Y = alkyl or aryl	Imino	Azomethines
X = NR Y = alkyl or aryl	Substituted imino	Schiff bases
$-C \equiv N$	Carbonitrile or cyano	Nitriles or Cyanides

IMPORTANCE OF WATER IN LIVING CELLS AND SYSTEMS

Water is most abundant molecule in living cells; it is absolutely essential for the functioning of living cells. The water molecule, consisting of two hydrogen atoms on one side of an oxygen atom; gives the molecule *polarity,* that is, a water molecule has positive and negative areas. This characteristic makes water an ideal solvent, or suspending medium, for other ionic or charged molecules. The hydrogen bonding (*i.e.,* the attraction between water molecules) is called *polar bonding.*

This polarity or water molecules accounts for four major characteristics that make t an essential part of living cells.

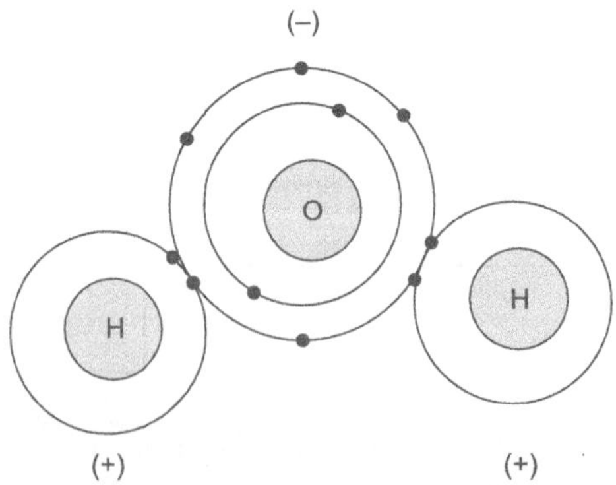

Fig. 2.6. A water molecule, showing two hydrogen atoms covalently bonded to an oxygen atom. The arrangement gives the molecule a more positive charge near the hydrogen protons and a negative charge near the oxygen electrons, thus providing an attraction between water molecules (polar bonding).

(Burton, p. 82)

1. Water is an excellent solvent (a liquid in which molecules of one or more other chemicals, called the solutes, are dissolved, thus forming a solution). Nutrients and waste materials may move in and out of a cell, crossing the cell membrane, because they are dissolved in water.
2. Its polarity makes the molecules adhere, producing surface tension and capillary action, hence water can move within and among cells and tissues.
3. It also serves as an excellent buffer against heat changes to protect the cell from fluctuation in temperature, because it can absorb and hold large amounts of heat.
4. Water enters into *hydrolysis* reactions (digestion of large macromolecules like cellulose by breaking apart the molecular units) and *dehydration* reactions (in which the macromolecules are synthesized with the removal of a water molecule.

SOLUTIONS, ACIDS, BASES, SALTS, pH

Solutions of various types result from the mixing of elements with elements, element with compounds, or compounds with compounds. It is these homogenous (evenly distributed throughout) mixtures, that form almost all existing substances. The most common solutions are those in which a compound is dissolved in the solvent water. Other solvents, such as alcohol, are more commonly used with organic compounds.

There are two types of compounds that have characteristics that set them apart from all others; these are acids and bases. One interesting feature of these two groups is that any acid will react with any base, almost without exception. Also, it is relatively easy to measure the acidity or alkalinity of solutions of acids and bases.

Acids

Acids are identified by using chemical indicators or dyes that exhibit a distinct colour in an acid solution.

All acids behave in a similar way because they all share a common feature: the ability to produce hydrogen ions when in solution. An acid molecule has at least one hydrogen atom. As an illustration, consider hydrogen chloride or hydrochloric acid, which has the formula HCl. Hydrochloric acid in pure form is a gas, but when bubbled into water, it releases a hydrogen ion (H^+) by donating an electron to form the chloride ion (Cl^-). The result is a cation (the positively charged hydrogen ion) and an anion (the negatively charge chloride ion), as illustrated below.

$$\underset{\text{Hydrogen chloride}}{HCl} + \underset{\text{Water molecule}}{H_2O} \longrightarrow \underset{\text{Hydrogen ion}}{H^+} + \underset{\text{Chloride ion}}{Cl^-} + \underset{\text{Water molecule}}{H_2O}$$

Note that the HCl is an electrolyte; that is, it is a substance with charged particles that can conduct electricity. All acids produce hydrogen ions, which are the particles that contribute to the sour taste of acids and also affect the colour of indicators. Indicator dyes are acidic or basic compounds that have distinctive colour when in acidic or basic solutions.

Bases

The bases have a hydroxide ion (OH^-) as their common feature. When dissolved in water, they release the hydroxide ion and a cation, and become electrolytes.

TABLE 2.5

Base	Formula	Location or use
Sodium hydroxide	$NaOH$	Caustic soda
Potassium hydroxide	KOH	Caustic potash
Ammonium hydroxide	NH_4OH	Household cleaners
Calcium hydroxide	$Ca(OH)_2$	Lime water
Magnesium hydroxide	$Mg(OH)_2$	Antacid drug

All acids will react with bases, and when they react they form one common product, water (H_2O).

$$HCl + NaOH \longrightarrow NaCl + H_2O$$

$$H_2SO_4 + Mg(OH)_2 \longrightarrow MgSO_4 + 2H_2O$$

$$H_2CO_3 + NaOH \longrightarrow NaHCO_3 + H_2O$$

General Equation : $\text{Acid} + \text{Base} \longrightarrow \text{Salt} + \text{Water}$

Salts

Salts are another general class of electolytic compounds. Chemical formulas are often written in a manner to assist in determining whether a substance is acid, base, or salt. For example, acids are written with H at the left, as in HCl and H_2SO_4. Bases have an OH at the right, such as NaOH and $Mg(OH)_2$. The formulas for salts have neither an H at the left nor an OH at the right. Sodium chloride, NaCl, and sodium hydrogen carbonate, $NaHCO_3$, are examples.

It is important for all living creatures, including microorganisms, to have and maintain the appropriate acid, base, and salt balance to metabolize and function properly.

pH (Potential of Hydrogen)

pH is a measure of acidity. Naturally, acids are "acidic", and bases are "basic" or "alkaline" in nature. The pH scale used to indicate acidity or alkalinity ranges from 0 to 14, with 7 as the neutral point. Pure water has a pH of 7. If an acid is added to the water, the pH decreases to between 0 and 7. Adding a base to water gives a pH between 7 and 14.

TABLE 2.6. The pH of Some Common Substance

Substance	pH
Household ammonia cleaner	11.9
Blood (human)	7.35-7.45
Pure H_2O	7
Milk	6.9
Black coffee	5
Orange juice	2.9
Gastric juice (stomach acid)	1.5

(Burton, p. 85)

The pH of a substance can be measured with a pH meter, with indicator solutions, or with chemically treated papers that turn a definite colour within certain pH ranges. These papers are called indicator papers.

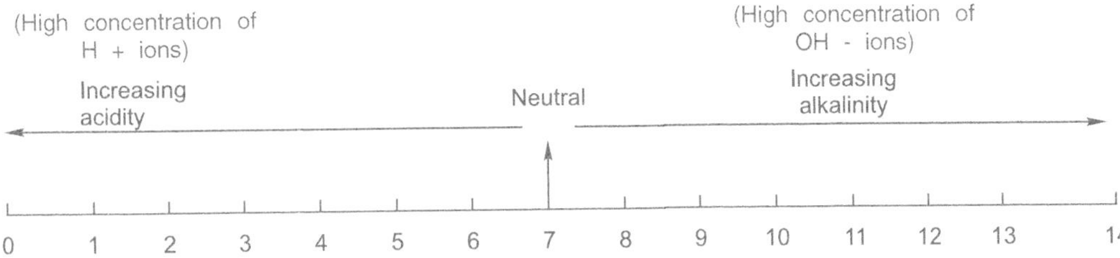

Fig. 2.7. The pH scale.
(Burton, p.85)

ORGANIC CHEMISTRY

Organic chemistry is that branch of the science of chemistry that deals primarily with the element carbon and its covalent bonds. This definition makes organic chemistry a broad and important branch of the science of chemistry.

Carbon Bonds

Carbon is the primary requisite for all living systems. The element carbon exists in three forms: diamond, graphite, and carbon or carbon black. These three forms have dramatically different physical properties, and it is difficult to believe that they are truly the same element. In addition to these unique physical differences, carbon has a valence of 4, which means that its atoms bond to four other atoms. For convenience, the carbon atom is illustrated with the symbol C and four bonds :

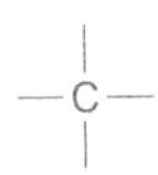

The uniqueness of carbon lies in the ability of its atoms to bond to each other to form a "multitude" of compounds. The variety of carbon compounds increases still more when atoms of other elements also attach in different ways to the carbon atom.

When two carbon atoms bond together there are three types of resulting covalent bonds : *single bonds, double bonds,* and *triple bonds.* Each line represents a bond between the carbon atoms, which is formed from a pair of shared electrons.

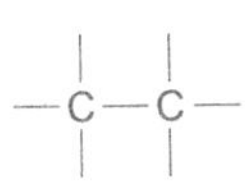

Single bond

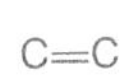

Double bond

Triple bond

When atoms of other elements attach to additional available bonds of the carbon atoms, stable compounds are formed. For example, if hydrogens are bonded to the available bonds, compounds called *hydrocarbons* are formed. Just a few of the many hydrocarbon compounds are shown below.

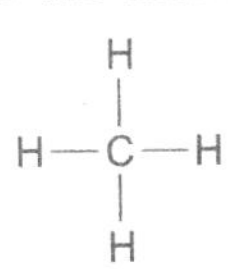

Methane

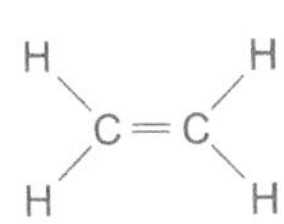

Ethylene

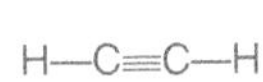

Acetylene

Fig. 2.8. Simple hydrocarbons.

When more than two carbons are linked together, longer molecules are formed. A series of many carbon atoms bonded together is called a *chain*. The long chain hydrocarbons are usually liquid or solid; the short-chain hydrocarbons, such as those shown above, are gases.

Cyclic Compounds

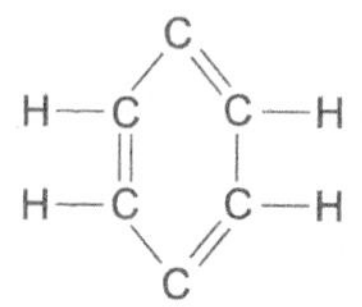

The benzene ring

Fig. 2.9

Carbon atoms may link to carbon atoms to close the chain, forming ring or cyclic compounds. An example is benzene, which has six carbons and six hydrogens bonded, as shown in the figure above.

BIOCHEMISTRY

Biochemistry is defined as the "Chemistry of living matter", and its biochemical changes or metabolism. In its broad aspects, biochemistry is the most comprehensive of all the branches of chemistry. It includes inorganic, organic, and physical chemistry to the extent to which each of these is related to the chemistry of living things, both plant and animal.

The basis of all forms of life is the material called protoplasm, which in chemical composition, physical organization, and function is enormously complex; in fact, it is the most complex physicochemical system with which the chemist has to deal. The problem of biochemistry in general is to relate the properties and functions of protoplasm to its physicochemical organization. The protoplasm of each different kind of cell in each kind of animal or plant is different and characteristic, yet the chemical composition, organization, and chemical processes in these many different forms of protoplasm are in many respects strikingly similar. The study of the biochemistry of one kind of protoplasm therefore is, in effect, the study of all kinds of protoplasm. The developments in animal biochemistry have been greatly aided by investigations of the chemical processes of plants and microorganisms and vice versa. Various chemical reactions first observed in microorganisms or plants have later been sought and found in higher animals. The reverse also is true.

The chief aims of biochemistry are to describe and analyze the chemical changes that occur in organisms. Investigations into the chemistry of living systems have shown that individual cells, whether of plants, animals, or microorganisms are fundamentally similar in function despite vast differences in structure. With its discoveries, biochemistry has greatly illuminated the functional aspects of organisms as they affect the fields of agriculture, medicine, nutrition, and other disciplines. But the effectiveness of biochemical knowledge in practical problems depends on our mastery of the fundamental principles of the chemistry of life.

DIVERSITY OF MACROMOLECULES

The large biochemical molecules, called macromolecules, are present in every cell in animals, plants, and microorganisms. These macromolecules are of different types and are classified as carbohydrates, lipids, proteins, and nucleic acids. The vitamins, enzymes, hormones, and high energy molecules, such as adenosine triphosphate (ATP), are also included among these biochemicals.

Humans obtain their nutrients from the foods they eat. The carbohydrates, fats, nucleic acids, and proteins contained in these foods are digested, and their components are absorbed by the blood and carried to every cell in the body. In the body cells, the components from the food biochemicals are absorbed and rearranged. In this way, the proper compounds necessary for cell structure and function are synthesized. Microorganisms also absorb their essential nutrients into the cell by various means (diffusion, osmosis etc.). These nutrients are then used in metabolic reactions as sources of energy and as building blocks for enzymes and structural macromolecules as well as for genetic materials.

TABLE 2.7. Approximate Composition of Materials from Plants and Animals

Materials (Plants & Animal)	Water (%)	Carbohydrate (%)	Protein (%)	Lipid%	Nucleic acid (%)	Ash* (%)
Mammalian muscle (lean)	72-80%	1%	17-21%	5%	0.3%	1%
Brain	79	1	8-10	10	0.2	–
Hen's egg	74	1	13	11	–	1
Cow's milk	87	5	3	4	–	0.7
Rat liver	69	4	16	5	0.3	1.4
Fish, Cod	83	0	16	0.4	–	1.2
Bananas	75	23	1.2	0.2	–	0.8
Rice	12	80	7.6	0.3	–	0.4
Peanuts (roasted)	2.6	24	27	44	–	2.7
Honey	20	80	0.3	0	–	0.2
Potatoes, White	78	19	2	0.1	–	1.0
Potatoes, Sweet	68	28	2	0.7	–	1.1
Spinach	93	3.2	2.3	0.3	–	1.5
Beer (4% alcohol)	90	4.4	0.6	0	–	0.2
Soybeans (dry)	7.5	35	35	18	–	4.7

*The value given as ash indicates the amount of material remaining when all organic matter has been oxidized. It is an indication of the amount of mineral elements present.

(De Bey, p. 7)

CARBOHYDRATES

Carbohydrates are composed of carbon (C), hydrogen (H), and oxygen (O) in the ratio of 1 : 2 : 1, or simply CH_2O.

Monosaccharides

The simplest carbohydrates are sugars; the simplest molecule of sugar is the monosaccharide (Greek, *mono* = one; *sakcharon* = sugar). The most important monosaccharide in nature is glucose ($C_6H_{12}O_6$), which may occur in the chain or alpha or beta ring configurations.

The main source of energy for body cells, glucose (dextrose), is found in most sweet fruits and in blood sugar. The glucose carried in the blood to the cell is oxidized to produce the high energy molecule ATP with its high-energy phosphate bonds. Thus, ATP molecule is the main source of energy used to drive most metabolic reactions. Other monosaccharides, with the aldehyde group on the first carbon, are galactose and ribose (a five-carbon sugar). An important monosaccharide that has a ketone group on the second carbon is *fructose*. Fructose, the sweetest of the monosaccharides, is found in fruits and honey.

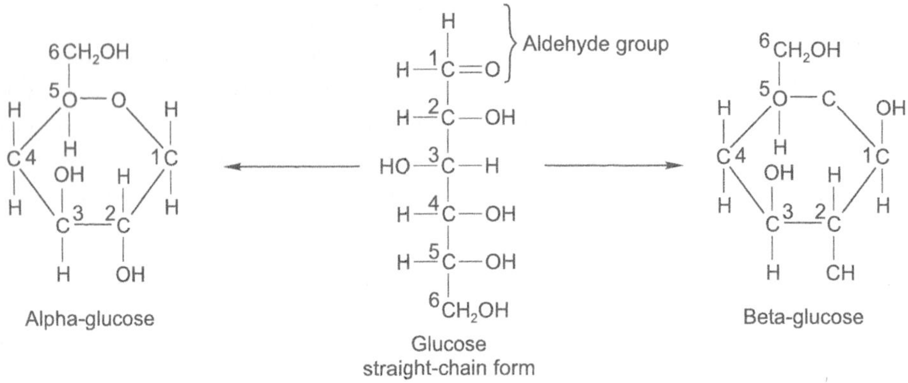

Fig. 2.10. Glucose. The straight-chain and alpha and beta forms may all exist in equilibrium in solution.

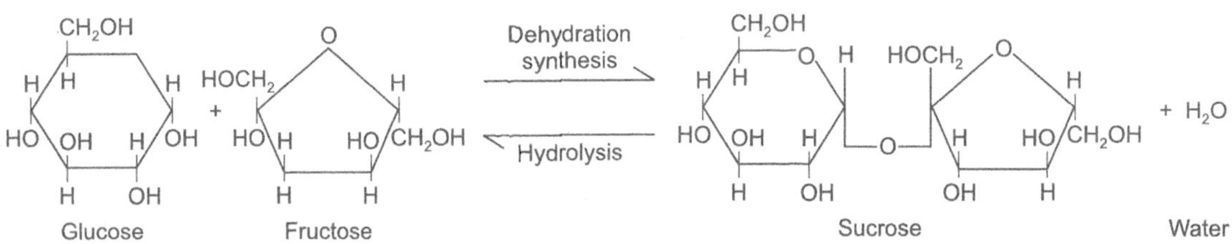

Fig. 2.11. The dehydration synthesis and hydrolysis of sucrose.

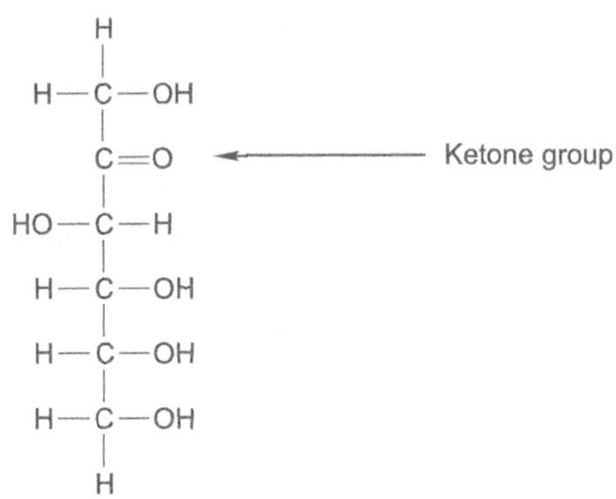

Fig. 2.12. Fructose in straight—Chain form.
(Burton, p. 88)

Disaccharides

Disaccharides are sugars that result from the combination of two monosaccharides. Sucrose (or table sugar) is a sweet *disaccharide* made up of a glucose molecule and a fructose molecule.

Disaccharides react with water in a process called hydrolysis. This causes them to break down into two monosaccharides. Maltose is the sugar in malt. Lactose is the sugar found in milk. Sucrose comes from sugarcane, sugar beets, and maple sugar. The synthesis of disaccharides from two monosaccharides with the removal of water is called *dehydration synthesis*.

Polysaccharides

Polysaccharides, such as starch, cellulose, and glycogen, are essentially made up of hundreds of repetitive glucose units, that is, they consist of long chains of monosaccharide, glucose, which are held together by glycosidic bonds. They are *polymers*, which means they consist of many similar subunits. These large polysaccharide molecules are made up of many monosaccharide molecules. Some of these molecules are so large that they are insoluble in water. In the presence of the proper enzymes or acids, polysaccharides may be hydrolyzed or broken down into disaccharides and then finally into monosaccharides.

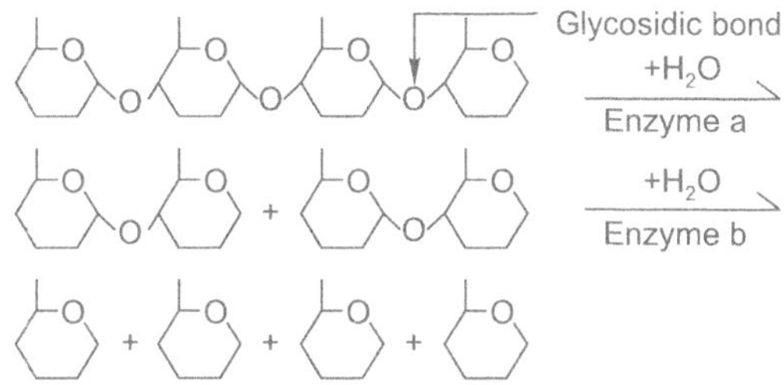

Fig. 2.13. The hydrolysis of starch. 1 Starch (polysaccharide), 2 Maltose (disaccharide), 3 Glucose (monosaccharide).

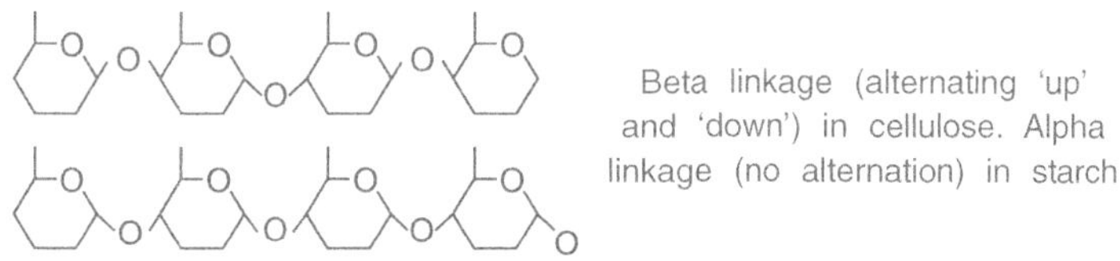

Fig. 2.14. The difference between cellulose and starch.

Polysaccharides serve two main functions. One is to store energy that can be used when the external food supply is low. The common storage molecule in animals is *glycogen,* which is found in the liver and in muscles. In plants, glucose is stored as *starch* and is found in potatoes and other vegetables and seeds. Some algae store starch; and bacteria contain glycogen granules as a reserve nutrient supply.

The other function of polysaccharide is to provide a "tough" molecule for structural support and protection. Many bacteria secrete polysaccharide capsules for protection against drying and phagocytosis. Plant and algae cells have cellulose cell walls to provide support and shape, as well as protection against the environment.

Cellulose is insoluble in water and indigestible for humans and most animals. Some protozoa, fungi, and bacteria have enzymes that will break the B-glycosidic bonds linking the glucose units in cellulose. These are the microorganisms (saprophytes) that are able to disintegrate dead plants in the soil and that live in the digestive organs of herbivores (plant eaters). Starch and glycogen are easily digested by animals because they have the digestive enzyme to hydrolyze the α-glycosidic bonds that link the glucose units into long, helical or branched polymers.

Fibers of cellulose extracted from certain plants are used to make paper, cotton, linen, and rope. These fibers are relatively rigid, strong, and insoluble because they consist of 100 to 200 parallel strands of cellulose.

When polysaccharides are combined with other chemical groups (amines, lipids, and amino acids), extremely complex macromolecules are formed that serve specific purposes. Glucosamine and galactosamine (amine derivatives of glucose and galactose respectively) are important constituents of the supporting polysaccharides in connective tissue fibers, cartilage, and chitin. Chitin is the main component of the hard outer covering of insects, spiders, crabs, and fungi. The main portion of the rigid cell wall of bacteria consists of aminosugars and short polypeptide chains that combine to form the peptidoglycan layer.

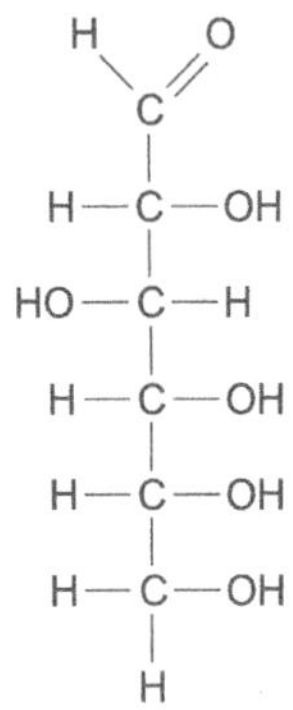

D-glucose
(dextrose)

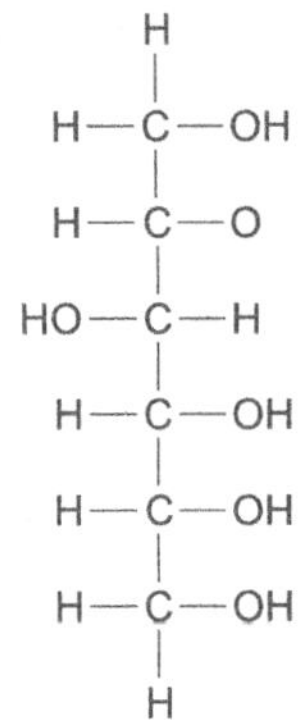

D-fructose
(levulose)

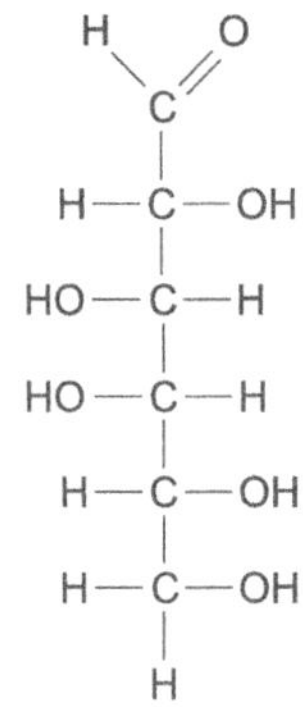

D-galactose

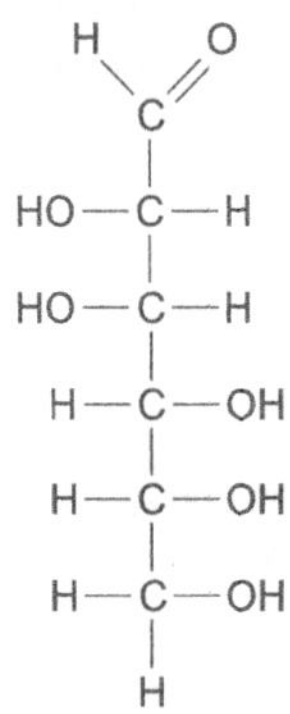

D-mannose

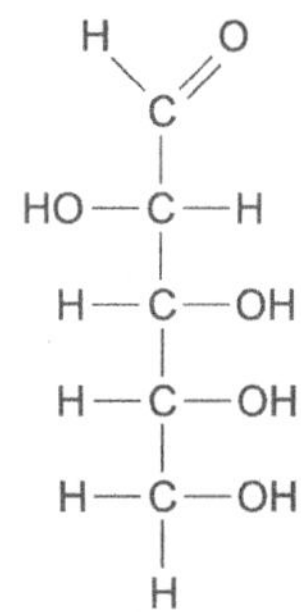

D-arabinose

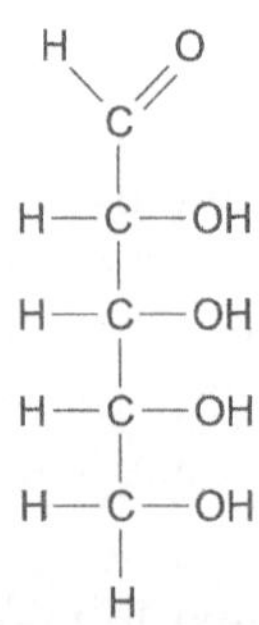

D-ribose

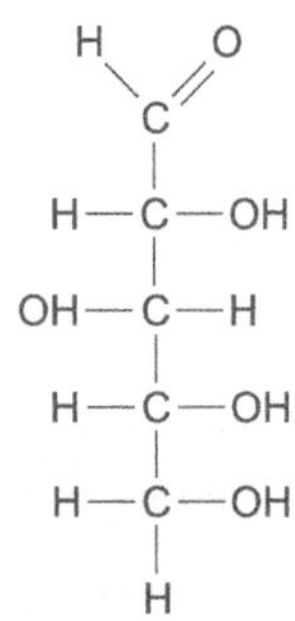

D-xylose

(De Bey, p. 16–17)

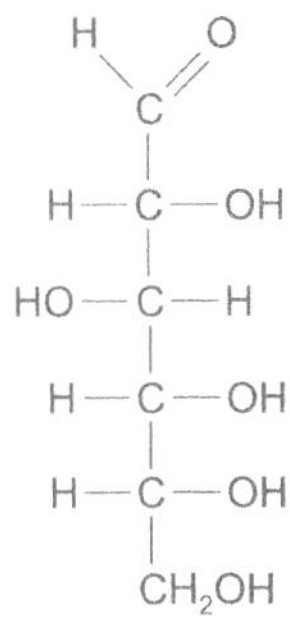

D-glucose

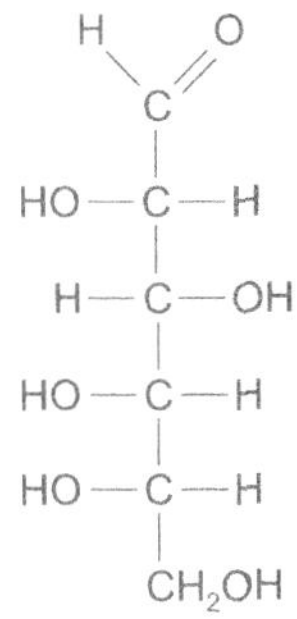

L-glucose

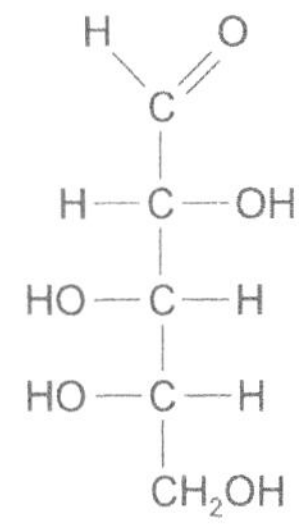

L-arabinose

The name *glucose* which is derived from a Greek word for sweet wine, reflects the fact that this sugar was originally isolated from grapes. The alternative name, *dextrose*, resulted from the observation that the common form of this sugar rotates polarized light to the right. The name *fructose* is derived from the Latin word for fruit, *fructus*; and the other common name for this sugar, *levulose*, indicates its levorotatory property. *Galactose* (a component of the disaccharide lactose) is found in milk and owes its name to the Greek prefix *galact*, meaning milk. *Lactose* is derived from the Latin prefix *lact*, meaning milk. The name *mannose* reflects the fact that its related hexahydric alcohol, *mannitol*, was found in the dried juice of the manna tree. *Arabinose* is produced from gum arabic, an exudate from certain types of acacia trees. Apparently the name *ribose* was coined by selection and rearrangement of the letters in arabinose, which is appropriate, since the two are isomers. *Xylose* is found in woody parts of plants; its name is derived from the Greek word, *xylon*, meaning wood. The name *sucrose* is derived from the French word for sugar, *sucre*, and maltose is named for its source, malt, which is formed in germinating cereals. The word *saccharide* is derived from words for sugar that are similar in Arabic, medieval Latin, and Greek.

TABLE 2.8. Solubility and Relative Sweetness of Sugars

Sugars	Solubility (g/100 ml of H_2O)	Sweetness
Sucrose	179	100
Glucose	83	74
Galactose	10	32
Fructose	Very soluble	173
Xylose	117	40
Lactose	17	16
Maltose	108	33
Saccharin*		

*Saccharin, which is not a sugar, is included for comparison.

(De Bey, p. 23)

Cellulose

Cellulose is the most abundant organic compound. It has been estimated that 50% of all carbon in the vegetation on the earth is tied up in cellulose molecules. Like starch and glycogen, cellulose is made exclusively of glucose units, but unlike them, the glucose units are β-linked. This β-linked polymer of glucose is extremely tough and indigestible, and for this reason is an effective structural material both within plants, and, in turn, for man who uses many materials which contain cellulose. The woody part of plants contain large amounts of cellulose. Cotton is 98% cellulose. Cellulose containing materials such as lumber, cotton, and rayon (which is a chemically modified cellulose) are important commercially.

Starch

Various kinds of starch are found in the cereal grains and in many kinds of tubers. The starch from each type of plant has a characteristic form of granule which could be identified microscopically. However, all forms of starch are quite similar chemically; all are composed totally of glucose. Most sources of starch yield two fractions, amylose and amylopectin. These two forms are very similar and separation is difficult.

The starches in the form of rice, potatoes, and wheat or cereal products supply about 70% of the world's food.

TABLE 2.9

Starch	Amylose Percent	Amylopectin
Potato	19-20	78-81
Wheat	24	76
Maize	21-23	77-79
Rye	17	80

(Kretovich, p. 75)

Glycogen

Glycogen is found in the liver and muscles of animals, and it is often called "animal starch". Glycogen is similar to the amylopectin fraction of starch in that it is made totally of glucose and is highly branched. Since animals do not synthesize starch, glycogen represents the only form in which they can store excess glucose. The total amount of glycogen in an animal is small; about 1% of the muscle and between 1.5 and 4% of the liver is glycogen. However, this small amount is extremely important in maintaining the correct level of glucose in the blood and in supplying the muscles with a source of energy.

Catalysis

A catalyst is a substance which accelerates a chemical reaction without itself appearing as a product of the reaction. Catalysts can be shown to accelerate both the forward and reverse components of a reaction. That is, they accelerate the attainment of equilibrium although they do not alter its position. In general terms catalysts of all types produce their effect by lowering the free energy of activation of the reaction which they catalyse. Virtually all cellular reactions are catalysed by enzymes.

Enzymes

Metabolic Enzymes : Enzymes are organic catalysts; that is, they are organic molecules that accelerate the rate of a biochemical reaction at certain temperatures without being used up in

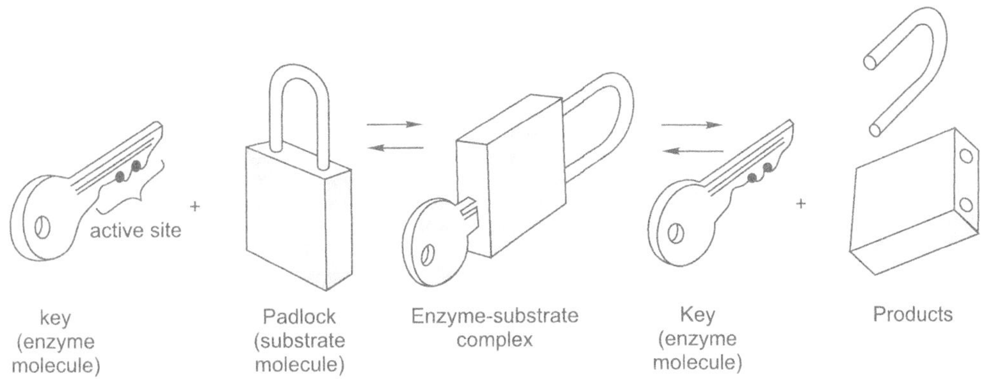

Fig. 2.15. The lock-and-key model for picturing the interaction of substrate and enzyme.

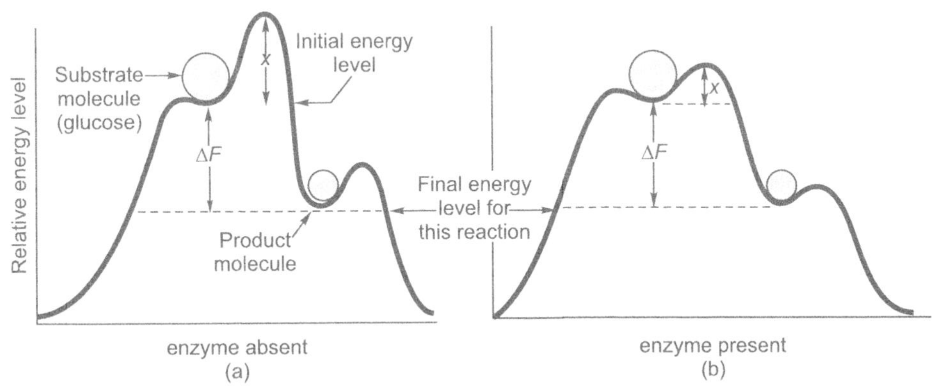

Fig. 2.16. The effect of an enzyme on the activation energy. The amount of activation energy is represented by X and the change in free energy by ΔF. The uncatalyzed reaction (on the left) requires a greater energy of activation. In reactions that proceed in a series of stepwise reactions, each reaction has its energy of activation. (De Bey, p. 92)

those reactions. These reactions would occur at the same temperature without enzymes but at a much slower rate. A *substrate* is a compound on which an enzyme exerts its effect. The enzyme must fit the combining site of the substrate, as a key fits a lock. The substrate (s) is (are) bound non-convalently to a region on the surface of the enzyme known as the active site or centre, where the catalytic reactions occur. Usually, an enzyme is specific, that is, it works on only one type of substrate, but occasionally it can attach to different substrates because the shapes of the combining sites on the substrate molecules are similar. All enzymes are proteins; some, however, require the presence of a small non-protein molecule before they can exert their catalytic effect. These small molecules were formerly classified as co-enzymes or as prosthetic groups dependent on whether they could be removed from their *apo-enzyme* by dialysis (the enzyme to which a co-enzyme or prosthetic group binds is known as the apo-enzyme, while a co-enzyme-apo-enzyme or prosthetic group-apo-enzyme complex is referred to as a holo-enzyme). However, all of the non-protein materials which are essential for the activity of various enzymes are designated as co-enzymes or co-factors. Co-enzymes act in one of two ways. Some, such as nicotinamide adenine dinucleotide (NAD^+), act as obligatory substrates in the reaction catalysed by their apo-enzyme. For example, in reactions catalysed by NAD^+ dependent dehydrogenases, the co-enzyme is the oxidizing agent, as in the reactions catalysed by FAD.

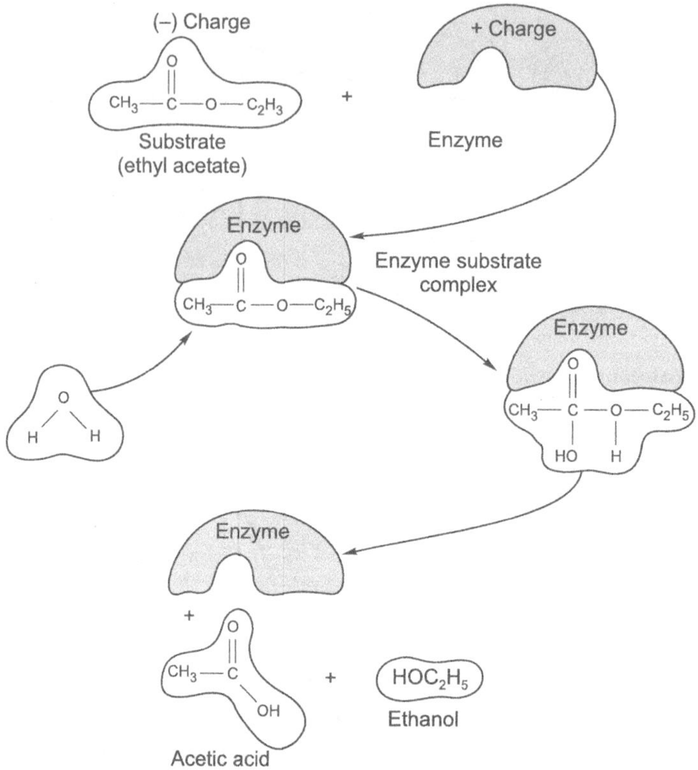

Fig. 2.17. A diagrammatic representation of an enzymatic reaction in which the enzyme makes the substrate more vulnerable to attack by the reacting molecule. In this reaction, the substrate (ethyl acetate) is bound to the enzyme by complementary shapes and by opposite charges. This binding facilitates the hydrolysis of the ester bond by a molecule of water.

(De Bey, p. 93)

$$\underset{\substack{\text{Ethanol} \\ \text{(ethyl alcohol)}}}{CH_3CH_2OH} + NAD^+ \xrightleftharpoons{ADH} \underset{\substack{\text{Ethanol} \\ \text{(acetaldehyde)}}}{CH_3CH{=}O} + NADH + H^+$$

ADH = alcohol dehydrogenase

$$\underset{\text{Succinate}}{^-O_2CCH_2CO_2^-} + Enz - FAD \rightleftharpoons {^-O_2CCH} = CHCO_2^- + \text{Fumarate } Enz - FAD\ H_2$$

Enz – FAD = Succinate dehydrogenase

The two oxidising co-enzymes differ in that one, NAD^+, is diffusible while the other, FAD, is not. Other co-enzymes, such as pyridoxal phosphate and thiamine pyrophosphate, act by enhancing the reactivity of the substrate of their apo-enzyme.

The three-dimensional shape of the protein enables it to attach to one or more substrates to accelerate a very slow reaction, without causing the enzyme to change in the process. The enzyme continues to move from substrate molecule to substrate molecule at a rate of several hundred each second, producing a supply of the end product, as long as this particular end product is needed by the cell. However enzymes do not last indefinitely; they finally degenerate and lose their activity. Therefore, the cell must synthesize and replace these important proteins.

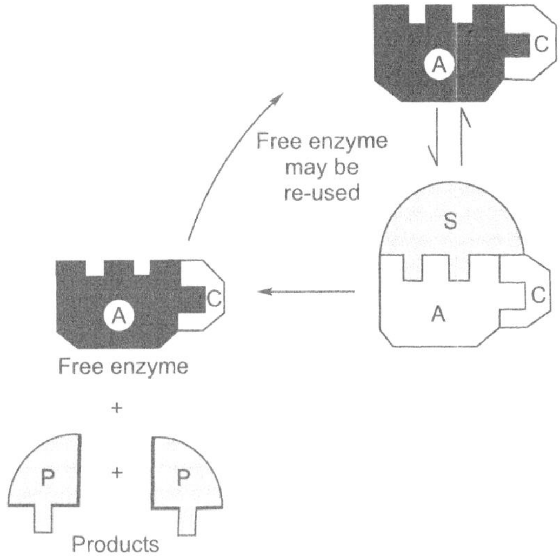

Fig. 2.18. A diagram showing the involvement of a coenzyme in an enzymatic reaction. The protein part of an enzyme (apoenzyme) combines with its coenzyme. The substrate molecule then attaches to the completed enzyme and is converted into the products of the reaction with the regeneration of the free enzyme. Not all enzymes require coenzymes.

(De Bey, p. 91)

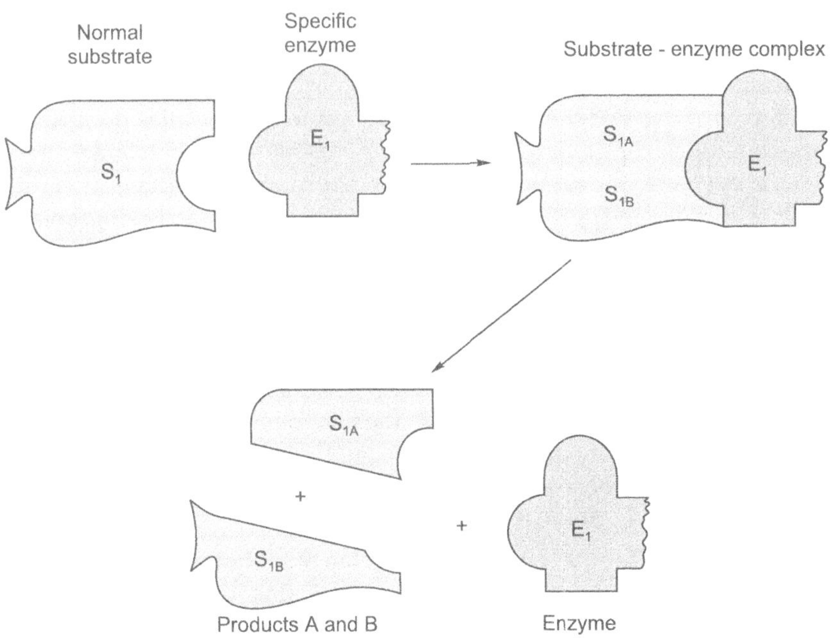

Fig. 2.19. Action of a specific enzyme breaking a substrate molecule.

Because there are hundreds of metabolic reactions continually occurring in the cell, there must be thousands of enzymes available to control and direct the essential metabolic pathways. At any particular time, all of the required enzymes need not be present; this situation is controlled by the genes on the chromosomes and the needs of the cell, which are determined by the internal and external environment. Those enzymes found inside the cell are called endoenzymes, and those secreted outside the cell to perform extracellular functions are exoenzymes. The digestive enzymes within phagocytes are good examples of endoenzymes. Exoenzymes are the enzymes secreted by saprophytes, which are then able to digest materials outside the organism. These nutrients digested outside are then absorbed into the organism.

Some enzymes require cofactors to activate them to perform their intended functions because they normally exist in the cell in an inactive state. These co-factors are usually mineral ions such as magnesium, calcium, or iron. Other enzymes function only in the presence of a coenzyme that acts as a carrier of small chemical groups (such as H_2) that are removed from the substrate. These coenzymes are small organic vitamin-type molecules such as flavin-adenine dinucleotide (FAD) and nicotinamide-adenine dinucleotide (NAD). The activity of these coenzymes is easily recognized in the citric acid cycle for the production of energy from glucose. Coenzymes, like enzymes, do not have to be present in large amounts because they are recycled through many reactions. However, the lack of certain vitamins from which the coenzymes are synthesized will halt all those reactions involving that particular coenzyme-enzyme complex.

Some enzymes may also be called hydrolases because they break down macromolecules by the addition of water in a process called *hydrolysis.* These hydrolytic processes enable saprophytes to break apart such complex materials as leather, wax, cork, wood, rubber, hair, some plastics, and even mechanical equipment. Some of the enzymes involved in the formation of large polymers, DNA, and RNA, are called polymerases. These polymerases are active each time the DNA of a cell is replicated and during the synthesis of RNA molecules.

Inhibition of Enzymes

Many factors affect the activity of enzymes. Any physical or chemical change may diminish or completely stop enzyme activity, because these protein molecules function properly only under optimum conditions. Optimum conditions for enzyme activity include a relatively limited range of pH (acidity) and temperature, and the appropriate concentration of enzyme and substrate. Extremes in heat and acidity can denature (or alter) enzymes by breaking the bonds responsible for their three dimensional shape, resulting in the loss of enzymatic activity. This explains why bacteria grow best at certain temperatures and pH, characteristic for each species of organism.

Although the mineral ions, calcium, magnesium, and iron, enhance the activity of enzymes by serving as cofactors, other heavy metal ions such as lead, zinc, mercury, and arsenic usually act as poisons to the cell. These toxic ions inhibit enzyme activity by replacing the cofactors, or sometimes replacing only hydrogen, at the combining site of the enzyme, thus inhibiting normal metabolic processes. Some disinfectants containing mineral ions are effective in inhibiting the growth of bacteria by this means.

Sometimes, a similar substrate can be used as an inhibitor to deliberately interfere with a particular metabolic pathway. It binds with the enzyme; thus the end product is not produced. For example, a chemotherapeutic agent, such as a sulfonamide drug, may bind with certain enzymes to prevent essential metabolites from being formed and thereby inhibits the growth of a pathogen.

Classification and Nomenclature of Enzymes

Prior to 1961 no systematic classification of enzymes was available and their nomenclature was confused. Although many enzymes had been named in a manner which was, at least in part, indicative of their catalytic function many had been named on the basis of their initial source. For example, while the name lactate dehydrogenase describes the overall catalytic function of the enzyme in question (although it does not indicate the nature of the electron acceptor involved) the name papain indicates only that this particular enzyme (a protease) was first isolated from papaya latex. However, in 1961 The International Union of Biochemistry proposed a system of classification and nomenclature which allows any enzyme to be identical unambiguously. The class suffix is *–ase*. Each enzyme is allocated to one of six major groups, corresponding to the type of reactions which it catalyses, and is numbered accordingly. These groups are as follows :

1. *Oxidoreductases* : These enzymes catalyse the transfer of H, O atoms or electrons from one substrate to another, *e.g.*, alcohol dehydrogenase :

$$RCH_2OH + NAD^+ \rightleftharpoons RCHO + NADH + H+$$

2. *Transferases* : These are enzymes that catalyse the transfer of a specific group from one substrate to another *e.g.*, kinases, such as glucokinase, catalyse the transfer of a phosphoryl group, $-PO_3^-$ or $-PO_2^-$, from ATP to an acceptor :

$$\text{Glucose} + ATP^{4-} \rightleftharpoons \text{Glucose} - 6 - \text{Phosphate}^{2-} + ADP^{3-} + H^+$$

3. *Hydrolases* : These are the hydrolytic enzymes, *e.g.*, esterases :

$$R^1COOR^2 + H_2O \rightleftharpoons R^1COOH + R^2OH$$

4. *Lyases* : These are enzymes that catalyse the cleavage of C–C, C–O, C–N, or C–S bonds other than by hydrolysis, *e.g.*, decarboxylases :

$$R^1CHR^2\ COOH \rightleftharpoons R^1CH_2R^2 + CO_2$$

5. *Isomerases* : These are enzymes that catalyse an intramolecular rearrangement, *e.g.*, racemases which catalyse reactions of the type :

$$\text{L-amino acid} \rightleftharpoons \text{D-amino acid}$$

6. *Ligases* : These are enzymes that catalyse the formation of C–O, C–S, C–N, or C–C bonds with the concomitant breakdown of ATP (or another nucleoside triphosphate), *e.g.*, the amino acid activating enzymes :

$$\text{Amino acid} + ATP^{4-} + \text{tRNA} \rightleftharpoons \text{aminoayl} - \text{tRNA} + \text{Pyrophosphate}^{4-}$$

Each group is then subdivided according to the nature of the principle substrate involved. For example the oxidoreductases are divided into 13 groups

1.1 Enzymes acting on CHOH groups as the donor
1.2 Enzymes acting on aldehyde or keto groups as the donor
1.3 Enzymes acting on CH—CH groups as the donor
1.4 Enzymes acting on CH—NH_2 groups as the donor
1.5 Enzymes acting on C—NH groups as the donor
1.6 Enzymes acting on NADH or NADPH as the donor
1.7 Enzymes acting on other nitrogenous compounds as the donor
1.8 Enzymes acting on sulphur groups as the donor
1.9 Enzymes acting on haem groups as the donor
1.10 Enzymes acting on diphenols and other related substances as the door
1.11 Enzymes acting on H_2O_2 as an acceptor

1.12 Enzymes using H_2 as a reductant

1.13 Other enzymes using O_2 as an oxidant

Further subdivisions are then made according to the nature of the other substrate. For example :

1.1.1 With NAD^+ or $NADP^+$ as acceptor

1.1.2 With a cytochrome as acceptor

1.1.3 With O_2 as acceptor

1.1.4 With other acceptors

Individual enzymes are then listed numerically in the appropriate class. For example :

E.C.1.1.1.1 has the systematic name: alcohol: NAD oxidoreductase, and the common name: alcohol dehydrogenase.

E.C.1.4.3.3. has the systematic name: D-amino acid oxidoreductase (deaminating), and the common name: D-amino acid oxidase.

TABLE 2.10. Types of Enzymes

S. No.	Enzymes	Reaction catalyzed
1.	Oxidoreductases	Oxidation and reduction
2.	Transferases	Transfer of a chemical grouping from one compound to another
3.	Hydrolases	Hydrolytic reactions
4.	Lyases	Nonhydrolytic addition or removal of groups
5.	Isomerases	Conversion of a substance into one of its isomers
6.	Ligases	Synthesis or, more specifically, the formation of various types of bonds.

(De Bey, p. 93)

Properties of Enzymes

1. All enzymes are proteins. The great majority of those which have been investigated are soluble, globular, proteins; others, however, are bound to membranes and are difficult to bring into solution without loss of activity. Thus, enzymes are de-natured by moderate temperatures, extremes of pH, by detergents and by a number of reagents such as urea and by non-polar solvents such as ethanol and acetone.
2. Enzymes are extremely efficient catalysts. It has been estimated, for example, that the proteolytic enzymes of the gut accelerate the cleavage of peptide bonds by factor of 10^5, reducing the time necessary for digestion of an average meal from 50 years to a few hours.
3. Most enzymes exhibit a high degree of both structural and stereochemical specificity. It appears that a few enzymes exhibit absolute structural specificity. For example, asparaginase catalyses the hydrolysis of only L-asparagine; the hydrolysis of L-glutamine, which differs from L-asparagine by only a single methylene group, is not accelerated.

Similarly fumarase catalyses only the reversible interconversion of fumaric acid and L-malic acid.

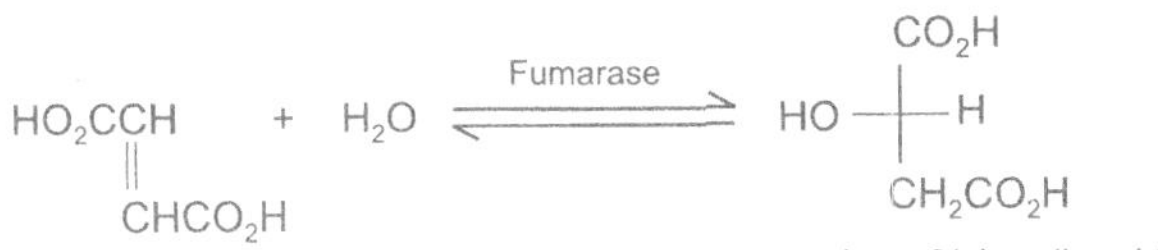

(Fumaric acid) L- or 2(s)-malic acid

Most enzymes, however, will catalyse the same reaction of a small group of compounds with homologous, or closely similar, structures.

4. Many enzymes require the presence of a non-protein co-enzyme in order to exert their catalytic effect.

Enzyme Inhibitors

Many compounds are known which decrease the rate of a particular enzyme catalysed reaction. Such compounds are known as enzyme inhibitors.

Enzyme inhibitors can be divided into two types, that is, those which act in an irreversible manner and those which act in a reversible manner.

Irreversible Inhibition of Enzymes

(*i*) *Active-site directed irreversible inhibitors* : A number of compounds are known which, when they react with certain enzymes, form a covalent bond to the active site and prevent the enzyme from functioning; such compounds are known as active site directed inhibitors.

The rate of inhibition is relatively slow and the process is not reversed by removing excess inhibitor by, for example, dialysis or chromatography.

Penicillins, as exemplified by penicillin G, is one of the irreversible enzyme inhibitors.

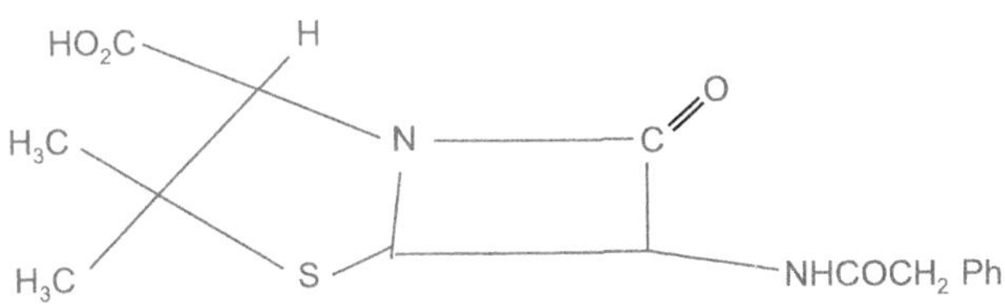

Penicillin G

The main morphological effect of penicillin G on penicillin sensitive bacterial cells is to cause them to undergo lysis.

(*ii*) *"Suicide" or "enzyme-activated irreversible inhibitors* : This type of inhibitor is chemically inert but after being accepted by the enzyme as a substrate, is converted via the catalytic reaction, to a reactive species which then inhibits the enzyme. A good example of an enzyme-activated irreversible inhibitor of a PLP (pyridoxal phosphate) dependent enzyme is D-3-fluoro-alanine.

$$FCH_2\ CH\ (\overset{+}{N}H_3)\ CO_2^-$$

D-3-fluoro-alanine

This compound shows antibacterial activity without host toxicity by blocking the biosynthesis of D-alanine, an essential constituent of bacterial cell walls.

Co-enzymes

Co-enzymes are small non-protein moieties, some of which act as obligatory second substrates of their apo-enzyme while others act as activators of the substrate of their apo-enzyme.

Adenosine triphosphate (ATP) acts as co-enzyme to a very large number of enzymes. The main roles of ATP as co-enzyme are as follows :

1. As a mediator of condensation reactions
2. As a mediator of the conversion of chemical energy to electrical, osmotic or mechanical energy, and
3. As a mediator of the elimination of water.

The flavin co-enzymes, flavin mononucleotide (FMN) and flavin adenine dinucleotide (FAD) are both derivatives of vitamin B_2 (riboflavin).

Pyridoxal phosphate (PLP) and thiamine pyrophosphate (TPP) are co-enzyme which modify the reactivity of the substrates of their apo-enzymes.

Co-enzyme A is the co-enzyme that is concerned with the reactions of acyl groups, particularly acetyl groups, in living cell. The functional part of the molecule is the thiol group and for this reason the co-enzyme is usually represented as CoASH.

Lipids

Lipids are a class of biochemical compounds consisting mainly of fats and oils. Most of them are insoluble in water but soluble in fat solvents such as ether, chloroform, and benzene. Lipids are essential constituents of almost all living cells. They may be classified into the following categories :

1. *Simple Lipids* (contain C, H, O)
   (*a*) – Fats and oils (butter, vegetable oils)
   (*b*) – Waxes (beeswax, lanolin)
2. *Compound Lipids* (Contain C, H, O, N and P)
   (*a*) – Phospholipids (in cell membranes)
   (*b*) – Glycolipids (in nerve cells)
3. *Derived Lipids* (Contain C, H, O)
   (*a*) – Steroids (sex hormones and cholesterol)
   (*b*) – Fat-soluble vitamins A, D, E, K.

The simplest lipids (fats and oils) consist of one molecule of glycerol and three fatty acid molecules joined together by the removal of three molecules of water (dehydration synthesis), which produces a triglyceride fat or oil.

If the hydrocarbon side chains have no double bonds between carbons, it is called a *saturated fatty acid* because every carbon atom is saturated with hydrogen. When there are double bonds between the carbons (C = C), it is an *unsaturated fatty acid.* Fats are liquids or low melting solids at room temperature, depending on the relative composition of the fatty acids. Unsaturated fatty acids with short carbon chains are usually liquids.

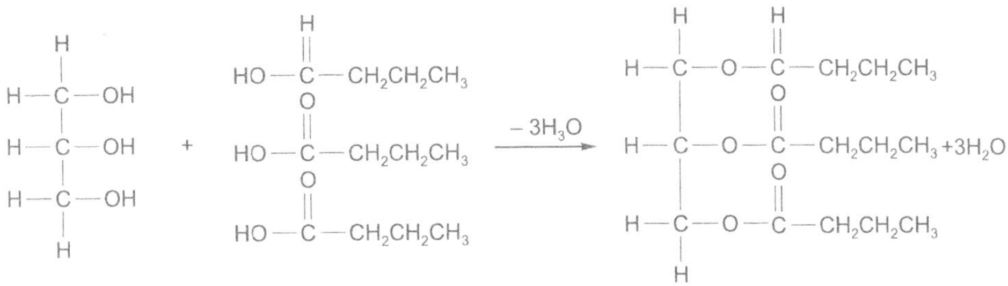

Glycerol + 3 butyric acids $-3H_2O$ ⟶ Tributyrin (a triglyceric fat)

The synthesis of a fat

A nutritious human diet should include at least the two essential fatty acids that the body cannot synthesize. These are linoleic acid ($C_{17}H_{31}COOH$), from fish liver oils and vegetable oils, and arachidonic acid ($C_{19}H_{31}COOH$) from egg yolk, liver, kidney tissues, and fish liver oils. However, different fatty acids are produced by and are necessary for the growth of bacteria and other microorganisms, depending on the species of the organism.

Saponification is the hydrolysis of a fat into glycerol (glycerine) and its three fatty acids. When NaOH or KOH (strong alkalies) are used to hydrolyze fats, the results are sodium or potassium salts of the fatty acids, which are soaps. Detergents, bile salts, and fat-digestive enzymes also saponify fats.

Waxes

Waxes are chemically different from fats because they are long chain or complex alcohols, other than glycerol, with fatty acid attached. Waxes are semisolid substances that serve as a protective coating on the surfaces of leaves, stems, fruits, insects, and some bacteria.

Compound Lipids

Compound lipids include the phospholipids of cell membranes and the glycolipids of nerve and brain cells. Even when an organism is starving, the amount of these nonfat lipids in the cells does not vary.

Derived Lipids

Derived lipids are classified as lipids because they are also soluble in fat solvents. These complex molecules of four interlocking carbon rings are called steroids. Examples of steroids include cholesterol, male and female hormones, adrenal cortex hormones, and vitamin D. Vitamin A, E, and K are not steroids but are fat-soluble vitamins.

Cholesterol is a necessary body metabolite. It is found in most eucaryotic membranes and in nervous tissues and aids in fatty acid adsorption. It is also essential for the production of sex hormones, vitamin D, and adrenal cortex hormones. However, an excess of cholesterol may be deposited in either the arteries, causing heart disease, or in the gallbladder, becoming gallstones.

Vitamin D helps maintain the calcium and phosphorus balance in the body. Good sources of vitamin D are fish oils and fortified foods. Vitamin A can be obtained from liver, eggs, butter, cheese, carrots, and green leafy vegetables; a lack of this vitamin can lead to night blindness or hardening of the mucous membranes.

Vitamin E (from plant oils, leafy vegetables, and eggs) reduces sterility in animals and the wasting effects of aging. Vitamin K, which is synthesized in the body, is necessary for normal blood clotting. Some colon bacteria (specially *Escherichia coli*) also secrete vitamin K, which is absorbed into the blood.

Proteins

Proteins are among the most essential chemicals in all living cells. Some are the structural components of membranes, cells, and tissues, whereas others are enzymes and hormones that chemically control the metabolic balance within both the cell and the individual organism. All proteins are polymers of amino acids; however, they vary widely in the number of amino acids present and in the sequence of amino acids, as well as in their size, configuration, and functions.

Proteins are synthesized by plants, algae, and bacteria from carbon dioxide, water, nitrates, and sulphates. Humans can synthesize proteins, but they must ingest eight essential amino acids to synthesize the other amino acids necessary in the building of proteins. Nonphotosynthetic microorganisms also must absorb their essential amino acids that are not synthesized or stored in living cells. All the amino acids (ingested and synthesized) must be present during the protein synthesis process in the cells. Any excess can be used to produce energy or eliminated with the soluble wastes.

Amino Acid Structure

There are 20 different amino acids. Each is composed of carbon, hydrogen, oxygen, and nitrogen; three of the amino acids also have sulphur atoms in the molecule. The general formula for amino acids is shown below.

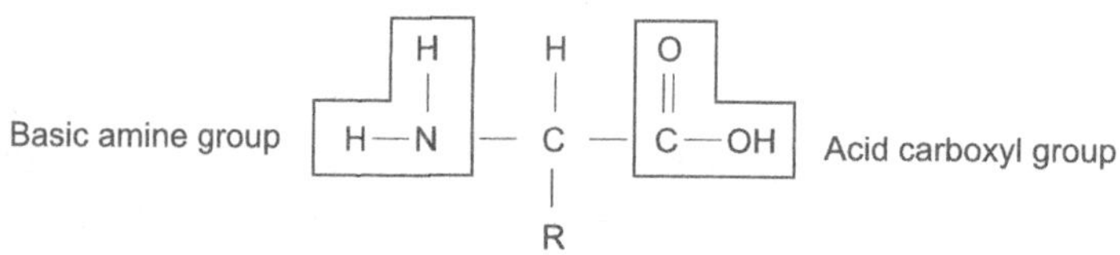

Fig. 2.20. The basic structure of amino acids.

TABLE 2.11

Elements	Percent
Carbon	51.0
Nitrogen	16.8-18.4
Hydrogen	6.9
Oxygen	21.7-23.0
Sulphur	0.7-1.3

Elementary composition for the proteins of wheat

(Kretovich, p. 1)

In Fig. 2.20, the "R" group represents any of the 20 groups that may be substituted into position to build the various amino acids. For instance, –H in place of the –R represents glycine and $-CH_3$ in that position results in a structural formula representing alanine.

The thousands of different proteins in the body are composed of a great variety of amino acids in various arrangements and amounts. The number of proteins that can be synthesized is virtually unlimited. They are not limited by the number of different amino acids, just as the number of words in the written language is not limited by the number of letters in the alphabet. The actual number of proteins produced by an organism and the amino acid sequence of those proteins are determined by the particular genes present on the chromosomes.

Protein Structure

When water is removed by dehydration synthesis, the amino acids are linked together by a peptide bond as shown below.

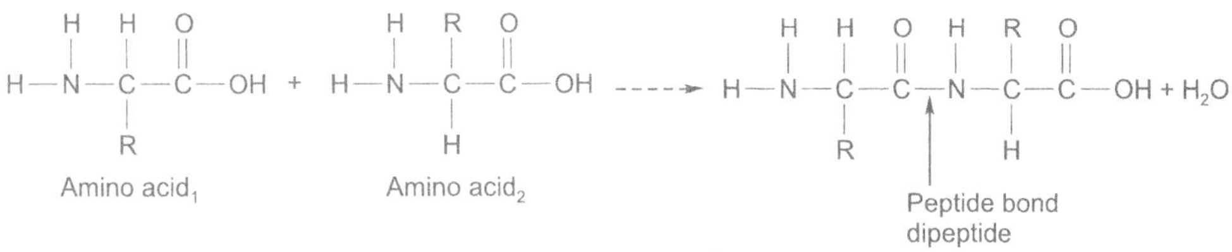

The formation of a dipeptide
(R = any amino acid side - chain group)

A *dipeptide* is formed by bonding two amino acids; three amino acids form a *tripeptide*. A long chain or polymer of amino acids is referred to as a *polypeptide*. Polypeptides are said to have *primary protein structure*, a sequence of amino acids in a chain. Some fibrous proteins are made of strands of polypeptides bonded to each by hydrogen bonds and wound together into a helix. Collagen, found in connective tissue, scars, and tendons is another fibrous protein, consisting of three polypeptides in a triple helix, forming a very strong structure.

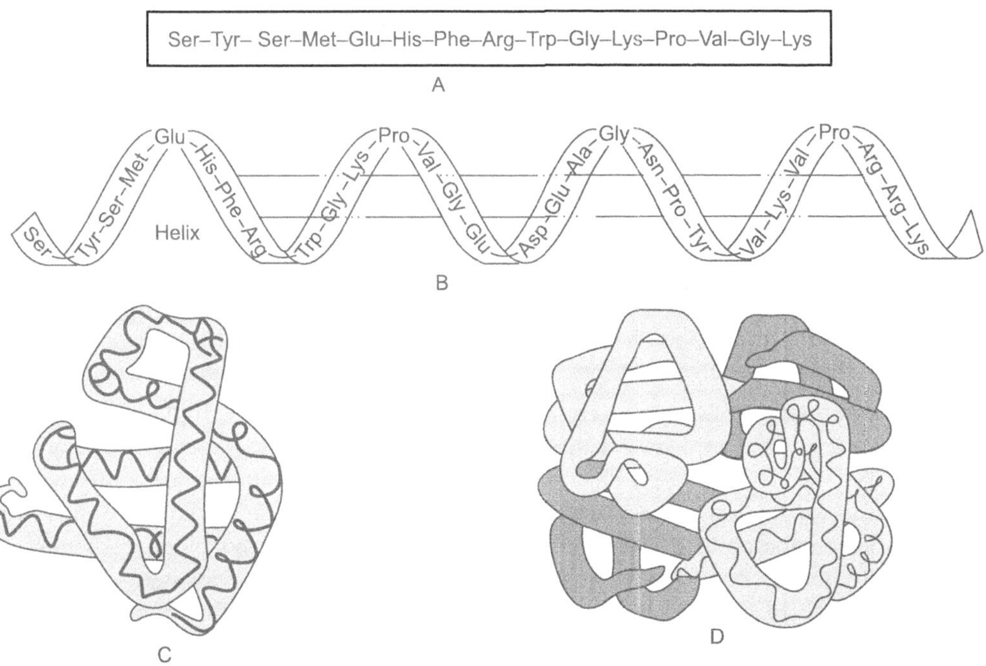

Fig. 2.21. Basic protein structure. (A) Primary sequence of amino acids. (B) Secondary helix. (C) Tertiary globular structure. (D) Quarternary structure with four polypeptide chains.

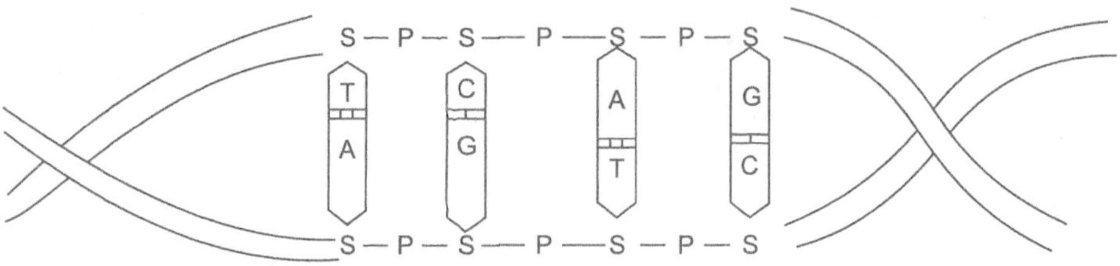

Fig. 2.22. The DNA double helix.

Enzymes : Enzymes are protein molecules produced by living cells as "ordered" by the DNA or genes on the chromosomes. Enzymes are essential as catalysts in the biochemical reactions of metabolism; in other words, they speed up these reactions. Almost every reaction in the cell requires the presence of a specific enzyme. All enzymes are proteins; many consist of protein only, whereas others require a nonprotein cofactor, such as metal ions (Ca^{2+}, Fe^{2+}, Mg^{2+}, Cu^{2+}) or organic compounds such as vitamin C. Although enzymes influence the direction of the reaction and increase its rate of reaction, they do not provide the energy needed to activate the reaction.

Enzymes are usually named by adding the ending *–ase* to the word indicating the compound or types of compounds on which an enzyme exerts its effect. For example, proteases, carbohydrases, and lipases are enzymes specific for proteins, carbohydrates, and lipids, respectively.

Poisons and toxins usually cause damage to the body by interfering with the action of certain necessary enzymes. For example, cyanide poison binds to the iron and copper ions in the cytochrome systems of the mitochondria of cells. As a result, the cells cannot use oxygen to synthesize ATP, which is essential for energy production, and they soon die.

Proteins, including enzymes, may be denatured by heat or certain chemicals. In a denatured protein, the bonds that hold the molecule in a tertiary structure are broken. With the bonds broken, the protein is no longer functional. Most polypeptide chains naturally twist into spirals or helices as a result of the charged side chains protruding from the carbon–nitrogen backbone of the molecule. This helical configuration is referred to as *secondary protein structure* and is found in fibrous proteins such as the keratin fibers of hair, nails, and skin, and the microtubules and microfilaments of cells.

Because a long coil can become entwined by folding back on itself, a polypeptide helix may become globular. In some areas the helix is retained, but other areas curve randomly. This globular, *tertiary protein structure* is stabilized, not only by hydrogen bonding but also by disulfide bond cross-links between two sulphur groups (S–S). This three—dimensional configuration is characteristic of enzymes, which work by fitting on and into specific molecules.

When two or more polypeptide chains are bonded together by hydrogen and disulfide bonds, the resulting structure is referred to as *quaternary* protein structure. For instance, haemoglobin consists of four globular myoglobins. The size, shape, and configuration of a protein is specific for the function it must perform. If the amino acid sequence, and thus, the configuration of haemoglobin in red blood cells is not perfect, the red blood cells may become distorted and assume a sickle shape (as in sickle cell anemia). In this state they are unable to carry oxygen, which is necessary for cellular metabolism. Myoglobin, the oxygen—binding protein found in skeletal muscles, was the first protein to have its primary, secondary, and tertiary structure defined.

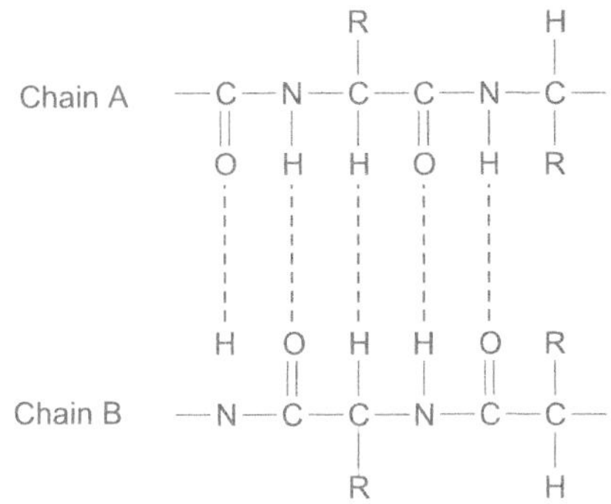

Hydrogen bonds between amino acid chains

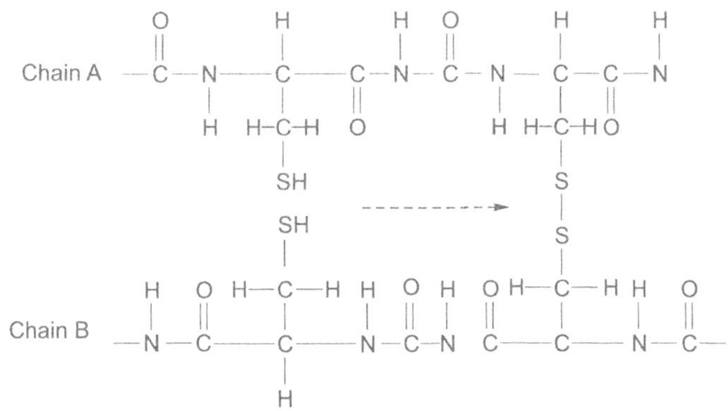

Formation of disulphide bonds between protein chains
(De Bey, p. 42)

TABLE 2.12. Molecular Weights of Proteins (Determined from the Sedimentation Rate in an Ultracentrifuge)

Protein	Molecular weight
Ribonuclease	12,700
Lactalbumin (from milk)	17,400
Myoglobin (from mouse)	16,900
Hordein (from barley)	27,500
Pepsin	35,500
Albumin (from egg)	40,000
Haemoglobin (from human blood)	63,000
Diphtheria toxin	74,000
Edestin (from hemp seed)	310,000
Urease (from soybeans)	480,000

(Kretovich, p. 26)

Importance of Proteins

Proteins play many vital roles in the living organism. Besides their participation in enzymatic catalysis, as oxygen carriers, and as structural materials, some other important functions of proteins are discussed on next page.

Since animals synthesize their body proteins from amino acids, and especially since they can synthesize only half of these amino acids, it is extremely important that their diets include adequate amounts of the ones that are not made in their own bodies. The major source of these essential amino acids is the proteins of the diet.

In addition to haemoglobin, the blood contains many other proteins. These are albumins which serve to maintain sufficient osmotic pressure to help water and other nutrients pass freely through the walls of the capillaries. The particular kind of globulin that has been called gamma globulin constitutes one of our major supplies of antibodies for resisting some diseases. The proteins fibrinogen and prothrombin are essential for the formation of the clump of red blood cells called a clot. Clotting is useful when it prevents excessive loss of blood, but clots can also form in the blood vessels of the heart or brain, causing a type of heart attack or stroke.

Many of the hormones which serve as important regulators of the processes of the body are proteins. Most of the hormones of the pituitary gland are protein-containing. Two of them, vasopressin and oxytocin, are octapeptides; others are either glycoproteins (compounds containing both protein and carbohydrate) or pure proteins. The widely used hormone, insulin, contains 51 amino acids and may be classed as either a large polypeptide or a very small protein.

The importance of proteins in our lives may be traced from the time a sperm cell (made up of nucleic acids) and proteins unites with an ovum. In animals, the fertilized egg develops through stages of differentiation which are largely influenced by a succession of enzymes, to the birth of the individual. Bird's eggs, which provide nutrition for the embryo birds, are especially rich in proteins. The gradual growth of a person is due to the increase in bone and muscle under the control of the proteinaceous hormones and with the help of enzymes. The production of new sperm and ova under the influence of other protein containing hormones leads to the perpetuation of the species. The continued life and health of the individual depends on sufficient nutrition by proteins (along with other foods) and on his ability to manufacture sufficient enzymes to catalyze the essential reactions and sufficient antibodies to resist infections. The failure of any of these protein-influenced systems or, in some cases, the overactivity of some of these processes leads to the death of the individual.

While the importance of other classes of compounds cannot be forgotten, the statement that protein is without doubt the most important of the known components of living matter, and it would appear that, without it, life would not be possible, is still true.

Nucleic Acids

Nucleic acids—deoxyribonucleic acid (DNA) and ribonucleic acid (RNA) are one of the major groups of macromolecules in living cells. DNA is the macromolecule that makes up the major portion of the chromosomes. Its important task is to carry the genetic information for cell of an organism. The information from the DNA must be carried to the rest of the cell for the cell to function properly; this is accomplished by RNA. Thus, RNA is found not only in the nucleus but in the cytoplasm as well.

Function

Nucleic acids have two very important functions. First, the nucleic acids determine precisely all the proteins that are synthesized, and those protein enzymes that are synthesized, control

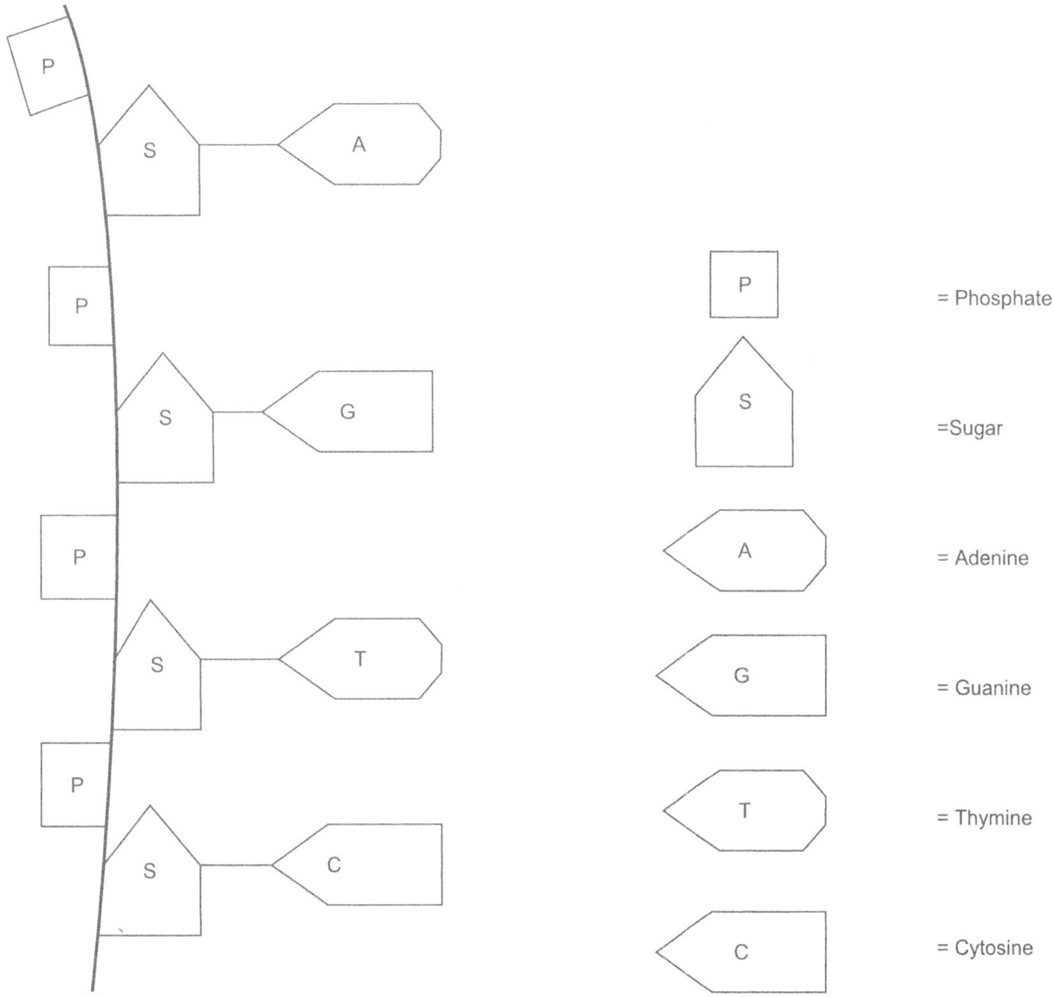

Fig. 2.23. One small section of a nucleic acid polymer. (Burton, p. 97)

the metabolism of the organism. DNA's second role is to act as the genetic "molecular blueprint of life". The genetic information must be passed from one generation to the next, from each parent cell to each daughter cell via the DNA.

Structure

In addition to carbon, hydrogen, oxygen, and nitrogen, DNA and RNA contain the element phosphorus, whereas the proteins contain sulphur. The building blocks of these nucleic acid polymers are the *nucleotides.* These are more complex monomers (single molecular units that can be repeated to form a polymer) than the amino acids, which are the building blocks of proteins. The nucleotides consist of three subunits: a nitrogen containing base, a five carbon sugar, and a phosphate group joined together, as shown Fig. 2.24.

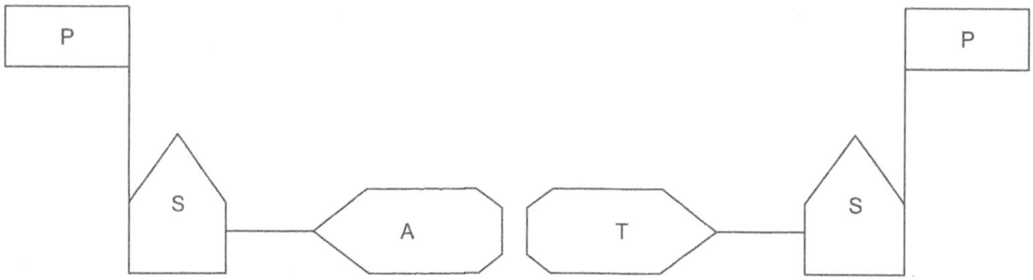

Fig. 2.24. Two nucleotides consisting of a nitrogen base, a five-carbon sugar, and a phosphate group. (Burton, p. 97)

As previously stated, there are two kinds of nucleic acids in cells: DNA and RNA. DNA contains deoxyribose as its five carbon sugar (a pentose), whereas RNA contains the sugar ribose. There are three types of RNA, which are named for the function they serve: messenger RNA (m RNA), ribosomal RNA (r RNA), and transfer RNA (t RNA). The five most important nitrogen containing bases are adenine (A), guanine (G), thymine (T), cytosine (C), and uracil (U). Thymine occurs in DNA and uracil in RNA. The other occur in both DNA and RNA.

The nucleotides join between their sugar and phosphate groups to form very long polymers, 100,000 or more monomers long.

DNA Structure

In 1953, James Watson and Francis Crick proposed a double stranded helical structure for DNA to indicate how it could copy (replicate) itself exactly to pass the genetic information to each daughter cell. For a double stranded DNA to form, the bases must bond together. It was found that adenine always bonds with thymine (with two hydrogen bonds) and guanine with cytosine (with three hydrogen bonds) because of the size and bonding attraction between the molecules. The bonding forces of the double-stranded polymer cause it to assume the shape of a double helix, which is similar to a right handed spiral staircase.

DNA Replication

When a cell is preparing to divide, all of the DNA molecules in the chromosomes of that cell must duplicate, thereby ensuring that the same genetic information is passed to both daughter cells. These process is called DNA replication. It occurs by the separation of the DNA strands and the building of complementary strands by the addition of the correct nucleotides. The duplicated DNA of the chromosomes can then be separated during ordinary cell division so that the same number of chromosomes, the same genes, and the same amount of DNA as in the parent cell are found in each daughter cell (except during meiosis, the reduction division to produce egg and sperm cells).

Protein Synthesis

In a normally functioning living cell, the DNA of the chromosomes controls all the metabolic activities. It accomplishes this by directing the synthesis of the protein enzymes that regulate the chemical reactions occurring in metabolism.

The translation and interpretation of the information carried by the DNA molecule to produce the proper proteins is a complex procedure. It is the sequence of the four nitrogen—containing bases (A, T, C, and G) that carries the code for the amino acid sequences of the protein to be synthesized in the cytoplasm of the cell. A diagrammatic representation of this process is shown in Figs. 2.25 and 2.26.

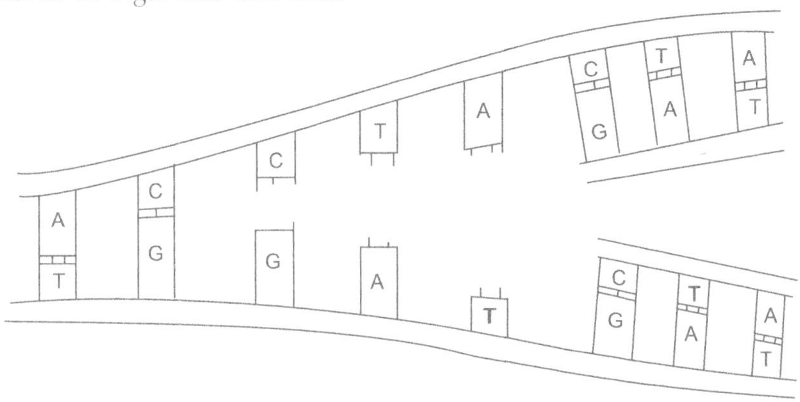

Fig. 2.25. Replication of DNA before cell division.

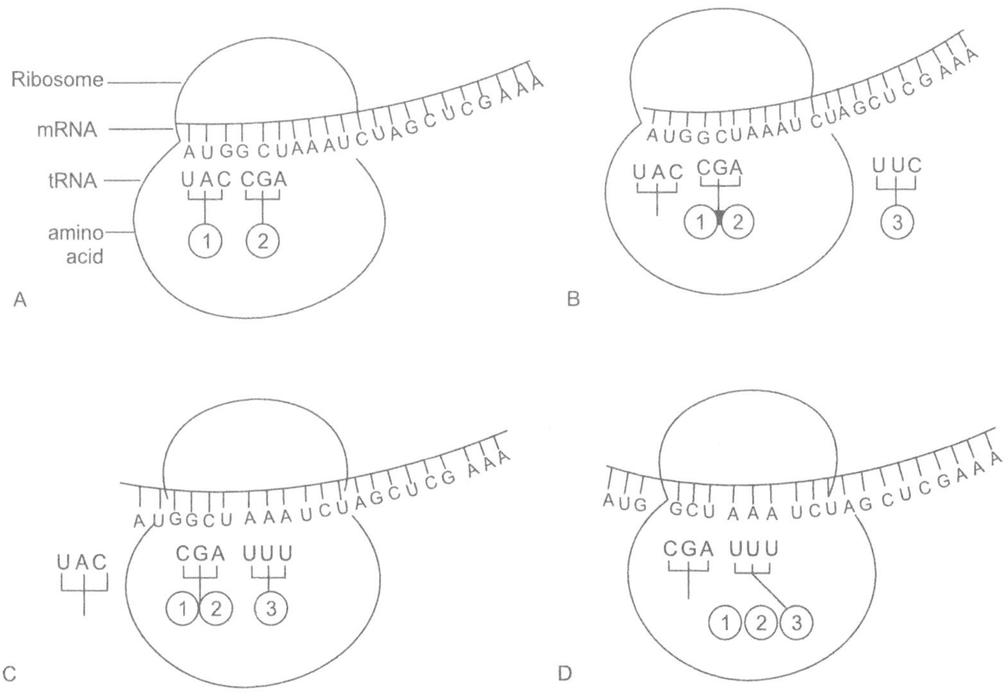

Fig. 2.26. Outline of protein synthesis. (A) Messenger RNA bound to ribosome. Amino acids 1 and 2 are bonded to their transfer RNAs. The t-RNA-amino acid 1 is bound to the peptidyl binding site. The t-RNA-amino acid 2 is bound to the amino acid binding site. (B) Amino acid 1 is released from its t-RNA and is joined to amino acid 2 through a peptide bond. (C) The ribosome moves along the m-RNA, and t-RNA 1 leaves the peptidyl binding site. The t-RNA 2 with the attached dipeptide moves to the peptidyl site and t-RNA-amino acid 3 binds to the amino acid binding site. (D) Amino acid 2 is bonded to amino acid 3 as in step B. The process repeats until the protein is completed.

You should remember that the chromosomes, which are located in the nuclear region of the cell, consist of many genes that carry the genetic information for the inherited traits of the organism. Each gene consists of several cistrons, which are the portions of the chromosomal DNA that code for one particular protein or enzyme.

When a cell is stimulated (by need) to produce a particular protein, such as insulin, the DNA of the appropriate cistron is activated to unwind temporarily from its helical configuration. This unwinding exposes the bases, which then attract the bases of free nucleotides, and a messenger RNA (mRNA) begins to build on one strand, the activated strand, of the opened DNA. Thus, the DNA has served as a template, or pattern, and has coded for a complementary mirror image of its structure in the RNA. This procedure is called transcription because the message from the DNA is transcribed onto the mRNA. After the mRNA has been synthesized over the length of the cistron, it is freed from the active DNA strand to carry the message to the cytoplasm and direct the synthesis of that particular protein, which is insulin in this case.

In eucaryotes, the freed mRNA passes through the pores of the nuclear membrane into the cytoplasm, and takes its position on the protein "assembly line", that is, the ribosomes containing ribosomal RNA (rRNA), which attract the mRNA. The ribosomes usually reside on the endoplasmic reticulum membranes in eucaryotic cells.

In procaryotes, the ribosomes may attach to the mRNA as it is being transcribed at the DNA genome, because no nuclear membrane is present; thus transcription and translation may occur simultaneously.

The base sequence of the mRNA is read or interpreted by the rRNA in groups of three bases, called *codons.* The sequence of a codon's three bases is the code that determines which amino acid goes in that position in the protein being synthesized. The amino acids must be transferred or carried by an activated *transfer* RNA (tRNA) from the cytoplasmic matrix to the site of protein assembly. The three-base sequence of the codon determines which tRNA brings its specific amino acid to the ribosome, because the RNA has an *anticodon*, three-base sequence, which is complementary to, or attracted to, the codon of the mRNA. For example, the tRNA with the anticodon base sequence UUU carries the amino acid lysine to the mRNA codon AAA. Similarly, the mRNA codon CCG codes for the tRNA anticodon GGC, which carries the amino acid proline at the opposite end of the tRNA molecule. This information system is called the *genetic code.* A chart can be prepared to show which sequence of three bases in the DNA codes for a particular codon in mRNA, which calls a particular anticodon on the tRNA carrying a specific amino acid.

The process of translating the message carried by the mRNA and decoded by the rRNA, which calls the particular tRNAs to bring the amino acids to be bound together in the proper sequence to make a specific protein (*e.g.*, insulin), is called *translation.* It should be noted that the living cell is constantly producing mRNA in its nucleus, which directs the synthesis of all the proteins, including metabolic enzymes necessary for the normal functions of that specific type of cell. Also the mRNA and tRNA are short-lived nucleic acids that may be reused many times, then destroyed and resynthesized in the nucleus. The rRNA is made in the dense portion of the nucleus called the nucleolus; the ribosomes last longer in the cell than the mRNA.

As the tRNA molecules attach to mRNA while it is sliding over the ribosomes, they bring the correct activated amino acids into contact with each other so that peptide bonds are made and a polypeptide is formed. As the polypeptide grows and becomes a protein, it folds into the unique shape determined by the amino acid sequence. This characteristic shape allows the protein to perform its specific function. If one DNA cistron base is out of sequence, the amino acid sequence will not be correct and the protein configuration will not allow the protein to function properly. For instance, some diabetics may not produce insulin properly because a mutation in the DNA of their chromosomes caused a rearrangement of the bases in the cistron that codes for insulin. DNA errors are the basis for most genetic and inherited diseases, such as phenylketonuria (P.K.U), sickle cell anemia, cerebral palsy, cystic fibrosis, cleft lip, clubfoot, extra fingers, albinism, and many other birth defects. Likewise, nonpathogenic microbes may mutate to become pathogens, and pathogens may lose the ability to cause disease by mutation, or a change in the DNA sequence.

The new science of genetic engineering attempts to repair the genetic damage in some diseases. As yet, the morality of manipulation of human genes has not been resolved by society. However, many genetically changed microbes are able to produce substances, such as human insulin, interferon, growth hormones, new pharmaceutical agents, and vaccines, that will have a substantial effect on the medical treatment of humans.

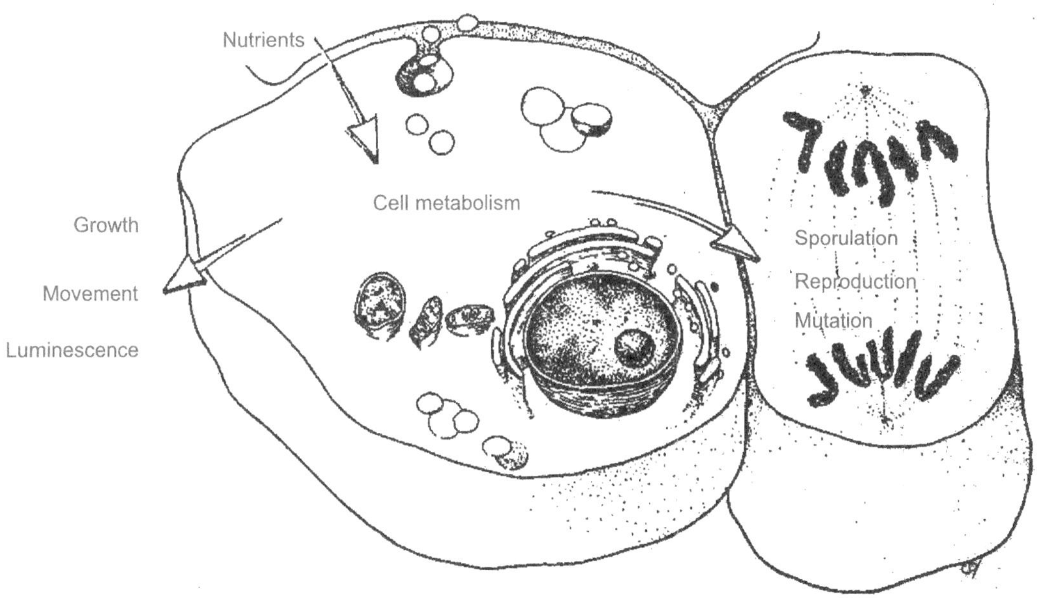

Fig. 2.27. Cell metabolism. The metabolic cycles within the cell enable it to use nutrients for growth, movement, luminescence, sporulation, reproduction, and mutation.

(Burton, p. 30)

REFERENCES

1. Burton, G.R.W. 1992. *Microbiology For The Health Sciences,* J.B. Lippincott Company, Philadelphia.
2. De Bey, H.J. 1968. *Introduction To The Chemistry of Life: Biochemistry,* Addison—Wesley Publishing Company, Reading, Massachusetts.
3. Embree, H.D. 1968. *Introduction To The Chemistry of Life: Organic Chemistry,* Addison—Wesley Publishing Company, Reading, Massachusetts.
4. Hanson, R.W. 1984. *Bio-Organic Chemistry.* Edward Arnold Publishers, London.
5. Kretovich, V.L. 1996. *Principles of Plant Biochemistry,* (English translation), Pergamon Press, Oxford.
6. White, E.H. 1965. *Chemical Background For The Biological Sciences,* Prentice—Hall of India (Private) Ltd., New Delhi.

Chapter 3

The Scientific Method, Living Organisms and the Groups of Microscopic Organisms

INTRODUCTION

Since earliest times man has been trying to solve the mysteries of nature to satisfy his curiosity. Why does iron rust ? What happens to coal when it burns? How does a mighty tree grow out of a tiny seed? Why to certain things float on water while others sink ? How do living things differ from non-living objects? His method of finding answers to all such questions has followed a set pattern; careful *observation* of the phenomenon; derivation of *inferences* from the observations; and experimentation for the *verification* of inferences. The knowledge thus gained is *empirical;* it merely describes facts. The foundations of science were laid by classifying and correlating various facts. Science may be defined as knowledge classified, correlated, and generalized into a system. In short, it is the *systematized knowledge.* A mere collection of facts is no more science than a pile of bricks in a house.

One description of science, therefore, is knowledge that has been accumulated by the use of scientific method under the direction of a scientist. Although science has been frequently called "organized knowledge" or the search for universal truth", it is questionable whether science can be defined accurately. When science is being defined, the alternatives are to define it either by its subject matter or by its methods. The purpose of science is to study the whole field of actual knowledge since science has no special topic of its own. However, any definition of science must necessarily stress its importance as a method of discovering facts that can be applied to the solution of a problem. Facts or truths are found by experimentation. They are then filtered into the literature and focused by publication of the results so that they are there for all the world to see, repeat, accept, or rejects. Thus, the approach and the methods of all sciences are alike. The one function common to all is the collection, organization, and interpretation of knowledge.

THE SCIENTIFIC METHOD

The two most important concepts in science are *observation* and *experimentation.* To observe and to experiment are only the beginning. These activities supply verifiable facts which can be applied to the solution of a problem. By such procedures, what appear to be unrelated phenomena are frequently found to be related, thus making possible the discovery of underlying principles. In other words, science only discovers facts, laws, forces, and elements already in existence which then must be organized in order to eventually offer some interpretation of them. *Organization* implies, first of all, the classification of facts collected by observation and experiment. If only a few facts were involved, the process of classification would be simple. However, the facts in science are so numerous that specialists in the field have had to resort to short cuts, the most obvious of which is a special vocabulary, for a single word is easier to use than a long description. By the recognition of similarities in facts that can be organized into groups, the content of a science is, thus, first named then classified.

The work of a scientist involves much more than the collection and classification of facts. The discovery, the naming, and the classification are merely his means to an end.

Many times great masses of information are obtained from a scientific investigation. From this information the scientist can reach certain logical conclusion. He offers these conclusions in the form of a *hypothesis,* as a possible solution to a problem. Therefore, a hypothesis is merely a provisional supposition—a good guess—regarding the causes or relations between observed facts. This information will be presented to other scientists who may test its validity, and whenever necessary, the conclusions can be reversed in the light of future discoveries. It is obvious that any problem may have more than one possible answer but only a single correct one.

When a hypothesis is supported by sound and convincing evidence through the work of many independent experimenters, and when such evidence proves to be unquestionably reliable and valid, then only is a *theory* proposed. A good theory must be reliable enough so that it can be used in future research. Theories that are universally true and have a high degree of predictability are often termed *laws.* Therefore, it can be seen that science is steady progress and not sudden change.

HOW A HYPOTHESIS IS SAID TO BE ESTABLISHED

Science may be defined in terms of either knowledge or method. *Scientific knowledge* is the total body of factual material which has been accumulated (by scientific *method*) relating to the events of the material world.

In order to satisfy their curiosity, scientists must continually pose questions about the world. The secret of success in science is to ask the right questions. According to Albert Einstein, "The formulation of a problem is often more essential than its solution, which may be mainly a matter of mathematical or experimental skill. To raise new questions, new possibilities, to regard old problems, from a new angle, requires creative imagination and marks real advance in science".

Scientific investigations may begin in response to observations made by scientists or in response to some internal "inductive" process on the part of scientists. Those aspects of knowledge which are described as scientific, must be capable of "refutation". This means that the facts of scientific knowledge must be testable and repeatable by other scientists. Thus, it is essential that all scientific investigations are described fully and clearly. If investigations yield identical results under identical conditions, then the results may be accepted as valid. Knowledge which cannot be investigated as described above is not scientific and is described as "metaphysical".

Facts are based on *observations* obtained directly or indirectly by the senses or instruments, such as light or radio telescopes, light or electron microscopes and cathode ray oscilloscopes, which act as extensions of our senses. All the facts related to a particular problem are called *data.* Observations may be *qualitative* (that is describe colour, shape, taste, presence and so on) or *quantitative.* The latter is a more precise form of observation and involves the measurement of an amount of quantity which may have been demonstrated qualitatively.

Observations provide the raw material which leads to the formulation of a hypothesis. A *hypothesis* is an assumption or question based on the observations, that may provide a valid explanation of the observations. A hypothesis has two functions:-

1. It should account for all the observed facts relevant to that problem.
2. It should lead to the *prediction* of new information. New observations (facts, data) which support the hypothesis will strengthen it. New observations which contradict the hypothesis must result in it being modified or even rejected.

In order to assess the validity of a hypothesis it is necessary to design a series of experiments aimed at producing new observations which will support or contradict the hypothesis. In most hypotheses there are a number of factors which may influence the observation; these are called *variables.* Hypotheses are objectively tested by a series of experiments in which each one of the hypothetical variables influencing the observations is systematically eliminated. The experimental series is said to be *controlled* and this ensures that only one variable of the problem is tested at a given time.

The most successful hypothesis becomes a *ruling hypothesis,* and if it withstands attempts at falsification and continues to be successful in predicting previously unexplained facts and relationships it may become a *theory.*

The trend throughout scientific study is to achieve higher levels of predictability (probability). When a theory has proved invariable under all circumstances, or such variations as occur are systematic and predictable, then it may be accepted as a *law.*

As knowledge increases and techniques of investigation improve, hypotheses and even well established theories may be challenged, modified and even rejected. Science is dynamic and controversial and the objective methods of science are always exposed to challenge.

THE SCIENTIFIC REASONING

Science begins with facts and from these, through scientific logic and reasoning, tries to draw a picture consistent with those facts. In other words, the scientist uses *inductive reasoning* to arrive at a conclusion from a set of observations. However, *deductive reasoning* is also a part of scientific method. This type of reasoning involves accepting a general idea to be a fact and then judging individual cases accordingly. To illustrate, a hypothesis is likely to be deduction that remains to be tested. Theories and laws, on the other hand, are obtained by inductive reasoning. The applications of laws and theories are usually deductive predictions. Therefore, science is essentially inductive knowledge for ideas, as well as for deductions, and must be tested experimentally.

The interpretations of observations and experiments can be simple or complex, but the conclusions should be consistent with the facts and at the same time as direct and simple as possible.

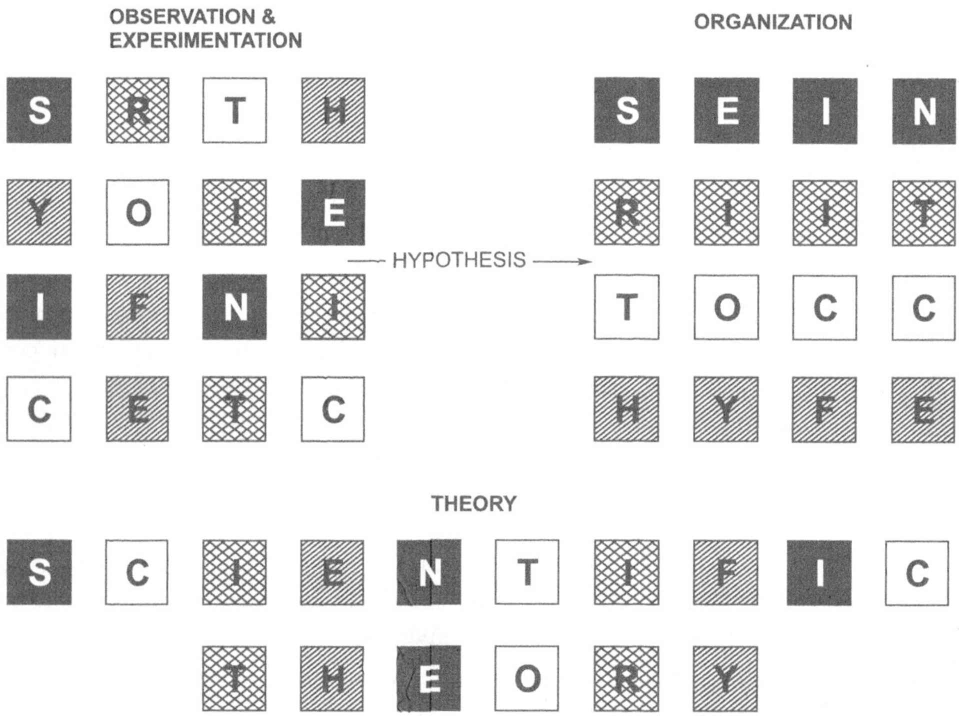

Fig. 3.1. Diagram of the scientific method.
(Braungart & Arnett, p. 4)

While trying to decide what makes one theory acceptable and another unacceptable, the scientist is forced to make some critical evaluations. All of his observations, therefore, are not going to be of equal importance in arriving at a theory, since a scientist must decide which facts are most pertinent. A judgement such as this is likely to be a personal opinion unless later experiment determines its validity.

Any scientific theory proposed should never close the door to further research on the subject. It should also be as simple as possible, capable of experimental testing, and consistent with all the pertinent and observed facts. If a concept in science cannot be tested, it cannot be refuted or supported. The only alternative is to ignore it.

CONCLUSIONS

Facts are not science, only knowledge. The method of science is to work upon these facts and find the general principles which will correlate a certain body of facts. The scientific method consists of the following four stages.

1. *Collecting the facts* by observation and experiments.
2. *Organising the facts.* When the facts are studied, many are found to be similar. The similar facts are classified and put together.
3. *Proposing a hypothesis* to account for the facts collected and organized. A scientific hypothesis is a brief statement of what seems to be an explanation of the facts as known at that time.

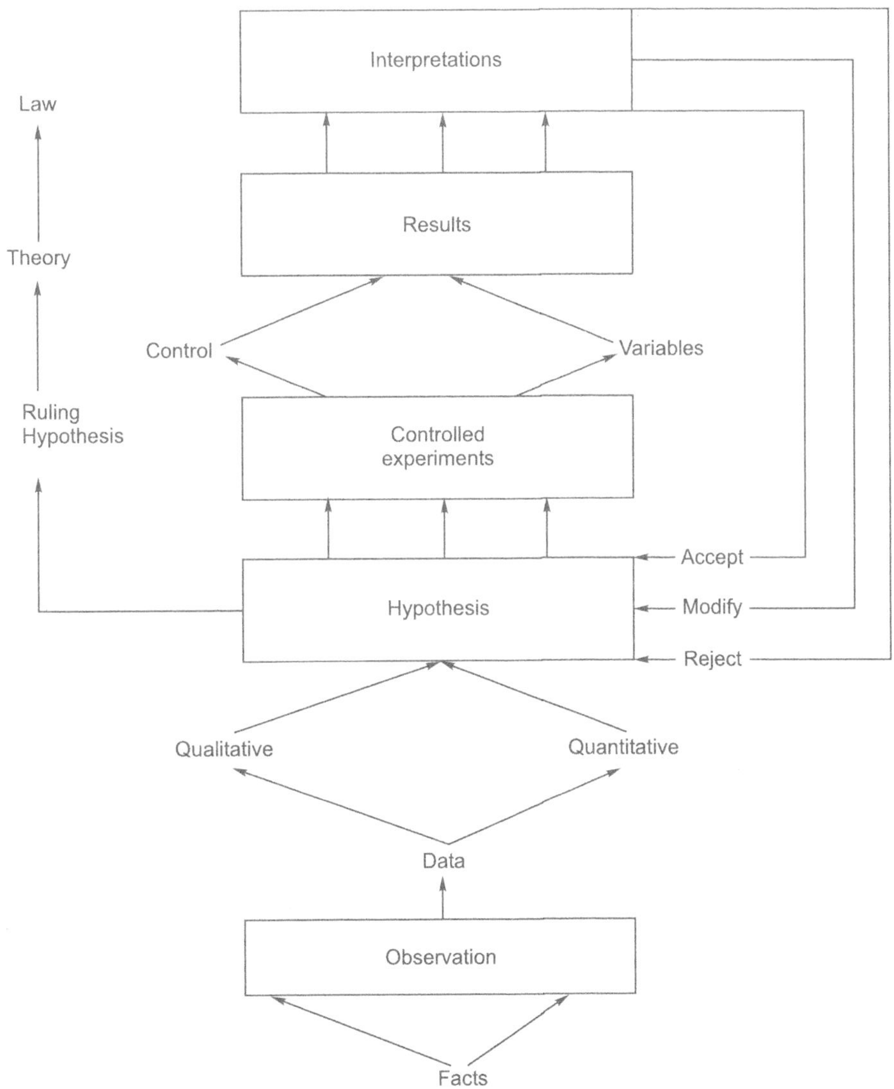

Fig. 3.2. Summary of scientific method.
(Soper, p. 937)

4. *Testing and verifying the hypothesis.* The hypothesis is tested by using it to *predict* what would happen under specified conditions if it be true. These *predictions* are verified by preforming new experiments. If the results of these experiments agree with the predictions, the hypothesis is true. It is then called *theory.* If the results do not confirm the predictions, the hypothesis is rejected. A new hypothesis is then worked out and similarly tested.

A theory is similarly submitted to the test of prediction and verification by experiments. If theory is confirmed by every fresh test it is termed a *law.* Otherwise the theory is also rejected.

THE LIVING ORGANISMS

"Life", in simple words may be defined as an organization of activities manifested by living beings (Tippo, 1961). All plants are alive and, in common with all other living things, possess certain characteristics which differentiate them from nonliving things. However, the following set of properties will serve as a working basis for distinguishing living organisms from lifeless materials.

(1) The Cellular Structure : Living organisms possess a characteristic type of structure which distinguishes them from non-living things. The bodies of plants and animals are composed of small structural units termed *cells.* These units contain a complex, sticky (viscous) fluid (the substance of life) known as *protoplasm,* which is composed of many organic and inorganic constituents, and an enclosing wall or membrane. The cell is composed of many different substances and contains tiny particles called *organelles* that have important functions. Living material of the cell consists of two parts, the cytoplasm (cytosol) outside the nucleus and the nucleoplasm inside the nucleus.

The cell is defined as the fundamental living unit of any organism because, like the organism, the cell exhibits the basic characteristics of life. The cell obtains food from the environment to produce energy and nutrients for metabolism. Metabolism is an inclusive term to describe all the chemical reactions by which food is transformed for use by the cells. Through its metabolism, the cell can grow and reproduce. It can respond to changes in its environment such as light, heat, cold, and the presence of chemicals. It can mutate (change) as a result of accidental changes in its genetic material, the DNA (deoxyribonucleic acid), which makes up the genes of the chromosomes, and thus become better or less suited to its environment. As a result of these genetic changes, the mutant organism may be better adapted for survival and development into a new species of organism.

(2) Reproduction : The formation of new individuals of the same kind is another characteristic usually associated with living things. It is one of the most fundamental of the criteria of life and one of the most universally apparent; not only does a species produce offspring of the same species, but the offspring usually resemble in a high degree their parents; that is, the individuality within a species is capable of being transmitted from parent to offspring.

In plants, the process of reproduction may be vegetative, in which an existing portion of the parent, or a new growth produced by it, becomes detached and forms a new individual.

Vegetative reproduction (propagation) in the higher plants, for example, may be brought about by suckers, shoots, bulbs, cuttings, runners, and other structures. Or reproduction may be sexual, involving male and female cells (gametes) which fuse to form a single cell (zygote) which gives rise to a new individual.

(3) Response : Living organisms have the property of *irritability*, that is, they respond to stimuli. It involves the sensitivity of protoplasm to changes in the environment and the capacity of specific reaction, usually in the form of movements, to those changes (the reactions of living organisms to stimuli consists chiefly of movements of the whole organism or its parts with reference to the exciting stimuli). Protoplasm is sensitive to a wide variety of the external features of its environment—to light, variations of temperature, moisture and drought, wounding, mechanical stimuli, such as pressure, tension, chemicals, gravity, etc. The reactions of protoplasm to these changing factors possess survival value, for the reactions tend usually to adjust the organism more or less favourably with reference to the particular stimuli involved.

Plants commonly undergo alterations in external and internal form and in their life processes when subjected to changes in temperature, lack of water, intense sunlight, or alterations in mineral composition of the soil or in length of day.

In animals, the response to stimuli is movement rather than growth. In plants, the reverse is generally true, although some microscopic water plants are as free swimming and active as microscopic animals. Higher types of plants, anchored to one spot by their roots, also may exhibit limited movement in response to such stimuli as light, touch, and changes in temperature. Some of the common examples are the opening and closing of flowers, the twining of tendrils, and the movement of leaves.

(4) Energy Transformation : Living systems accumulate and expand energy, which is derived, directly or indirectly, from the sun and is stored in food materials. An organism sets free the energy stored in food by the process of *respiration*. It is this energy which is utilized in assimilation, in the construction of complex compounds from simple ones, and in many other processes. Such energy is necessary to the maintenance of life itself. The term *metabolism* is usually applied to the overall chemical processes of the living body necessary for life and growth. Some metabolic processes break down the materials of organism into simpler constituents (catabolism); others build them up (anabolism). Growth occurs only when the building up predominates over the breaking down.

(5) Growth : Living organisms are capable of growth. Growth is a complex of phenomena. This involves not only an increase in size of living organisms and their parts, but includes processes of division of labour, of coordination of activities, and of development according to a definite plan for every kind of organism. A seed, for example, is alive, but it grows only when supplied with water, oxygen, and a suitable temperature.

As mentioned above, growth is generally accompanied by increase in physical size. However, nonliving things such as a ball of snow rolling down hill or a crystal in a solution also exhibit increase in size. When such things increase in size, they do by a process of *accretion*-that is, building up from outside. Increase in size, but not growth, is also exhibited by dead seeds and by a piece of gelatin placed in water. On the other hand, true growth centers around the process of *assimilation*—that is, the utilization of foods, chiefly proteins and sugars, in the construction of the living body. Assimilation involves the synthesis, or putting together, of living from nonliving materials. It is basic to growth in living objects as distinguished from mere increase in size, which may also occur in the world of inanimate matter.

The basic plan of growth for each kind of living thing is determined by the hereditary characteristics received from its parents. Although the fundamental architecture of growth is determined by heredity, it is susceptible to constant and considerable modification by factors of the external environment.

(6) Organic Composition : A final aspect of all living things is that they are constructed of *organic* compounds. All organic compounds contain the element carbon, which possesses the capacity to form long chains composed of many atoms. It also unites readily with other elements, such as hydrogen, oxygen, and nitrogen. Examples of organic compounds are sugars, starch, cellulose, proteins, fats, and alcohols.

Among the simplest organic compounds are those of carbon and hydrogen. Examples of those, known as *hydrocarbons,* are methane (marsh gas, the chief component of natural gas), lubricating oils, and rubber.

The most essential constituents of living organisms are proteins, composed of the elements carbon, hydrogen, oxygen, nitrogen, and sometimes sulphur and phosphorus. One hypothesis of the origin of life is that proteins were formed spontaneously during the early history of earth. Proteins were preceded by simple organic compounds, such as hydrocarbons, formed by high-temperature chemical reactions. Ammonia, a compound composed of nitrogen and hydrogen, was another compound formed during such reactions. These substances, dissolved in the warm seas of the primeval world, combined to form particles of simple proteins. Some of these particles possessed a kind of organization which enabled them to build more of their own substance and to reproduce themselve. When such self-duplicating protein particles arose, life began. Combinations of such particles led gradually to the organization of primitive living cells. These, through the process of evolution, gave rise to the infinite variety of living things that now inhabit the earth.

Support for this hypothesis, has been obtained by bringing methane, ammonia, hydrogen, and water together in the laboratory under conditions believed to be similar to those prevailing on the primitive earth. When these materials were held together for a time, they combined to form a number of the units (amino acids) of which proteins are composed.

PROCARYOTES AND THE EUCARYOTES

Much evidence exists to indicate that almost 4000 million years ago the first bit of life to appear on earth was a very primitive cell similar to the simple bacteria of today. Bacterial cells exhibit all the characteristics of life, even though they do not have the complex system of membranes and organelles found in the more advanced single-celled organisms. These less complex cells, which include bacteria and cyanobacteria (blue-green algae), are called *procaryotes* or procaryotic cells. The more complex cells, with a nucleus and many organelles, are called *eucaryotes* or eucaryotic cells; these include such organisms as protozoa, fungi, green, brown, and red algae, and all plant and animal cells, including those that make up in the human body.

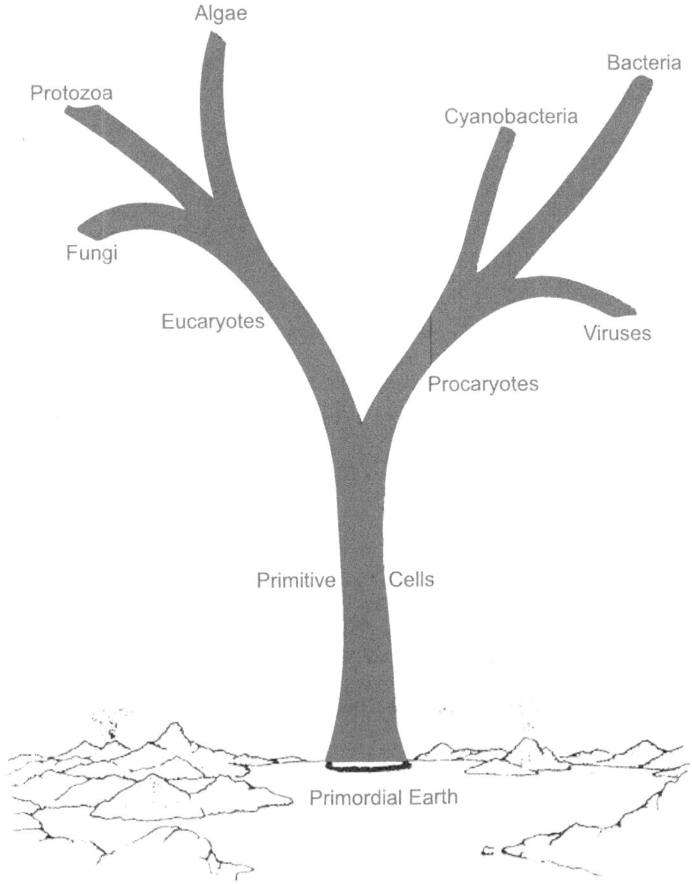

Fig. 3.3. Family tree of microorganisms. The primitive cells are divided into eucaryotes (organisms with a true nucleus), such as fungi, protozoa, and most algae, and procaryotes (organisms without a bound nucleus), such as bacteria, cyanobacteria, and viruses.

(Burton, p. 3)

Viruses appear to be the result of regressive, or what might be termed reverse, evolution, because they are composed of only a few genes protected by a protein coat, a few enzymes, and little other material. Viruses depend on the energy and metabolic machinery of the host cell to live and reproduce. Therefore, because they are not truly viable cells, they are usually placed in a completely separate category and are not classified with the simple procaryotic cells (*Burton,* 1992).

Eucaryotes appeared first in the late Pre-Cambrian period, about 2000 million years ago, and probably evolved from procaryotes.

The cells of procaryotes (*pro,* before; *karyo,* nucleus) lack true nuclei. In other words, their genetic material (DNA) is not enclosed by nuclear membranes, and lies free in the cytoplasm. The cells of eucaryotes (*eu,* true) are much more complex and are characterised by a true nucleus, that is genetic material enclosed by membranes (the *nuclear envelope*) to form a definite easily recognisable structure.

Differences between Procaryotic and Eucaryotic Cells

Eucaryotic cells are divided into plant and animal types because animal cells do not have a cell wall, whereas plant cells have a simple cell wall, usually consisting of cellulose. Procaryotic cells have a complex cell wall consisting of proteins, lipids, and polysaccharides. Eucaryotic cells are filled with membrane bound organelles, such as the endoplasmic reticulum, whereas procaryotic cells have only a few cytoplasmic membranes (mesosomes and photosynthetic membranes), which arise from the plasma membrane. The cytoplasmic ribosomes (involved in protein synthesis) are larger and more dense (80 s) than those found in procaryotes (70 s). The fact that 70 S ribosomes are found in the mitochondria and chloroplasts of eucaryotes may indicate that these structures are derived from parasitic procaryotes during their evolutionary development.

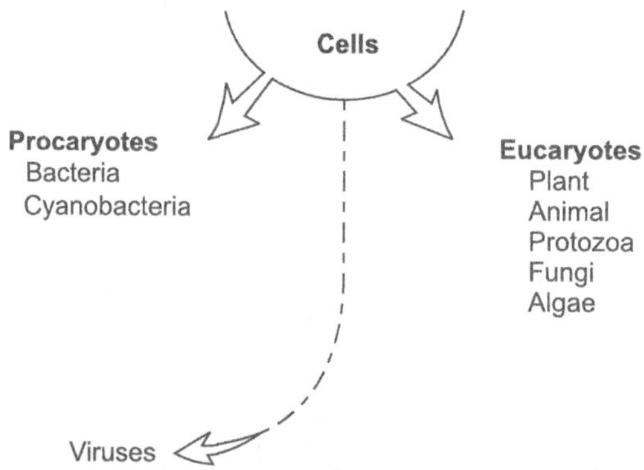

Fig. 3.4. Typical cells: procaryotes, the less complex cells, and eucaryotes, the more complex cells. Viruses are not true cells. (Burton, p. 30)

Many other fundamental differences exist between procaryotes and eucaryotes, the important of which are summarised in Table 3.1.

TABLE 3.1. General Organization of Procaryote and Eucaryote Cell

S.No.	Feature	Procaryote	Eucaryote
1.	Cell size	Average diameter 0.5-5 μm L (1 μm = 10^{-6} m)	Upto 40 μm diameter common; 1000–10,000 times volume of procaryotic cell.
2.	Form	Unicellular or filamentous	Unicellular, filamentous or truly multicellular.
3.	Genome	Naked DNA molecule in cytoplasmic matrix	"True nucleus" in nuclear envelope.
4.	Chromosome	Circular DNA molecule. No true nucleus or Chromosomes. No nucleolus.	Paired chromosomes in nucleus, diffuse in interphase, compact during mitosis and meiosis. Circular DNA molecule in mitochondria and chloroplasts.
5.	Chromosome composition	DNA	Linear DNA associated with proteins and RNA to form chromosomes within a nucleus. Nucleolus in nucleus.

6.	Chromosome number	Single	Two of a kind; often numerous kinds; single DNA molecule in mitochondria and chloroplasts.
7.	Chromosome division	DNA replication	Mitosis and miosis after DNA replication in mitochondria and chloroplasts.
8.	Protein synthesis	On ribosomes in cell matrix. Ribosomes 70 s (smaller). No endoplasmic reticulum involved. (Many other details of protein synthesis differ, including susceptibility to antibiotics, *e.g.*, procaryotes inhibited by streptomycin)	On ribosomes in cell matrix and on rough endoplasmic reticulum; ribosomes 80 S (larger) 50 to 60 S in mitochondria and chloroplasts, may be attached to endoplasmic reticulum. In nucleus, Chloroplasts and mitochondria.
9.	Organelles	Few Organelles. None are surrounded by an envelope (2 membranes). Internal membranes scarce; if present usually associated with respiration and photosynthesis	Many organelles. Envelope bound organelles present, *e.g.*, nucleus, mitochondria, Chloroplasts. Great diversity of organelles bounded by single membranes, *e.g.*, Golgi apparatus, lysosomes, vacuoles, microbodies, endoplasmic reticulum.
10.	Cell Walls	Generally present; rigid and contain polysaccharides with amino acids. Murein is main strengthening compound.	Generally present on plant cells but absent in animal cells. Cell walls of green plants and fungi rigid and contain polysaccharides. Cellulose is main strengthening compound of plant walls, chitin of fungal walls.
11.	Flagella	Simple, lacking microtubules. Extracellular (not enclosed by cell surface membrane); 20 nm diameter; composed of flagellin. Unknown energy source for movement.	Complex, with "9+2" arrangement of microtubules; intracellular (surrounded by cell surface membrane), 200 nm diameter; composed of tubulin. ATP energy source for movement.
12.	Microfilaments and microtubules	Absent	Probably universal
13.	Centriole	Absent	Present in animal cells, rare in plant cells.
14.	Glycocalyx	Generally present	Generally present on algae, protozoan, and animal cells.
15.	Glycolytic enzymes	Glycolytic (fermentative) enzymes in cell matrix.	Glycolytic enzymes in cytoplasmic matrix.

16.	Oxidative enzymes	On cell membrane and its extension (mesosome)	On membranes, including cristae of mitochondria and in peroxisomes.
17.	Hydrolytic enzymes	On cell membranes	In lysosomes
18.	DNA and RNA synthesis	On DNA molecule	On Chromosomal DNA, in mitochondria and Chloroplasts on DNA molecule.
19.	Respiration	Mesosomes in bacteria, except cytoplasmic membranes in cyanobacteria	Mitochondria for aerobic respiration
20.	Photosynthesis	No Chloroplasts. Takes place on membranes which show no stacking	Chloroplasts containing membranes which are usually stacked into lamellae or grana
21.	Nitrogen fixation	Some have the ability.	None have the ability

Histones (Proteins)—The chemical substances predominant in the nucleus are the nucleic acids (DNA and RNA) and proteins. Histones appear to be most closely associated with chromosomes, and contain the basic amino acids lysine, histidine, and arginine.

Internal Membranes—In addition to the membrane at the periphery of the cell, most procaryotes possess internal membranes, which can often be seen in electron micrographs. These internal membranes may be simply extensions or invaginations of the cell membrane, or they may be much more complicated. Many bacteria possess a special membranous structure called a *mesosome* (meaning "middle body") which is probably associated with the formation of the bacterial cross wall and may have other functions as well.

In photosynthetic procaryotes, the photosynthetic membranes often form an extensive internal membrane system.

Ribosomes—In electron micrographs of thin sections, small, dark particles can often be seen within the cytoplasm. These particles, which are part of the protein—synthesizing machinery of the cell are called ribosomes, because they are ribonucleic acid—containing bodies. They are composed of about 60% ribonucleic acid (RNA) and 40% protein, and are about 20 nm in diameter.

Nanometer—For very small dimensions the unit *nanometer* (nm) is often used. The prefix *nano* means that the unit following should be multiplied by 10^{-9}. The older unit *millimicron* (mμ) is identical with the nanometer. Thus, 0.001 μm (1 mμ) equals 1 nm.

Sedimentation Constant—is a physical parameter of a particle and is a function of both molecular size and shape, given in units called Svedbergs (abbreviated as S). Smaller subunits have low S, while larger ones have high S. A major difference between procaryotes and eucaryotes is in the size of the ribosomes. Procaryotes have ribosomes with a sedimentation constant of 70 S, which are composed of two subunits having constants 50 S and 30 S. Cytoplasmic ribosomes of eucaryotes have a sedimentation constant of 80 S, and these are composed of two subunits with sedimentation constants of 60 S and 40 S. The ribosomes of the mitochondria and chloroplasts of eucaryotes are 70 S, however.

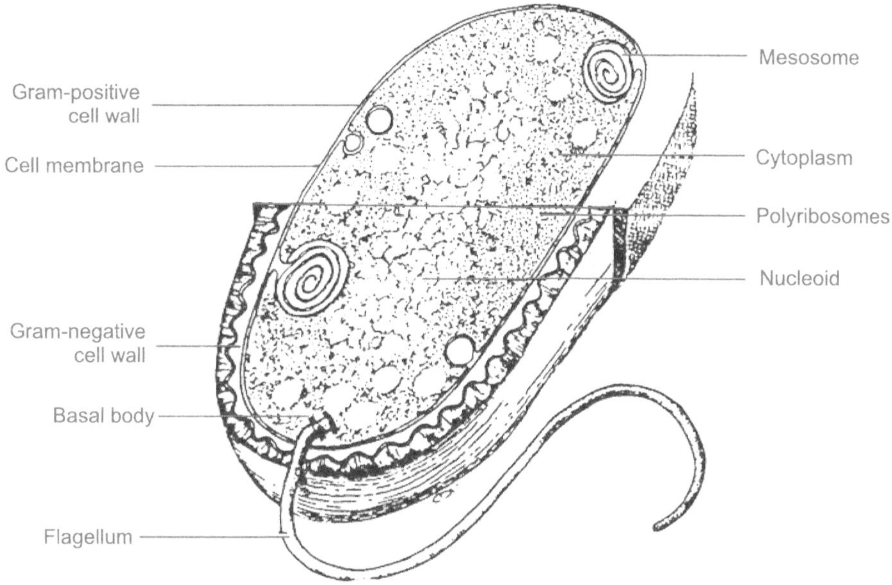

Fig. 3.5. An illustrated composite procryotic bacterial cell showing the difference between gram-negative and gram-positive cell walls. (Burton, p. 35)

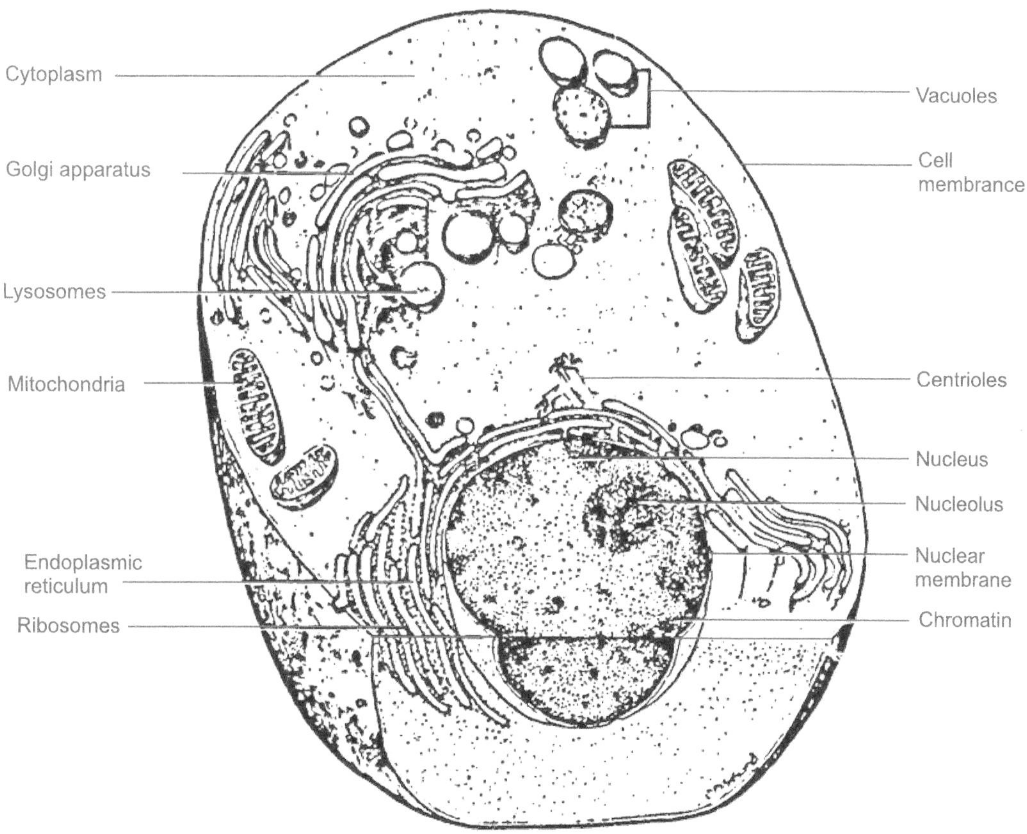

Fig. 3.6. A typical eucaryotic animal cell (Burton, p. 31)

The ribosomes from procaryotic microorganisms are thus slightly smaller and lighter in weight than those of eucaryotes.

Flagella—Flagella (singular flagellum) are long filamentous structures that are attached to one end of the cell and move in a whip—like manner to impart motion to the cell.

Bacterial flagella are composed of protein subunits, the protein is called flagellin. These flagella are long, thin appendages free at one end and attached to the cell at the other end. They are so thin (about 20 nm) that a single flagellum can never be seen directly with the light microscope, but only after staining with special flagella stains. They are also readily seen with the electron microscope. Eucaryotic flagella however, are large enough to be seen with the light microscope, and it is possible to observe their movements. The manner in which the flagellum moves determines the direction of cellular movement. The maximum rate of movement of flagellated microorganisms range from 30 to 250 nm/second.

The flagella of eucaryotes are quite different in structure from those of procaryotes in some other respects also, even though the same word is used for both. Without exception, flagella of eucaryotes contain two central fibres surrounded by nine peripheral fibres, each of which is doublet in nature. The whole unit is surrounded by a membrane and the fibres are embedded in an organic matrix. Each fibre is in reality a *microtubule,* which is composed of a large number of identical molecules of a protein called *tubulin.* Microtubules are an important feature of the eucaryotes. Attached to each of the outer doublet fibres are molecules of another protein called *dynein,* and it is this protein that is involved in the conversion of chemical energy (from ATP) into the mechanical energy of flagellar movement. The flagellum is anchored to the cell at a structure called a *basal body* or *kinetosome.*

Cilia (*singular cilium*)—are similar to eucaryotic flagella in fine structure but differ in being shorter and more numerous.

Glycocalyx—The cytoplasm of every cell is bounded by a *plasma membrane,* so thin and fragile that it was not visualized until the advent of the electron microscope. The plasma membrane is covered by protective coatings: in bacteria and plants by a cell wall, in animal cells by a galactoprotein *glycocalyx.* The glycocalyx is helpful in cell adhesion and migration. In many bacteria and plant cells a glycocalyx is present in addition to the cell wall. Bacteria stick to surfaces by strands of polysaccharides of the glycocalyx.

CLASSIFICATION

Early classification systems recognized only the plant and animal kingdoms, with the primary criterion for separation being that animals were motile and plants were stationary. Classification systems, as expounded by Aristotle and other ancient philosophers, were developed centuries before the beginnings of evolution theory. Microorganisms were considered to be little plants and animals when they were discovered in the late seventeenth century. When the Swedish botanist Carl Linneaeus established the first comprehensive classification system he placed all microbes into a single genus, which he appropriately named *Chaos.*

NATURAL SYSTEMS

Although a classification system should reflect the branching of groups that is a natural consequence of evolution, evolutionary affinities among microorganisms are difficult to understand because there is no fossil record of most microorganisms. Therefore, the lack of a fossil record makes direct examination of the ancestors of today's microorganisms for the most part impossible.

One of the earliest classification systems that attempted to reflect evolutionary relationships was developed by a German zoologist, Earnst Heinrich Haeckel, who was a disciple of Darwin and adopted his theories. Haeckel proposed a three kingdom classification system that divided living organisms into animals, plants, and protists. The protista were defined by a lack of tissue differentiation; therefore, in this system all microorganisms were classified as *protists* and, conversely, all protists as microorganisms, Haeckel's system has been superseded by other theoretical systems but still retain its practical appeal because Haeckel's kingdom Protista encompasses all the organisms (bacteria, algae, fungi, and Protozoa).

THE COMPONENTS OF THE FIVE-KINGDOM SYSTEM

Of all the classification schemes used in recent years, the one that now has gained widest acceptance is the *five-kingdom system* proposed by Robert H. Whittaker in 1969.

According to Whittaker's proposed classification, evolutionary divergence occurs along three principal modes of nutrition :

(*a*) synthetic to evolve plants,

(*b*) adsorptive to evolve fungi, and

(*c*) ingestive to evolve animals.

In this system microorganisms exclusively comprise three of the five kingdoms: in addition to the kingdoms of Plants and Animals, there are the kingdoms Monera, which includes all bacteria; Protista, which includes the algae and protozoa, and Fungi, which is composed of the fungi. Whittaker's system assumes that the Monera (bacteria) represent the most primitive form of life and that evolutionary processes led to the development of the Protista (algae and protozoa) which then underwent further evolution to form the remaining three kingdoms.

According to this scheme of classification, therefore, the living world is divided into five kingdoms :

(1) Plantae

(2) Animalia

(3) Protista

(4) Fungi, and

(5) Monera (Procaryotes)

The relationships between these five kingdoms suggests the descent of all organisms from a common ancestor, the first living cell.

Whittaker's system is based on profound differences between eukaryotic and procaryotic forms of life, including their cellular organization—unicellular, unicellular-colonial, or multicellular and their nutrition—absorptive, ingestive, photosynthetic, or combination of these. Brief descriptions of these kingdoms are given below.

(1) Plantae : Many of the species in this kingdom are well known and easily recognized green plants—garden vegetables, grasses, shrubs, mosses, and most algae. In general, plants are multicellular, eucaryotic, and have rigid cell walls. Most plants are photosynthetic, containing one or more types of chlorophyll, and, with few exceptions, they are nonmotile. Other properties of the plant kingdom members include differentiation of their tissues into organized structures such as roots, stems, and leaves, and complex life cycles, generally involving both asexual and sexual reproduction.

(2) Animalia : The animal life, included in the kingdom Animalia ranges from organisms without backbones, such as sponges and worms, to highly developed forms with backbones, such as mammals. Animal cells are eucaryotic and enclosed only by a flexible membrane. They lack cell walls. Animals are multicellular, motile most of their lives, and they exhibit ingestive nutrition. Some show tissue differentiation. Animals may reproduce sexually and/or asexually.

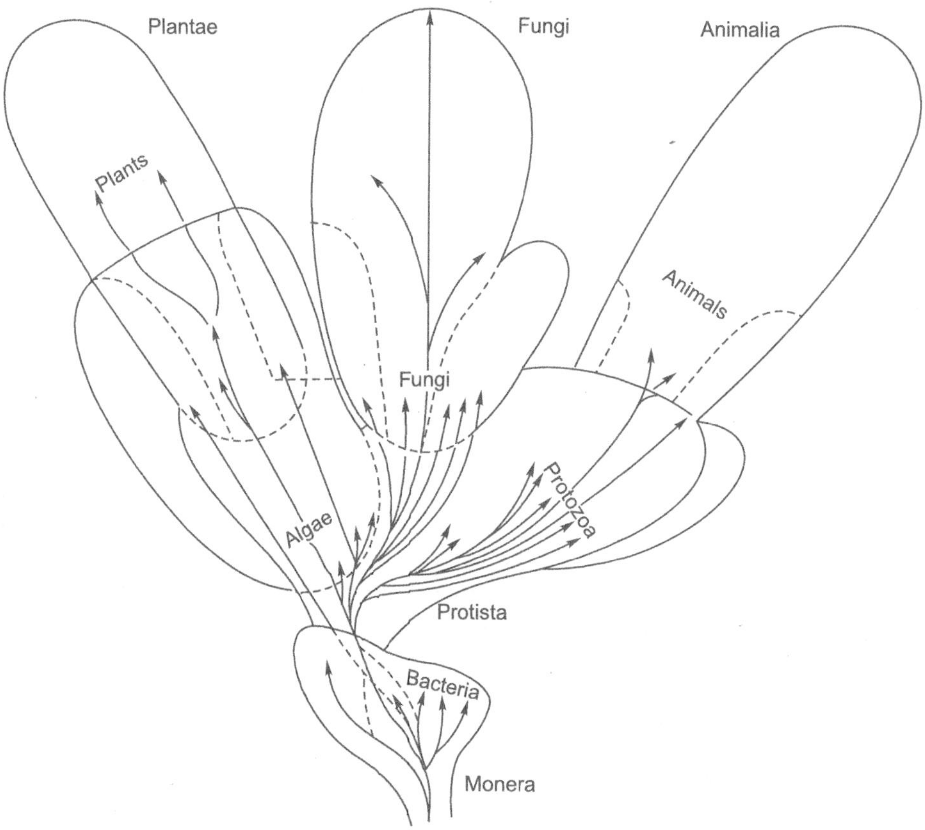

Fig. 3.7. Whittaker's classification is a five-kingdom system based on three levels of organization, the prokaryotic (kingdom Monera), eukaryotic unicellular (Kingdom Protista), and eukaryotic multicellular and multinucleate (Kingdom Fungi, Kingdom Animalia, and Kingdom Plantae). (Atlas, p. 34)

(3) Protista : Various protozoa and unicellular and colonial forms of algae make up the Protista. Unlike typical animals and plants, protists are biologically and biochemically independent. There is no dependence upon other cells, such as exists in the tissue and organs making up higher plants and animals. The Protista is considered by some a transitional kingdom; in an evolutionary sense it falls between procaryotic organisms of the Monera kingdom and the more structurally sophisticated members of the animal and plant kingdoms.

Others view Protista as a group for any organism that does not readily fit into another kingdom.

Protists are eucaryotic and obtain their nutrients in various ways, including absorption, ingestion, photosynthesis, and combination of these modes. They exhibit both asexual and sexual reproduction.

(4) Fungi : This kingdom includes the familiar mushrooms, molds and yeasts, as well as the less familiar slime molds, rusts, and smuts. Fungi are eucaryotic, possess cell walls, and may be unicellular or multicellular. Like animals, they are not photosynthetic. Fungi obtain their nutrition from the environment by absorption.

(5) Monera : The kingdom Monera consists of unicellular and colonial microorganisms with procaryotic cellular organization. All of the functions and activities of life are carried out within the confines of these extremely small but far from simple cells. The monera includes a wide variety of bacteria, such as the blue-greens and other photosynthetic organisms, agents of disease such as the rickettsia and the chlamydia, and numerous beneficial and harmless forms. Most organisms in this group have absorptive nutrition, although some of them are photosynthetic. Reproduction is mainly asexual, but a few species also reproduce sexually.

THE THREE-KINGDOM SYSTEM

Determining the validity of classification systems that aim at reflecting the process of evolutionary separation of species has been greatly facilitated by the development of an understanding of how genetic information is stored at the molecular level. Carl Woese in 1980 proposed a new classification systems with three primary Kingdoms—Archaebacteria, Eubacteria, and Eukaryotes based upon molecular–level genetic analysis. Both *eubacteria* and *archaebacteria* have prokaryotic cells, which are simply organized, lack a nucleus, and are structurally quite different from eukaryotes, which have a nucleus and various other organelles. The archaebacteria and eubacteria, however, are genealogically as distinctly related to each other as each is to eukaryotes. Woese proposed that the archaebacteria, eubacteria, and an *urkaryote*, which was the original eukaryotic cell (German *ur* = prototype), all evolved from a common simple ancestor, the progenote. According to Woese's theory, bacteria began to live within urkaryotes. Eventually these prokaryotic bacteria and their urkaryotic host cells became mutually dependent. Over time the prokaryotic partners in this mutually beneficial relationship developed into cellular organelles (mitochondria and chloroplasts), giving rise to the eukaryotic cell. This system is receiving great deal of attention from microbiologists but has yet to gain wide support from general biologists.

THE BERGEY'S MANUAL

John H. Bergey (an American bacteriologist) published a system of bacterial classification in 1923 in a book form under the title *Manual of Determinative Bacteriology*. The manual has been revised at intervals with great improvements and additions. Seven subsequent editions have been titled *Bergey's Manual of Determinative Bacteriology*. The seventh edition was prepared under the editorship of Professors Robert S. Breed, E.G.D. Murray, and Dr. Nathan R. Smith,

Dr. Bergey having died in 1937. This tremendous task has been undertaken cooperatively by 94 specialists whose names appear on the title page of the *Manual.* The eighth edition of *Bergey's Manual* was called *Bergey's Manual of Systematic Bacteriology,* and published in 1974.

In 1984, however, the Bergey's Manual Trust began publication of a four volume set of *Bergey's Manual of Systematic Bacteriology,* which describes the taxonomic status of particular groups of microorganisms.

Volume-1. Deals only with the Gram-negative bacteria of general, medical, and industrial importance.

Volume-2. The Gram-positive bacteria other than the filamentous actinomycetes.

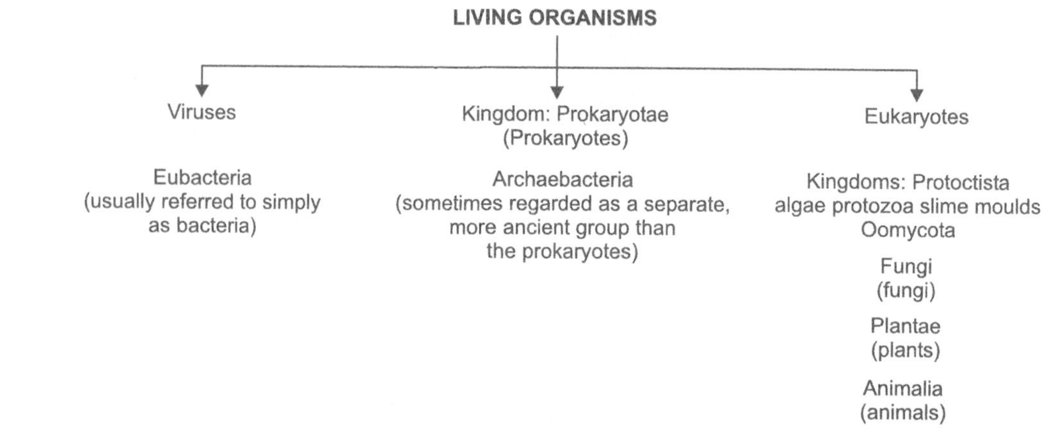

Fig. 3.8. Summary classification of living organisms.

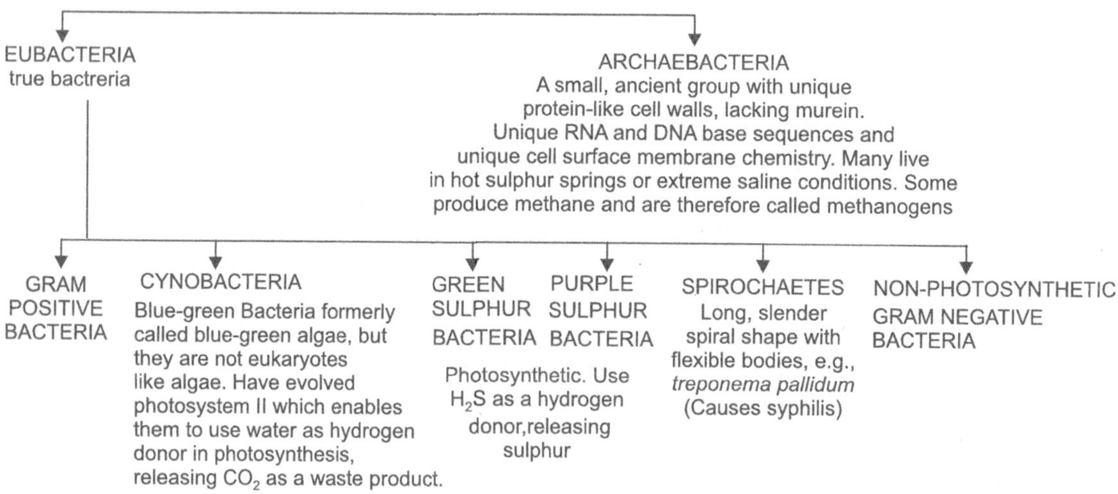

Fig. 3.9. Classification of Prokaryotes.
(Soper, p. 5)

Volume-3. Archaebacteria, the cyanobacteria, and the remaining Gram-negative bacteria including the chemolithotrophs.

Volume-4. The Gram-positive, filamentous bacteria.

TABLE 3.2. Characteristics of the Five Kingdoms

S.No.	Characteristics	KINGDOMS				
		Plantae	**Animalia**	**Protista**	**Fungi**	**Monera**
1.	Cellular Organization	Eukaryotic and multicellular	Eukaryotic and multicellular	Eukaryotic, unicellular, and some colonial forms	Eukaryotic, multicellular and unicellular	Prokaryotic and unicellular
2.	Cell wall	Present	Absent	Present with algae	Present	Present
3.	Differentiation of tissues	Present	May be present or absent	Absent	Absent	Absent
4.	Mode or types of nutrition	Primarily photosynthetic	Ingestive and some absorptive	Absorptive, ingestive, photosynthetic, and combinations	Absorptive	Absorptive
5.	Reproduction	Generally both sexual and asexual	Generally both asexual and sexual	Asexual and sexual	Asexual and sexual with most	Asexual and rarely sexual
6.	Motility (ability to move)	Mostly nonmotile	Motile	Both motile and nonmotile	Generally nonmotile	Both motile and nonmotile
7.	Microbial representatives	None	None	Single celled algae and protozoa	Molds and yeasts	Bacteria

(Wistreich & Lechtman p. 38)

Aside from its being the last unified volume, several very significant features characterized the eighth edition of *Bergey's Manual.* The cyanobacteria were included as a division of the kingdom prokaryotae, recognizing that the blue-greens are properly considered bacteria and not algae. The bacteria were placed with a single kingdom (the Prokaryotae) that is separated into two divisions (division I, the Cyanobacteria; division II, the Bacteria). In the latest edition, four divisions of the kingdom Prokaryotae are recognized.

1. *Gracilicutes* : bacteria with a typical Gram-negative cell wall,
2. *Firmicutes* : bacteria with a typical Gram-positive cell wall,
3. *Tenericutes* : bacteria lacking a cell wall,
4. *Mendosicutes* : bacteria lacking peptidoglycan in their cell walls.

Proposed Higher Taxa for Bacteria (according to *Bergey's Manual of Systematic Bacteriology*)

Kingdom : Prokaryotae

Division-I : Gracilicutes (Gram-negative, typical bacteria)

Class-1. Scotobacteria (Nonphotosynthetic bacteria)

Class-2. Anoxyphotobacteria (Nonoxygenic photosynthetic bacteria)

Class-3. Oxyphotobacteria (Oxygenic photosynthetic bacteria).

Division-II : Firmicutes (Gram-positive, typical bacteria)

Class-1. Firmibacteria (Simple, Gram-positive bacteria)

Class-2. Thallobacteria (Branching, filamentous bacteria)

Division-III : Tenericutes (Bacteria lacking a cell wall)

Class-1. Mollicutes

Division-IV : Mendosicutes (Bacteria with defective cell walls or lacking peptidoglycan)

Class-1. Archaebacteria

The manual includes 33 sections, each of which consists of one or more closely related genera. Those sections having numerous genera usually are subdivided into families.

VOLUME-1

Section-1 The spirochetes

Section-2 Aerobic/microaerophilic, motile, helical/vibroid, gram-negative bacteria.

Section-3 Nonmotile (or rarely motile), gram-negative, curved bacteria.

Section-4 Gram-negative aerobic rods and cocci.

Section-5 Facultatively anaerobic gram-negative rods.

Section-6 Anaerobic gram-negative straight, curved, and helical rods.

Section-7 Dissimilatory sulphate- or sulphur-reducing bacteria.

Section-8 Anaerobic gram-negative cocci.

Section-9 The rickettsias and Chlamydias.

Section-10 The mycoplasmas

Section-11 Endosymbionts

VOLUME-2

Section-12 Gram-positive cocci.

Section-13 Endospore-forming gram positive rods and cocci.

Section-14 Regular, nonsporing, gram-positive rods.

Section-15 Irregular, nonsporing gram-positive rods.

Section-16 The mycobacteria

Section-17 Nocardioforms

VOLUME-3

Section-18 Anoxygenic phototrophic bacteria.

Section-19 Oxygenic photosynthetic bacteria.

Section-20 Aerobic chemolithotrophic bacteria and associated organisms.

Section-21 Budding and/or appendaged bacteria.

Section-22 Sheathed bacteria

Section-23 Nonphotosynthetic, nonfruiting gliding bacteria.

Section-24 Fruiting gliding bacteria: The myxobacteria.

Section-25 Archaebacteria (archaeobacteria).

VOLUME-4

Section-26 Nocardioform actinomycetes

Section-27 Actinomycetes with multilocular sporangia.

Section-28 Actinoplanetes

Section-29 Streptomycetes and related genera

Section-30 Maduromycetes

Section-31 *Thermonospora* and related genera

Section-32 *Thermoactinomyces*

Section-33 Other genera

The *Oxygenic Photosynthetic Bacteria* are classified as follows :

Group-I *Cyanobacteria*

Subsection I Order. Chroococcales

Subsection II Order. Pleurocapsales

Subsection III Order. Oscillatoriales

Subsection IV Order. Nostocales

Family I Nostacaceae

Family II Scytonemataceae

Family III Rivulariaceae

Subsection V Order Stigonematales

Group-II Order Prochlorales Family-I. Prochloraceae other Texa.

This is the first time cyanobacteria have actually been integrated into *Bergey's Manual.* In the 8th edition, they were a separate division; prior to that time, they were not included with bacteria at all and were called blue-green algae. However, Roger Stanier suggested that while their photosynthetic systems resembled those of eukaryotic organism, all their other attributes were those of prokaryotes. His suggestion is now generally accepted by bacteriologists (although not necessarily by all biologists).

TABLE 3.3. Taxonomy of the Cyanobacteria

Sub-section	Order	Typical Genus	Characteristics
I	Chroococcales	*Syne-chococcus*	Unicellular; divide by binary fission or budding; coccoid to rod-shaped.
II	Pleurocapales	*Xenococcus*	Reproduce by formation of multiple baeocytes by internal fission of a single cell (Baeocytes are small cells produced internally by multiple fission of a single parental cell)
III	Oscillatoriales	*Spirulina, Oscillatoria*	Filamentous; fission in one plane; have trichomes but no heterocysts or akinetes (Trichomes are nonsheathed chains of cells)
IV	Nostocales	*Nostoc, Anabaena*	Same as subsection III but can make heterocysts
V	Stigonematales	*Chlorogleopsis*	Multiseriate trichomes or branches; fission in more than one plane

Prochloron is the only genus included in group II. The photosynthetic system of the members of *Prochloron* is intermediate in composition between those of the cyanobacteria and the eukaryotic cells. None of these organisms have been grown in culture, and the only known examples are found as extracellular symbionts associated with marine invertebrates. *Prochlorothrix* is included as "other taxa".

REFERENCES

1. Atlas, R.M. 1989. *Microbiology, Fundamentals and Applications,* Macmillan Publishing Company, New York.
2. Birge, E.A. 1992. *Modern Microbiology, Principles and Applications,* Wm. C. Brown Publishers, Dubuque (U.S.A.).
3. Braugnart, D.C. and R.H. Arnett. 1965. *An Introduction to Plant Biology,* The C.V. Mosby Company, Saint Louis.

4. Burton, G.R.W. 1992. *Microbiology for the Health Sciences,* J.B. Lippincott Company, Philadelphia.
5. Fuller, H.J. and O. Tippo. 1961. *College Botany,* Oxford & IBH Publishing Co., Calcutta.
6. Giese, A.C. 1979. *Cell Physiology,* W.B. Saunders Co., Philadelphia.
7. Holt, J.G. (ed.) 1984–1989. *Bergey's Manual of Systematic Bacteriology* (4 volumes), Williams & Wilkins Co., Baltimore.
8. Soper, R. (ed.) 1990. *Biological Science* 1 & 2, Cambridge University Press, Cambridge.
9. Whittaker, R.H. 1969. New Concepts in Kingdoms of organisms. *Science,* 163 : 150–160.
10. Wilson, C.L. and W.E. Loomis. 1957. *Botany,* Holt, Rinehart & Winston, Inc. New York.
11. Wistreich, G.A. and M.D. Lechtman. 1980. *Microbiology,* Macmillan Publishing Co., Inc. New York.
12. Woese, C.R. 1981. Archaebacteria, *Scientific American,* 244(6): 98–122.

Chapter 4

PURE CULTURE METHODS FOR THE STUDY OF MICROORGANISMS

INTRODUCTION

The ability to determine the characteristics of a microorganism depends in large part on being able to grow *pure cultures* (cultures containing only that organism) of that microbe for study. To cultivate, or *culture* miroorganisms it is necessary to establish a suitable environment, one in which the particular microbe can survive and reproduce. For each type of microorganism there are minimal nutritional requirements, tolerance limits for a variety of environmental factors, and optimal conditions for growth. By understanding the growth requirements of a given microbial species, it usually is possible to establish the necessary conditions *in vitro* (within glass or plastic culture vessels) to support the optimal growth of that microorganism.

Cultures are routinely grown in the clinical microbiology laboratory to aid in the determination of the cause of a patient's disease. Water quality testing laboratories culture microorganisms to determine the safety of the water supply. Various industries grow pure cultures of microorganisms in huge vessels called fermentors to produce numerous products of economic value.

CALL TO MONITOR ATMOSPHERIC POLLEN

With approximately 15 per cent of the Indian population suffering from some allergy or the other, an aerobiologist has spelt out the need for carrying out daily atmospheric pollen monitoring to help sufferers take preventive measures. The practice of publicising day-to-day pollen monitoring results is being presently carried out in the US and many other European countries according to the Indian Aerobiological Society president S.N. Agashe.

Prof Agashe is here to attend the three-day 8th national conference on "Impact of airborne microbes" being held at the Bharati Vidyapeeth's Yashwantrao Mohite college here.

Although some plants flower throughout the year, their pollen percentage in the atmosphere varies from month to month and even daily. This variation in the pollen count influences manifestation of allergy symtoms and, so allergy patients are aware of the allergy concentration in the atmosphere, they can take preventive measures including medication, the expert has pointed out.

Prof Agashe also urged the aerobiologists to take the initiative in compiling pollen calenders, which indicate the occurrence of different types of airborne pollen at different times of the year, for the benefit of clinicians and as an aid to allergy diagnosis.

While compiling the pollen calender, however, only those pollens are to be considered from amongst the numerous types of pollens, which are known to cause allergic disorders, he pointed out.

The conference has been organised in view of the immense impact of airborne microbes not only on human and animal health but also on inanimate objects like sculptures and paintings. The aerobiologists are concerned about bio-pollution or the increasing population of unwanted microorganisms in the atmosphere.

The growth of a microorganism in pure culture requires that all other microbial species be eliminated. Obtaining and maintaining pure cultures requires the elimination of other microorganisms from the growth medium (*sterilization*), the separation of the microorganism being cultured from a mixture of microbes (*isolation*), the movement of the microorganism from one place to another without contamination (*aseptic transfer*), and the maintenance of the culture in a favourable environment (*incubation*).

Eearly Culture Methods

Credit for developing the methodoogy to grow pure cultures generally is given to Robert Koch (1843–1910), a German physician, and his co-workers. Koch, like Louis Pasteur (1822–1895), intensely believed in the possibility of microorganisms being the causative agents of many diseases that afflict humans. According to his postulates, it is essential to study organisms in pure culture. In search for ways in which to accomplish this task, Koch soon determined that the most practical method was to place microorganisms onto a solid surface.

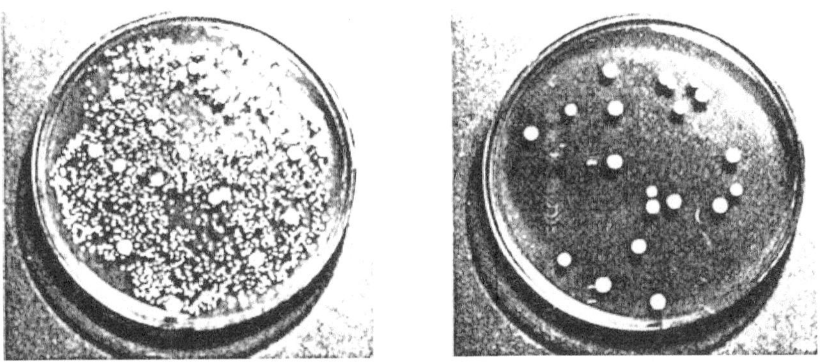

Fig. 4.1. Bacterial colonies growing on two types of agar media in petri dishes. The larger colonies are *Staphylococcus aureus* and the smaller colonies *Micrococcus luteus*.

(Atlas, p. 113)

Microorganisms grown on a solid medium tend to stay more or less in one place, depending on the extent of their motility. Each cell will eventually give rise to a large cluster of cells that form a mound on the surface of the medium. Each cluster of cells is called a *colony*, and unless the colonies overlap as they enlarge, the cells in a colony are all descendents of one original cell. If a new culture is *inoculated* (started) from a single truly isolated colony, it is of necessity a pure culture. Prior to Koch, microorganisms were grown in liquid broths. Koch, between 1881 and 1883 tried growing bacteria on solid substrates, either on the cut surface of a potato or on gelatin (a protein extracted from bone or skin of animals) supplemented with various nutrients, but both gave less than ideal results. There were two main difficulties. Many bacteria are capable of degrading starch in potato or protein in the gelatin and therefore tend to destroy the solid surface necessary for colony formation. Moreover, Koch also found it difficult to maintain a pure, uncontaminated culture when working around the various unprotected cultures that soon accumulated in his laboratory. Ultimately Koch used a substance called *agar* as a solidifying agent. The suggestion for using agar originated with the New Jersey born wife of one of the investigators at Koch's Institute, Frau Fanny Hesse, who had seen her mother using agar to make jellies. The use of these solidified media permitted the isolation and identification of microorganisms.

Agar is a mixture of complex polysaccharides (chains of sugars) produced by several types of algae. Originally, it was extracted primarily from the genus *Gelidium*, but now the genus *Gracilaria* has predominated. There are two major families of polysaccharides found in agar, agarose (the fraction that will solidify by itself) and agaropectin (the fraction that will not solidify by itself).

Agar has the property of being solid until it is heated to a temperature greater than 90°C (about 100°C), at which time it lilquifies and forms a suspension with water. It will stay liquid down to about 40°C, at which time it again solidifies. Agar has yet another advantage over other solidifying agents. Despite its polysaccharide nature, relatively few bacteria are able to break it down into its component sugars; therefore, it serves as an essentially inert substrate for bacterial growth. Agar is still the primary solidifying agent used today, although various other chemicals, primarily silica derivatives, are used for special purposes.

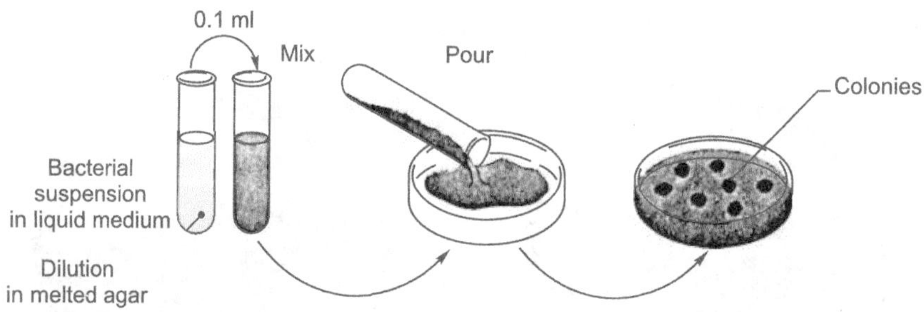

Fig. 4.2. The pour plate technique for enumerating microorganisms. A known volume of a microbial suspension is mixed with liquefied agar medium and poured into a petri plate. After incubation the numbers of colonies that develop are counted, and the concentration of microorganisms in the original suspension is calculated.

(Atlas, p. 115)

The other major development in Koch's laboratory was made by Richard Julius Petri, who in 1887, described the use of a new type of culture dish for growing bacteria on semisolid media. The basic design has become known as the *Petri dish* and is used in virtually all microbiological laboratories in essentially the same design described by Petri. The Petri dish was invaluable because it permitted cultures to be stacked in a small space and transported from one place to another without contamination.

CULTURE MEDIA AND THEIR INGREDIENTS

A culture *medium* (plural media) is any substance or mixture of substances used to grow one or more kinds of microorganisms in the laboratory. Bacteriologic media frequently are classified as complex or defined according to their chemical composition. A complex medium is one that contains such chemical composition. A complex medium is one that contains such chemically ill-defined substances as yeast or beef extract, while a defined medium can have 10 to 20 different chemical substances present, the identity of each being known precisely. For example, many media contain beef extract (a complex mixture of proteins, carbohydrates, lipids, and other biochemical constituents obtained by extracting the water soluble components from beef tissue), peptones (an enzymatic digest of protein–containing amino acids and other nitrogen–containing compounds, as well as vitamins and other compounds), and yeast extract (an aqueous extract of yeast cells containing vitamins and other growth factors). In the clinical laboratory we often incorporate blood into media that are designed for the culture of human pathogens. Such complex culture media will support the growth of many different types of heterotrophic organisms but not those of others that have very specific growth requirements.

Some examples of these media are given below, in which all ingredients are listed per liter (also see Agrawal and Hasija, 1986, and Fairbrother and Taylor, 1965.

Luria Brother :		
(Complex Medium)		
NaCl	8	g
Yeast extract	6	g
(Water – soluble portion of yeast Cell lysate)		
Tryptone	10	g
(from casein digested with the enzyme trypsin)		
Pseudomonas Medium :		
(Defined Medium)		
K_2HPO_4	2.65	g
KH_2PO_4	2.08	g
NH_4Cl	1.0	g
$MgSO_4 \cdot 7H_2O$	0.5	g
$CaCl_2$	0.5	mg
Ferric ammonium citrate	5.0	mg
Succinic acid	8.0	g

Methanobacterium Medium :
(Defined Medium)

K_2HPO_4	15	mg
KH_2PO_4	300	mg
$(NH_4)_2SO_4$	300	mg
NaCl	610	mg
$MgSO_4 . 7H_2O$	1,200	mg
$CaCl_2 . 2H_2O$	800	mg
Na_2CO_3	400	mg
$MnSO_4$	5	mg
$FeSO_4$	1	mg
$CoCl_2$	1	mg
$ZnSO_4$	0.1	mg
$CuSO_4$	0.1	mg
$Al\ K\ (SO_4)_2$	0.1	mg
H_3PO_4	0.1	mg
$Na_2Mo\ O_4$	0.1	mg
Cysteine HCl	60	mg
Resazurin	1	mg
Nicotinic acid	0.05	mg
Folic acid	0.02	mg
Pyridoxine HCl	0.1	mg
Biotin	0.02	mg
Thioctic acid	0.05	mg
Thiamine HCl	0.05	mg
Riboflavin	0.05	mg
Vitamin B_{12}	0.001	mg
Calcium pantothenate	0.5	mg

The term *synthetic* is sometimes used instead of the word *defined* in order to imply that the substances present have been chemically purified rather than extracted from natural products.

Materials used in complex media are generally water soluble portions of cells or tissues for various types of organisms. Meat extracts from different animals and/or organs and water–soluble portions of disrupted yeast cells commonly are used. Regardless of the ingredients, a culture medium must fulfil all the nutritional needs of the organisms to be grown on it. These needs are the same as the general nutritional requirements for plants or animals. An organism must have a source of energy to grow as well as various chemical elements to synthesize the molecules of which it is composed, and for the maintenance of transport activities across the cytoplasmic membrane. Microorganisms generally have many other specific inorganic nutritional requirements. In order of decreasing quantity, these elements generally include carbon, hydrogen, nitrogen, oxygen, phosphorus, potassium, sodium, sulphur, magnesium, calcium, and iron. *Trace elements* such as molybdenum, zinc, manganese, and copper also are required but in such small quantity that the normal contaminants in standard chemicals will often suffice to supply them.

Some other growth media are mentioned below.

Nutrient Broth :

(General medium for cultivation of heterotrophic bacteria)

Beef extract	3 g
Peptone	5 g
Water	1000 g

Glucose – Inorganic Salts Medium :

(For growth of *E. Coli*)

Glucose	5 g
K_2HPO_4 (Potassium dibasic phosphate)	7 g
KH_2PO_4 (Potassium monobasic phosphate)	2 g
Mg SO_4 (Magnesium sulphate)	0.1 g
$(NH_4)_2SO_4$ (Ammonium sulphate)	1 g
Water	1000 g

Glucose – Yeast Extract Medium :

(for growth of bacteria with growth factor requirements)

Glucose	5 g
K_2HPO_4 (Potassium dibasic phosphate)	7 g
KH_2PO_4 (Potassium monobasic phosphate)	2 g
Mg SO_4 (Magnesium sulphate)	0.1 g
$(NH_4)_2$ SO_4 (Ammonium sulphate)	1 g
$FeSO_4$ (Ferrous sulphate)	0.1 g
Yeast extract (contains numerous vitamins, amino acids, and other growth factors)	5 g
Water	1000 g

The nutritional requirements of bacteria are diverse. Some need essentially no organic material for growth, whereas others require many amino acids, nucleic acid bases, and even vitamins to grow; therefore, the media necessary to grow microorganisms range from the simple to the very complex.

In a growth medium, generally phosphorus is supplied in the form of phosphate, sulphur in the form of sulphate, and nitrogen as ammonium ion.

Carbon can be supplied in a wide variety of forms depending on the organism to be grown. *Autotrophs* can take carbon dioxide from the air and fix (reduce) it into carbohydrate compounds that can be used to provide the majority of compounds they need for metabolism. Often autotrophic microorganisms are grown in a medium containing simple inorganic compounds. Green plants, algae, and several ecologically important groups of bacteria are autotrophic.

In contrast to autotrophs, *heterotrophic* organisms require organic compounds as a source of cellular carbon and energy. Some heterotrophic microorganisms have simple nutritional requirements and are able to reproduce by using a single carbon source as their growth substrate. Such organisms have the metabolic capability of converting a particular organic carbon compound, such as glucose, into all the macromolecular constituents that compose the structures of those organisms. Other heterotrophic microorganisms have more complex growth requirements which must be met before they can be grown *in vitro.* The nutritionally most demanding microorganisms are termed *fastidious;* some such fastidious microorganisms are important human pathogens, and their culture in the clinical microbiology laboratory requires the use of various specialized media. Fastidious microorganisms are able to reproduce only under very restricted conditions. In some cases, it is impossible to grow such microorganisms in pure culture because we have been unable to define and to create the necessary growth conditions.

For the growth of heterotrophic microorganisms, an organic substrate serves as the source of essential carbon and energy. Normal laboratory culture media often contain proteins and/or carbohydrates as growth substrates. For the growth of specific heteroptrophic microorganisms, individual growth substrates may be included in the culture medium.

A few bacteria have the capability of reducing gaseous nitrogen (N_2) into ammonium ion and, therefore, do not require the addition of any nitrogenous compounds to the medium. This process, called *nitrogen fixation,* is analogous to carbon dioxide fixation and is extreamely important environmentally. Certain other bacteria are capable of reducing nitrate or nitrite ions to ammonium ions in order to supply their metabolic needs.

Not all microorganisms can be grown by using defined media. The nutritional requirements of many microorganisms are simply not known. These microorganisms are able to reproduce in nature where their nutritional needs are met, but we do not understand their growth requirements well enough to define the appropriate laboratory conditions needed for their growth. For example, we are typically able to culture less than 1 percent of the microorganisms that are present in a natural water or soil sample.

In some cases the natural environment in which the microorganism reproduces can be established in the laboratory without actually defining the growth requirements. For example, soil extract is often added to growth media to support the reproduction of soil microorganisms. Some microorganisms, such as viruses or other obligate intracellular microbial species can reproduce only within cells. Generally such microorganisms have metabolic capabilities and rely on the metabolism of the host cell to meet their nutritional needs. Such organisms can be grown in *tissue culture* or by using other methods that provide the necessary cells within which these microorganisms can reproduce. In these methods the host cells are cultivated *in vitro* (within glass) and the microorganisms are grown *in vivo* (within a living system) inside of the host cells. For example, some viruses are cultivated by inoculating them into eggs containing developing chicken embryos. Other viruses are cultivated in human and other animal cells that are grown as a tissue culture.

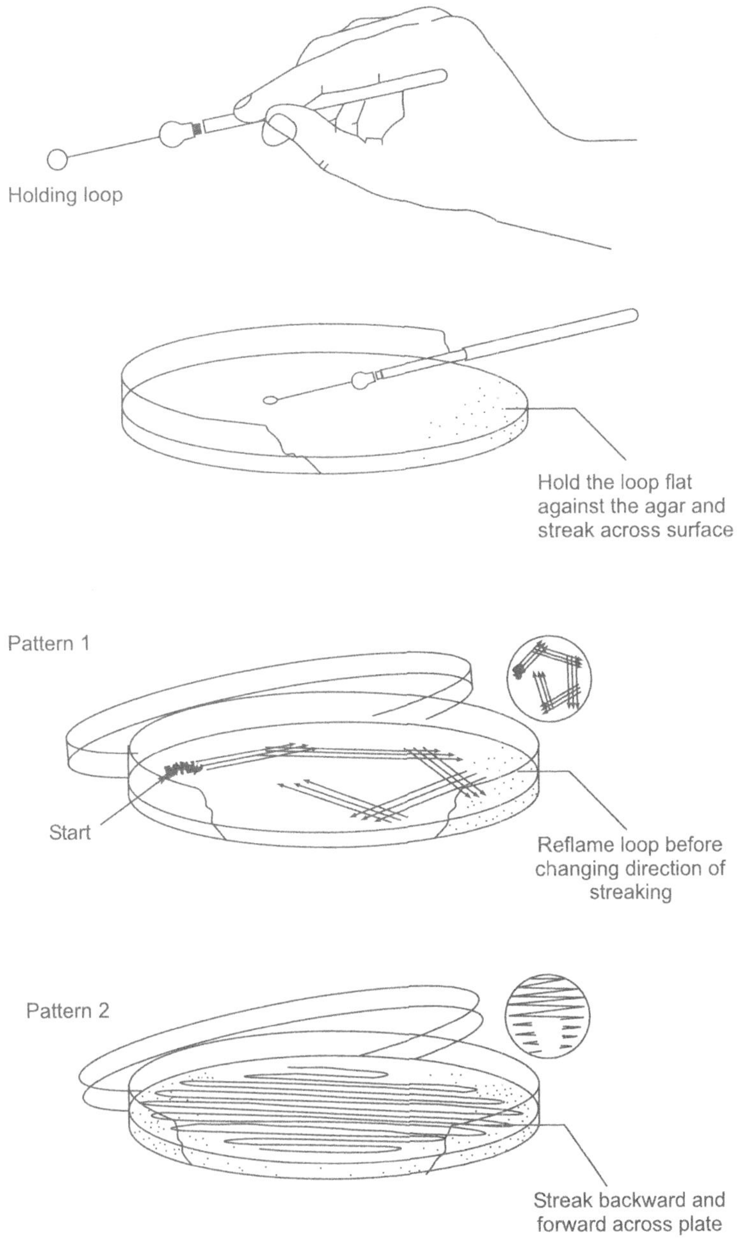

Fig. 4.3. Streaking for the isolation of pure cultures, showing two different streaking patterns; in this procedure a culture is diluted by drawing a loopful of the organism across a medium untill only single cells are deposited at a given location. The multiplication of each isolated cell results in the formation of a discrete colony.

(Atlas, p. 113)

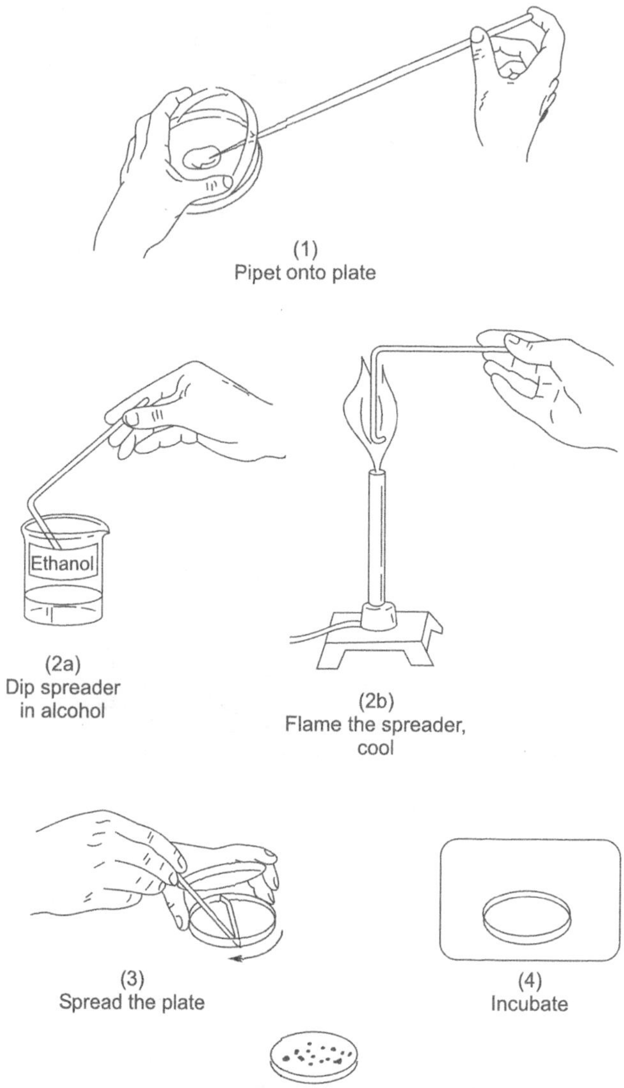

Fig. 4.4. The spread plate technique for enumerating microorganisms. (1) Aseptically apply a known volume of a suspension to a suitable solid medium; (2) sterilize a spreading rod by dipping in alcohol and flaming; (3) use the sterile rod to spread the suspension over the surface of the medium; (4) incubate; (5) count colonies and calculate number of microorganisms in the original suspension.

(Atlas, p. 114)

ENRICHMENT CULTURE

Technique

By taking into account the metabolic capabilites of specific microorganisms, it is possible to design growth media that will select for the growth of particular microorganisms based on their nutritional requirements. In the *enrichment culture* technique employed to isolate specific groups of microorganisms, the culture medium and incubation conditions are designed to favour the growth of a particular microorganism. For example, in order to isolate microorganisms capable of metabolizing petroleum hydrocarbons, one can design a culture medium containing a hydrocarbon as the sole source of carbon and energy. By doing so, one establishes conditions whereby only microorganisms capable of metabolizing hydrocarbons can grow. Because other microorganisms cannot reproduce in this medium, one thereby selects for hydrocarbon–utilizing microorganisms.

Similarly, a culture medium that favours the growth of a particular pathogen can be used to increase the numbers of that organism before attempting to isolate it in pure culture. Such enrichment culture techniques are sometimes used in the clinical microbiology laboratory to increase the likelihood of isolating a particular disease–causing microorganism, which might otherwise be obscured by the occurrence of larger number of nonpathogenic microorganisms in a sample.

TABLE 4.1. Some Culture Conditions that are Employed for Enrichment Cultures

Carbon Source	Oxygen	Nitrogen Source	Inorganic Sulphur Source	Light	Organism Enriched for
Nonfermentable organic carbon	+	N_2	–	–	*Azotobacter*
Nonfermentable organic carbon	–	NO_3^-	–	–	Denitrifying bacteria
Nonfermentable organic carbon	–	–	SO_4^{--}	–	*Desulfovibrio*
Hydrocarbon	+	NH_4^+	SO_4^{--}	–	Hydrocarbon oxidizers
–	+	NH_4^+	SO_4^{--}	–	*Nitrosomonas*
–	+	NO_2^{--}	SO_4^{--}	–	*Nitrobcater*
–	+	–	S	–	*Thiobacillus*
–	+	N_2	SO_4^{--}	+	Cyanobacteria
–	–	NH_4^+	H_2S	+	Green sulphur bacteria

(Atlas, p. 118)

Selective Culture Media

Microbial reproduction is not only affected by the qualitative composition of the growth medium but also by the concentrations of the various components in the medium. Although a nutrient may be essential for growth in low concentrations, it may be toxic at higher concentrations. For example, some autotrophic microorganisms require trace amounts of growth factors but cannot grow in the presence of high amounts of organic matter. Similarly, many microorganisms may use hydrogen sulphide at low concentrations, although at higher concentrations hydrogen sulphide is very toxic to most microoganisms. Accordingly, in order to achieve maximal growth rates, it is necessary to define carefully the growth medium.

COMPACT INVERTED MICROSCOPE FOR INVESTIGATION OF TISSUE CULTURE, LIVING PROTOZOA, etc.

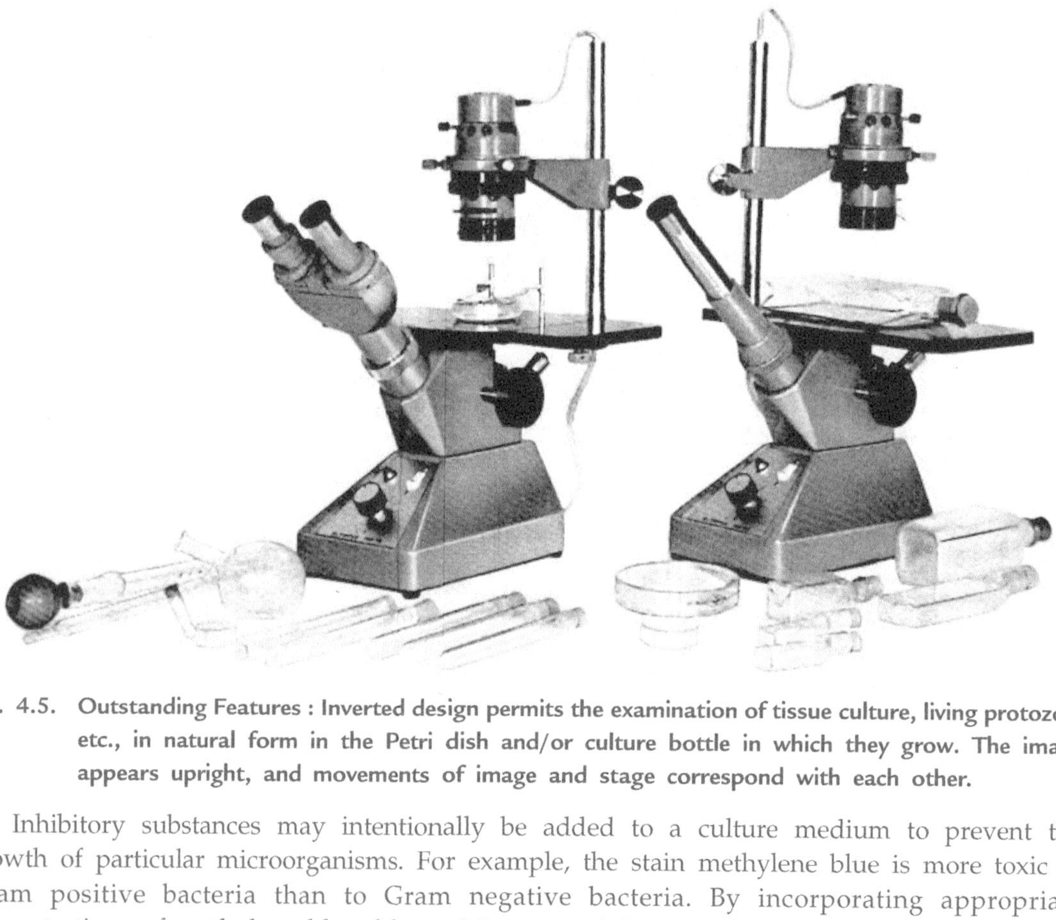

Fig. 4.5. Outstanding Features : Inverted design permits the examination of tissue culture, living protozoa, etc., in natural form in the Petri dish and/or culture bottle in which they grow. The image appears upright, and movements of image and stage correspond with each other.

Inhibitory substances may intentionally be added to a culture medium to prevent the growth of particular microorganisms. For example, the stain methylene blue is more toxic to Gram positive bacteria than to Gram negative bacteria. By incorporating appropriate concentrations of methylene blue (about 0.5 per cent) into a culture medium, the growth of Gram positive bacteria may be inhibited, while not interfering with the growth of Gram negative bacteria. The medium eosin methylene blue agar is frequently used for the selective culture of Gram negative bacteria, such as *E. coli.* The methylene blue in this medium inhibits the growth of Gram positive bacteria, and other constituents of the medium favour the growth of microorganisms found in association with animal intestinal tracts.

Various antibiotics can be similarly incorporated into a medium to prevent the growth of specific types of microorganisms. Antibiotics are substances produced by microorganisms that inhibit or kill other microorganisms; these substances are used in medicine to inhibit the growth of pathogenic microorganisms. Such culture media are frequently used in microbiological laboratories in order to select for the growth of specific microbial populations. For example, by incorporating an antibacterial antibiotic into a medium, fungi can be grown free of bacterial contamination. By incorporating the right inhibitory compound into a growth medium, the selectivity of that medium for a particular type of microorganism is established. Also, lowering the pH of the medium by adding acid inhibits bacterial growth, permitting the selective culture of fungi. Selective media are used in many areas of microbiology, including clinical microbiology where such media are employed for the isolation of pathogenic microorganisms.

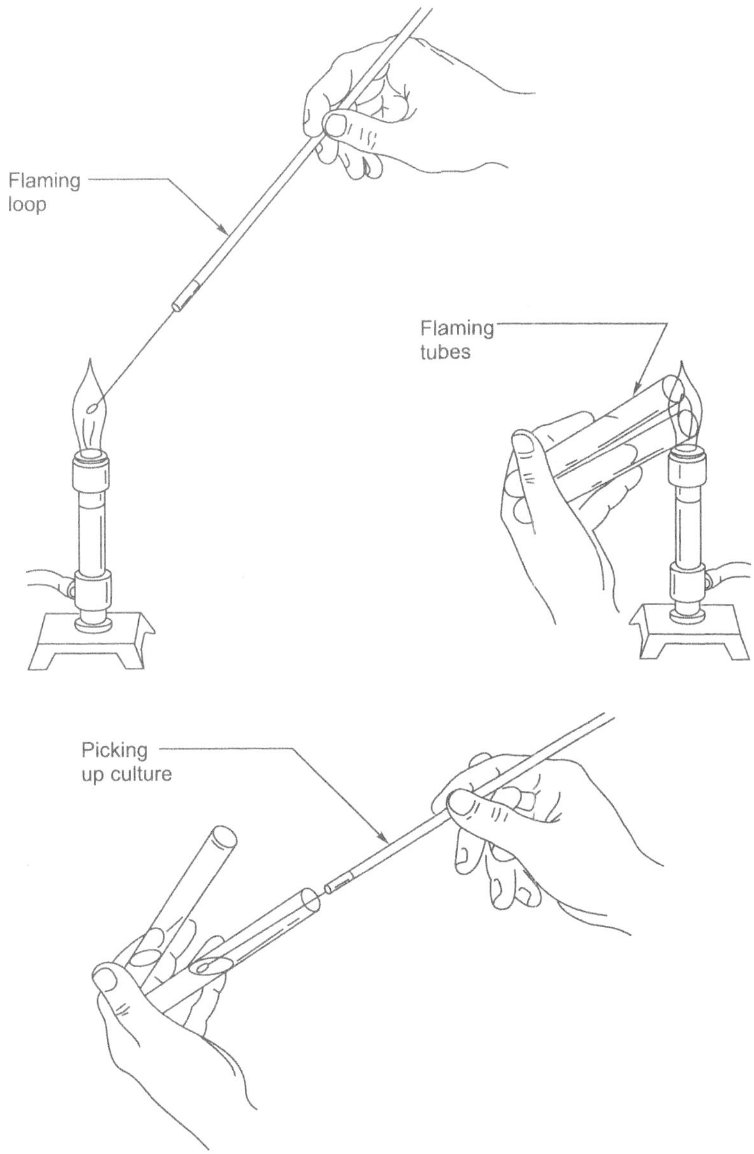

Fig. 4.6. Steps in aseptic transfer of bacteria.
(Atlas. p. 112)

SUMMARY

A culture medium must contain the nutrients that are required by the organism for growth. In some cases, it is necessary to include a variety of growth factors, such as vitamins and amino acids. In other cases, microorganisms are capable of growing on simple media. In general, microorganisms require a source of carbon, nitrogen, phosphorus, iron, magnesium,. sulphur, sodium, potassium, and chloride. Oxygen is also required by some microorganisms, but oxygen is toxic to others. Parameters that have a great influence on microbial growth and death rates include organic and inorganic nutrient concentrations, concentrations of organic and inorganic inhibitory substances, temperature, oxygen concentrations, water availability, and pH. The combined effects of these factors generally determine the ability of a microorganism to survive and grow in a natural system. It is also these parameters that are manipulated in controlled situations to regulate the rates of microbial growth. The modification of environmental parameters to control the rates of microbial growth is applied in many areas of microbiology. Understanding the factors that control the growth of a microorganism provides the basis for cultivating that microorganism.

Most microorganisms can be grown in pure culture on an appropriate medium or within cultivated cells. Microorganisms are transferred from one medium to another using aseptic technique, that is, methods that prevent contamination of the microbial culture and also protect against the uncontaminated spread of the organism being transferred.

Microorganisms can be isolated usign several different techniques. Streak plate, spread plate, and pour plate methods are all designed so that individual microbial cells are sufficiently separated so that when they reproduce they develop well isolated colonies. Colonies are macroscopic aggregations of *clones* of microbial cells (the progeny of a single individual asexually). In some cases specialized selective or differential media are employed to isolate specific microbial species. Enrichment cultures that favour the selective multiplication of specific types of microorganisms are also used to isolate microorganisms.

These methods have many applications, including the isolation and culture of pathogens that is essential for the diagnosis and initiation of appropriate treatment of various diseases.

METHODS FOR CULTURING ALGAE

Although it is possible to make many studies on algae collected from nature, cultures are essential for most detailed studies. A *unialgal* culture is one that has only one kind of alga, even though it may be contaminated with bacteria, whereas a pure (or *axenic*) culture is free of all contaminants. For studies of the taxonomy or life cycle of an alga, unialgal cultures may be satisfactory, but for physiological and biochemical studies, pure cultures are essential. In recent years the development of culture methods for algae has been considerably advanced, and it is now possible to culture virtually any alga, even some of the seaweeds that earlier had proved difficult to grow.

Algae are placed into artificial cultivation for a number of reasons, including the following :

(1) To trace the life cycle of a species and to learn the morphological and general phases of the life cycle.

(2) To study manner of gamete-formation, gametic union, and the ecological factors related to reproductive phases.

(3) To determine morphological and reproductive characteristics as aids in establishing taxonomic criteria and in developing phylogenetic relationships.

(4) To provide specimens for cytological and ultra microscopical examination of cell structure, organelles, and cell division.

(5) To study the physiology of algae (and to apply information to other plants); photosynthesis (its mechanism and its products); elaboration of ergastic substances; respiration; nutrition requirements; response to light; and temperature. Culture for physiological studies may involve either organic or inorganic media or a combination of both.

(6) To study the elaboration of extrametabolites; endo and exotoxins; and antibiotics.

(7) To determine the chemical composition of algae for both scientific and practical purposes.

(8) To study the production and productivity rate, and their relationships to limnology, to the food chain of aquatic animals, and as food for domestic animals, and for man.

(9) To study and test algicides, their effects on different species and their use in various concentrations.

(10) To study relationships of algae to water quality or sanitation.

Isolations Methods

Because of their large cell size, many algae can be isolated directly from natural materials with capillary pipettes. A sample of water containing the alga of interest is placed in a dish and observed under a dissecting microscope. The capillary pipette held by a rubber suction bulb is carefully lowered into the water, and the tip is placed adjacent to an algal cell. The cell is then drawn into the capillary, and after microscopic examination of the capillary to make certain that only one cell is present, the contents of the capillary are expelled into the well of a depression slide, which contains the culture medium. The well is covered to prevent evaporation, and the slide is incubated in the light. After the alga has reproduced, the contents of the well can be transferred to a large volume of culture medium to allow further growth.

In many cases the alga of interest may not be present in sufficient qunatity in the natural material to allow its direct observation and isolation. If a sample of natural material, such as pond water or soil, is added to a culture medium, and then incubated, a variety of forms may develop that were in the minority in the original material, and those of interest can be isolated by the capillary pipette method.

In some cases one can use selective culture media to prepare enrichment cultures for species requiring particular environmental conditions. For instance, media of low pH can be used to select algae from acidic natural habitats such as bogs. Motile algae that are positively phototactic can be isolated by illuminating from only one end a tube filled with natural water, and removing material from the illuminated end for culture and further study. Likewise, high temperatures of incubation can be used to enrich for thermophilic algae.

Algal Culture Media and Growth Conditions

All the various techniques and media employed attempt to do one of the two things: either to duplicate natural conditions as nearly as possible, or to provide a set of artificial conditions which are suitable for growth and which can produce desired results. Flasks, tubes, bottles, and Petri plates with appropriate media are used for minute or microscopic algae, whereas basins, glass vats, or tanks must be employed for macroscopic algae, especially the larger marine forms. The culture vessels, both large and small, may be connected (either singly or in series) with a continuous flow nutrient supply and gas supply (O_2, CO_2, etc). The culture vessels or tubes may be vertical or horizontal and so equipped that they receive nutrients at constant temperatures and under controlled illumination. In many instances culture vessels are maintained in a shaker device to provide a more nearly equal exposure of plants to medium and to light, and to prevent setting out of cells. Such a vibration is often desirable because agitation somewhat duplicates water movement in the natural habitat.

Natural water are usually more or less deficient in certain algal nutrients and hence cannot be used directly as culture media if large algal populations are desired. For freshwater alge a variety of culture media containing only inorganic compounds can be devised, which will permit the growth of many species. Although many algae will grow well in simple media, it is often advantageous to supplement the basal salts medium with trace elements or organic growth factors.

In general, a temperature of 20°C (28°C commonly applield also), and a light source of from 800 to 1200 (or up to 2000) foot candles is used. For a host of fresh water species, ordinary room temperature and the day light from a north window are suitable for successful growth. Such cultures may be maintained in jars or batteries of lake water, or in soil water culture media. Many marine algae also require vitamins, and these must also be added. Any of the algal culture media can be made semisolid by adding agar. In most cases it is desirable to keep the agar concentration lower (1.0 to 1.2 per cent agar) than for bacteria and fungi since algae usually form colonies better on less rigid agar.

After the agar mixture is sterilized, Petri plates are poured and allowed to cool. Then by a micropipette or sterile transfer loop or needle, the surface of the agar is streaked with inoculum from an alga culture (preferably from a unialgal culture). This permits the development of unicellular, colonial, or prostrate "creeping" algae in an isolated condition.

An added carbon source is essential to achieve high cell yields. The most important carbon source of these autotrophs is CO_2, which is present in normal air in only trace amounts. By adding $NaHCO_3$ or by aerating cultures with air enriched with CO_2, much heavier cell yields will usually be obtained. Addition of organic carbon sources such as acetate will often increases cell yield, but can only be used with bacteria–free cultures since acetate is also an excellent bacterial nutrient.

Depending upon the objective of the culture study, types of cultures are established. For some purposes a mixture of algae (a gross culture) may be grown together in a balanced aquarium. In others, a plant may be isolated and placed in a medium to form *unialgal* culture, either free from or with bacteria. Some cultures are established by isolating a single cell or a spore (zygospore, zoospore, or other) to initiate an *axenic* culture. Such axenic culture is especially necessary for studies to determine and isolate physiological or genetic clones or races.

In continous–flow cultures the supply of nutrients, the temperature, and light can be regulated so that the growth rate and population development can be measured at a desired time, the cells or plants being all at approximately the same stage. This is referred to as a *synchronous* culture.

Soil–Water Culture Medium

The simplest method to grow selected species is the soil–water extract method. Essentially, this method takes advantage of the nutrient elements and vitamins in soil. A layer (about one – fourth inch) of garden loam is placed in an Erlenmeyer flask, or about four centimeters in a test tube. The vessel is then nearly filled with sterile water, plugged with cotton and pasteurized in a steamer for three hours. This allows sufficient time to kill algae and other organisms present (except for bacterial spores). After pasteurization and after the flasks have been cooled, desired amounts of algal material are added, the nature of the specimen determining the amount to be introduced. One ml (or less) of a suspension of unicells or colonial forms is sufficient.

There are various modifications of the soil–water method. One involves pasteurizing the soil in water and then pouring off the liquid to be used as a basic medium to which various selected ingredients are added. Another involves using the soil–water to make up an agar medium for Petri plate or slanted tube cultures if a surface growth of an alga is desired.

Mineral Salt Solutions for Algal Culture

(1) **The "A–Z" Solution**

$Al_2(SO_4)_3$	1.0 gram
$Ca(NO_3)_2.6H_2O$	1.0
$CuSO_4.5H_2O$	1.0
H_3BO_4	11.0
KBr	0.5
KI	0.5
$MnCl_2.4H_2O$	7.0
$NiSO_4.6H_2O$	1.0
$SnCl_3.1H_2O$	0.5
TiO_2	1.0
$ZnSO_4$	1.0
LiCl	0.5

When dissolved in one liter of water this stock solution is then further diluted so that the above amounts of minerals are distributed through 18 liters of water for a stock supply. This solution is suitable for a great variety of algae (fresh–water) and for other plants.

(2) **Chu No. 10 Solution**

Chu's solution (with ferric citrate recommended in place of his original ferric chloride) is (in mg per liter); sometimes used with 10% soil extract medium –

Ca	4.0	N	5.25
Cl	13.3	Na	7.5
Fe	0.18	P	0.89
HCO_3	12.0	Si	4.6
K	2.2	SO_4	20.0
Mn	0.01	Mg	5.0

Or

$Ca(NO_3)_2$	0.0040%	$Fe\ Cl_3$	0.00008%
K_2HPO_4	0.001%	$Mg\ SO_4.\ 7H_2O$	0.0025%
Na_2CO_3	0.002%	$Na_2Si\ O_3$	0.0025%

This solution may be prepared as follows (in mg per liter) :

$Ca(HCO_3)$	160.0
$MgSO_4$	25.0
$NaSi\ O_3$	20.0
K_2HPO_4	5.0
Ferric citrate and citric acid each	1.0
$MnSO_4$	0.03

All the solutions listed above are used as stock, and from each the following proportions are added to one liter of water or nutrient medium :

$Ca(HCO_3)_2$ (as $CaCO_3$)	10 mg
$MgSO_4$	25 ml
$NaSiO_3$	20 ml
K_2HPO_4	5 ml
NH_4Cl	20 ml
Ferric citrate and Citric acid	100 ml
$MnSO_4$	0.3 ml

The above combination is suitable for diatoms, green algae, and also for blue–green species when organic substances are added.

(3) **Culture Medium for Blue–Green Algae** (Cyanobacteria)

$CaCl_2.2H_2O$	0.0359
Citric acid	0.003
Ferric citrate	0.003
KCl	0.0086
$MgCl_2.\ 6H_2O$	0.0209
$NaCO_3$	0.02
$NaNO_3$	0.0413
Na_2PO_4	0.0082
$Na_2Si\ O_3$	0.250
Na_2SO_4	0.0146

(in gram per liter)

(4) **Synthetic Medium for Dinoflagellates (*Katodinium* and *Gyrodinium*)**

NaCl	2.4 gm
$Mg\ SO_4.\ 7H_2O$	0.9 gm

KCl	0.07 gm
Ca (with Chlorine)	30.0 mg
$NaNO_3$	5.0 mg
$NaSiO_3.9H_2O$	1.0 mg
Nitrilotriacetic acid	20.0 mg
B_{12}	1.0 mg
D – ribose	5.0 mg
Tris	0.1 gm
Biotin	1.0 μg
Thiamine HCl	0.01 mg
Pyridoxine HCl	2.0 μg
Folic acid	2.0 μg
K_2HPO_4	0.1 mg
Urea	0.1 mg
DL – alanine	5.0 mg
Na_2 fumarate	5.0 mg
$(NH_4)_2SO_4$	0.1 mg
H_2O	100 mg
pH adjusted	7.6 – 7.8

(5) A medium suitable for *Volvox*, and probably most volvocoids is suggested as follows:

One kg of garden soil in one liter of water is autoclaved (15 pounds pressure for one hour). After this cools the water is decanted. To 30 ml of the soil water one ml of 5% KNO_3 and 300 ml of distilled water are added to produce a culture liquid.

The Culture of Marine Algae

The culture of marine algae usually requires the preparation of artificial sea water. Natural sea water for culture purposes has been found unsuitable because some salts precipitate during sterilization in the autoclave. In general, it has been found necessary to increase the amounts of certain salts in natural sea water to produce suitable media. Further, modifications of techniques have been found necessary to make natural sea water successful as a culture medium.

An artificial sea water medium, especially useful for the culture of phytoplanktons, is as follows (in micromoles) :

NaCl	410,000
$MgSO_4.7H_2O$	24,000
$MgCl_2.6H_2O$	22,000
$CaCl_2.2H_2O$	10,000
KNO_3	1,000

K_2HPO_4	100
$Na_2SiO_3.9H_2O$	100
$FeCl_3.6H_2O$	1.5
H_3BO_3	185.0
$MnCl_2.4H_2O$	7.0
$ZnCl_2$	0.8
$CoCl_2.6H_2O$	0.02
$CuCl_2.2H_2O$	0.0002
Sodium ethylenediamine tetracetate	30.0

Johansen (1940) has described the various methods for collection, cultivation, preservation, fixation, and staining of all the classes of algae separately.

Purification of Algal Cultures

The development of bacteria–free cultures from unialgal cultures is easy with some algae and difficult with others. If the alga has a mucilaginous sheath to which bacteria can adhere, purification is especially difficult since agar–streaking methods are usually ineffective with nonmotile forms. Several methods for purification can be used. *Washing methods,* useful for unicellular algae, involve the washing of single cells with sterile culture medium; the cell is passed successively from one drop of culture medium to another with a capillary pipette, and the process is repeated until the algal cell is free of contaminants. Motile organisms that are positively phototropic can be induced to wash themselves by providing unilateral illumination and allowing the cells to move through a sterile medium toward the light. *Streaking methods* on agar or gelatin media can be used for those organisms that will form colonies on semisolid medium and are hardly enough to withstand the streaking process. Algae that glide across the surface of agar can often cleanse themselves by creeping movement, through the mechanical action of the cells moving over the agar gel. Finally, to rid algal cultures of bacteria, *antibiotics* can be used since many antibiotics are active against bacteria but are ineffective against the eucaryotic algae. After several passages through antibiotic–containing medium, the alga may then be transferred to antibiotic–free medium for maintenance and further study.

CONCLUSIONS

Whereas the preparation of media for the successful culturing of algae, especially marine species, is often tedious and difficult, such studies are highly rewarding. Some of the objectives of culture methods outlined above apply especially to marine algae, for in the sea there is an impressive number of genera for which life histories are still unknown or imperfectly so. Accordingly, the taxonomic position of many species is uncertain and the duplication of names for two different phases of the same plant await clarification. In addition to these aspects, the many details involving cytology of life cycle stages, and the anatomy of reproductive organs, pose a host of problems.

METHODS FOR THE STUDY OF FUNGI

Fungi, like all other organisms, should be studied as far as possible in their natural habitat, but the mycelia of many of the smaller species are so intermingled in these circumstances that the stages of their development cannot with certainty be separated, and the use of pure cultures becomes essential.

Cultivation of fungi is usually not especially difficult: nutritional requirements are generally simple, and adequate culture media can easily be prepared. An exception to this statement must be made for the rusts, which are obligate parasites and live only upon or within the living tissues of their hosts. For example, *Puccinia graminis tritici* lives only on wheat, some species of grasses, and some species of mulberry. These host plants when killed will no longer support growth of the fungus. However, many species, which in the past were considered to be obligate parasites, have since been cultured on nonliving media. Many fungi grow rather slowly even under optimal conditions. However, the only real problems in culturing slow–growing fungi are the desiccation of the agar medium during the long incubation period and the possibility of contamination by more rapidly growing, airborne fungi. Petri dishes should be wrapped in plastic film or placed in sealed jars during incubation, thus simultaneously preventing both desiccation and entry of contaminating spores.

ISOLATION AND CULTIVATION

The isolation and cultivation of fungi in pure culture constitutes an essential procedure in studies involving their structure, developmental history, pathogenesis, and physical activities. However, it should be borne in mind that fungi may be quite abnormal when grown on substrata entirely unlike those encountered in nature, and react abnormally when entirely separated from the associative effects of other organisms. These influences might be expected to be more pronounced among pathogenic species, having restricted host relations, than among saprophilous species. Therefore, in nature pure cultures seldom occur and, if they do, are not maintained for appreciable periods.

The common obligate saprophytes and such facultative saprophytes as *Pythium* and *Phytophthora* grow readily in culture. Many coprophilous forms develop freely on sterilised dung or on dung agar to which scraps of grass or filter paper have been added. Aquatic species may sometimes be obtained by leaving dead insects floating in pond water, and less often, by placing pieces of twig, taken from ponds, and still bearing bark, in about five hundred times their volume of cool, clear water. A search through dying material of *Spirogyra* or *Oedogonium* may yield some of the smaller forms.

In the laboratory, fungi are most conveniently grown in Petri dishes; tall species may be grown in beakers covered with half a Petri dish, or in conical flasks plugged with cotton wool. Cultures may also be made in test tubes containing a suitable medium; these cultures are useful for retaining a stock of the fungus, but they are inconvenient for purposes of examination.

Isolation

Isolation of a tall species is usually easy, as a platinum wire can be thrust among the spores, or as in *Mucor*, a sporangium can be pinched off with sterilised forceps: subsequent transfer to a medium usually gives a clean culture. When a mixture of low–growing species is to be sorted out, the material is examined and the spores distinguished; either a single spore is isolated, or a few spores taken up in sterilised water in a sterilised pipette, and spilled on to prepared media. Where bacteria are likely to be troublesome the medium is slightly acidified with citric acid. As soon as growth is viasible to the naked eye, the colonies are cut out with a hot scalpel, and separately transferred to fresh media; it may be necessary to make several transfers before a clean culture is obtained.

Cultivation

The substrata upon which fungi are cultivated in the laboratory are called media. Many different kinds of media have been compounded or synthesized. Although these media may contain all the essential food substances, namely, carbohydrates, proteins, fats, amino acids, vitamins, minerals, air, and water, none of them constitutes an ideal substratum. Their deficiency is evidenced by the fact that certain fungi fail to grow or even complete their life cycle in these artificial media. The ideal medium is, therefore, most nearly like the substratum on which the fungus occurs in nature.

KINDS OF MEDIA

(1) Natural Media : A natural medium is one which is composed entirely of complex natural materials of unknown compositions. Among the natural substances so used are the following : plant parts, malt, yeast, peptone, manure, bread, wort, fruit, and vegetables. Many of these substances are used in the form of extracts, infusions, or decoctions.

Natural media have many advantages. They are cheap and easy to prepare. In many instances it is necessary only to add water to the base material and autoclave. More important yet is the fact that many fungi grow well upon a wide variety of natural media.

(2) Semisynthetic Media : A semisynthetic medium is one which is composed in part of natural materials of unknown composition. Such media are made by adding compounds of known composition to one or more natural materials. The widely used potato–glucose (dextrose) medium is an example of this type. The addition of agar to an otherwise synthetic medium introduces a natural material of unknown composition.

(3) Synthetic Media : A synthetic medium is one of *known composition and concentration.* Such media may be simple or complex but must contain the essential elements in utilizable form.

The chief value of synthetic media is for nutritional studies. However, growth and reproduction are frequently poorer on a synthetic medium than on one containing some natural material.

For rough cultures of *Mucor* and *Rhizopus* it is only necessary to scatter a little dust upon a slab of moist bread. This is placed under a bell jar, and, at ordinary temperatures, sporangia will be present in about a week. If the bread is left for a longer period, the Mucors disappear and Aspergillaceae and *Fungi imperfecti* may be obtained. Dung treated in similar fashion often yields a number of interesting forms, including *Mucor, Piptocephalis, Pilobolus*, various Ascomycetes and later, Basidiomycetes.

A combination of the liquor from boiled potatoes with agar will support the growth of some parasites and most of the saprophytes found on plant material, coprophilous species usually flourish best on an appropriate dung agar.

On the basis of their composition, media may be regarded as of two kinds: *synthetic* (mineral) and *non-synthetic* (organic), and they may be either liquid or semisolid. Agar in the proportion of about 2% is used almost exclusively in making the media semisolid.

Fungi generally grow best in media abundantly supplied with carbohydrates. Another requisite is that the pH be within the range of 5 to 6.

Most kinds of organic media contain meat extract and peptone as essential constituents. The meat extract supplies mainly the ash constituents, organic nitrogen, and certain organic acids. Peptone supplies decomposable proteins that can be cleaved by many fungi to yield essential amino acids. When media containing these materials are enriched by the addition of plant substances extracted by boiling, like those from potatoes, corn meal, or fresh beans, they constitute satisfactory substrata for the cultivation of a large number of different species of fungi.

Mineral nutrient media have a more limited usage than organic media. They are employed especially in physiological studies of the suitability of a single carbohydrate, amino acid, or mineral. A mixture of mineral salts in liquid cultures or jellified with agar furnishes the basis for such media.

The nature of the substrate is also a vital factor in the production of sporophores in culture. Sawdust may be used as a basic material supplemented by readily available nutrients. As additional factors there should be a plentiful supply of moisture in the substratum, an atmosphere of high relative humidity, and exposure to diffuse light of moderate intensity.

STERILISATION

If the Petri dishes are to be used, the medium is poured into flasks which are then plugged. These may be placed, with the required number of dishes, in a cold autoclave. Heat is applied, and, when steam issues freely from the valve, the latter is closed and a pressure of thirty pounds is maintained for 20–30 minutes. The gas is then turned out, and, as soon as the gauge has fallen to zero, the valve of the outlet pipe is cautiously unscrewed. The Petri dishes when removed from the autoclave are placed flat on paper and at once poured. This is done by lifting the lid of the dish sufficiently to allow the entrance of the neck of the flask containing the medium, pouring in enough to make a layer 5 mm deep, and shutting down; it is sometimes desirable to pour in less, in order to get a shallow layer of medium which will allow of the examination of the culture through the bottom of the dish. The plates are allowed to cool, and, when the medium has set, may be inoculated. If time permits it is better, before inoculation, to incubate the plates at 20°C for three days or so. This dries off the water of condensation and allows time for contaminations introduced during pouring to become visible; contaminated dishes should be rejected.

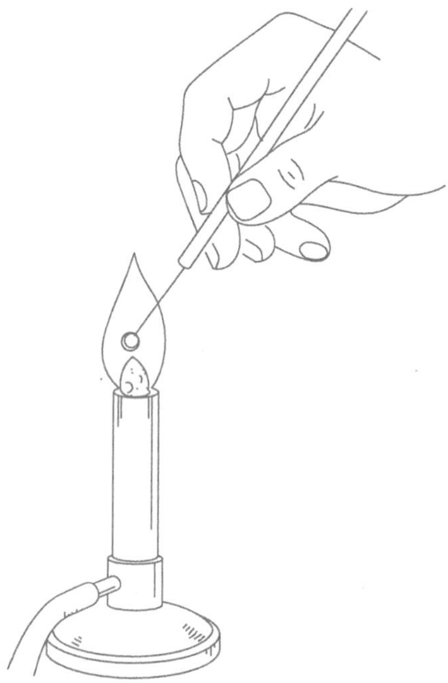

Fig. 4.7. The technique for flaming the bacteriological loop.
(Burton, p. 148)

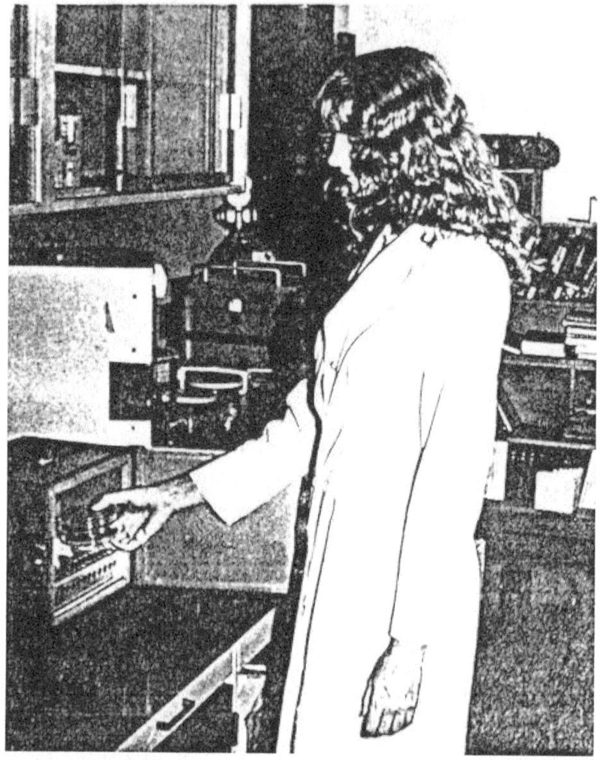

Fig. 4.8. Photograph of a technician loading a hospital Incubator
(Atlas, p. 115)

If a dry oven is available, the dishes are sterilised in hot air at about 140°C. The medium, after sterilisation in an autoclave, is left there till required, and is poured as cool as possible, to avoid water of condensation and current of air.

INOCULATION

It is usually most convenient to inoculate by means of a piece of platinum wire set in a glass rod. The wire is heated in a flame, and, as soon as it is cool, pushed rapidly among the spores of the fungus, and, as quickly, through a narrow space between the side and lifted lid of the prepared dish, into the centre of the medium. It is best to avoid opening a dish more than once for this purpose; when an inoculation has been made, the dish should be put aside and opened no more until the fungus is to be examined. When possible a number of cultures of the same fungus should be made at the same time.

Some fungi, such as *Phytophthora* and *Basidiobolus,* often do not form spores in large numbers. For their transfer, a piece of the medium, bearing the fungus, is cut out with a hot scalpel and rapidly carried to a fresh dish. The lid of the latter is opened only sufficiently to permit the introduction of the scalpel and its load, and, if the portion of inoculum be dropped on the way, or touch the outside of the dish, it is rejected, the scalpel reheated, and a fresh portion transferred. Extreme care should be taken to avoid the introduction of anything except the fungus and the piece of substratum.

TUBE AND FLASK CULTURES

When tubes and flasks are to be prepared for cultures, the medium is made up as for the dishes, but it is usual to sterilise in the flask or tube. Accordingly, the quantity of the medium poured into each vessel. The receptacles are loosely plugged with cotton wool, and sterilisation is carried out. When removed from the autoclave, tubes are placed in a slanting posititio n until cool, so that the medium sets in a slope. It is well to incubate for several days, in order to dry the plugs. Inoculation is carried out as with dishes, but, before the plug is removed, its upper part is passed through a flame; the plug is lifted between the fingers so that it stands out from the back of the hand, and is so held whilst inoculation is performed. The lower part of the plug should be passed through a flame before being replaced.

Common Culture Media for Fungi

(1) Potato – Dextrose Agar (PDA) :

Potato (Peeled)	200 g
Dextrose (glucose)	20 g
Agar	20 g
Water	1000 ml

(2) Corn – meal Agar (CMA) :

Ground corn (maize)	25 g
Peptone (if desired)	20 g
Glucose (if desired)	20 g
Agar	20 g
Water	1000 ml

(3) Malt Extract Agar (MA) :

Malt extract	20 g
Agar	15 g
Water	1000 ml

(4) Oat Agar (OA) :

Oat meal	40 g
Agar	15 g
Water	1000 ml

(5) Waksman's Medium :

Glucose	10 g
Peptone	5 g

Potassium dihydrogen	
Phosphate (KH_2PO_4)	1 g
Agar	25 g
Water	1000 ml

(pH is made 4 by the addition of normal H_2SO_4 or H_3PO_4)

(6) Rabbit Dung Agar (RDA) :

Agar	20 g
Water	1000 ml

Place 3 or 4 pellets in each 1 oz bottle, but do not crush them. Pour on the water agar. Inoculate on the pellets.

(7) Nutrient Agar :

Beef extract	30 g
Peptone	10 g
Agar	15 g
Water	1000 ml
Carbohydrate (if desired)	10 g

(8) Hansen's Medium for yeasts :

Peptone	1 g
Maltose	5.9 g
Potassium dihydrogen Phosphate (KH_2PO_4)	0.3 g
Magnesium sulphate ($MgSO_4.7H_2O$)	2.5 g
Water	1000 ml

(9) Richard's Solution :

Potassium nitrate	10 g
Potassium dihydrogen phosphate (KH_2PO_4)	5 g
Magnesium Sulphate ($MgSO_4.7H_2O$)	2.5 g
Ferric Chloride ($FeCl_3$)	0.02 g
Sucrose	5 g
Water	1000 ml

(10) Cazpek (DOX) Agar :

Sodium nitrate	2.3 g
Potassium phosphate (K_2HPO_4 or KH_2PO_4)	1 g
Potassium Chloride (KCl)	0.5 g
Magnesium Sulphate ($MgSO_4.7H_2O$)	0.5 g
Ferrous sulphate ($FeSO_4$)	0.01 g
Sucrose	30 g
Agar	15 g
Water	1000 ml

Phosphate should be dissolved separately in distilled water and should be mixed with rest of the constituents in the end.

Maintenance of Cultures

A fungus is kept alive by transfer at sufficiently frequent intervals. Not only must it be kept alive but also as a result of manipulation and examination each fungus will come to be recognized as having a distinctive appearance. The frequency of subculturing will depend upon several factors, including –

(1) the peculiarities of the fungus,

(2) the culture medium, and

(3) the temperature and humidity of the storage cabinet.

Fungi that produce sclerotia, as *Rhizoctonia solani* and *Sclerotium rolfsii,* will remain alive for several years even though the medium has dried up.

The result of tranfer of mycelium or spores from old cultures that are thoroughly desiccated may indicate that the culture is dead. It may be possible, however, to revive such an old culture by flooding its surface with a liquid medium and then incubating for a day or two before again attempting to subculture.

If the temprature of the culture cabinet can be maintained at about 20°C, most species of fungi need be transferred only two or three times a year.

Implications

Certain principles underlie the techniques of isolation and cultivation of fungi. Following routine directions and procedures may cause failure to isolate or to grow a specific fungus artificially. Failure may be attributable to use of the improper kind of substrate. Fungi should be expected to respond in a most nearly normal manner if food and environmental influences approximate those that they encounter in their natural habitat. It is of course impossible in many studies to duplicate these conditions artificailly, but a knowledge of these facts about a given fungus may prevent loss of time in attempts to isolate it and misinterpretation of results of its responses when grown in culture.

Fungal Mounting Media

(1) Lactophenol :

Phenol (Pure crystals)	20 g
Lactic acid	20 g or 16 ml
Glycerol	40 g or 31 ml
Water	20 ml

Add a small amount of cotton blue, which is the most popular stain for the microscopical examination of fungi.

(2) Andre and Hoyer's Fluid :

Gum arabic or gum guiac	15 g
Chloral hydrate	100 g

Glycerol	10 g
Water	25 ml
(3) Glycerine Jelly :	
Gelatin	1 g
Glycerol	7 g
Water	6 ml
Phenol 1%	6 ml

Slide Preparation

A drop of lactophenol is placed on a slide and a small tuft of the fungus, picked up with a needle, is put into the drop. The slide is then examined under the low power of the microscope and the material is teased and spread with two needles and a cover glass is applied and examined under the high power of the microscope.

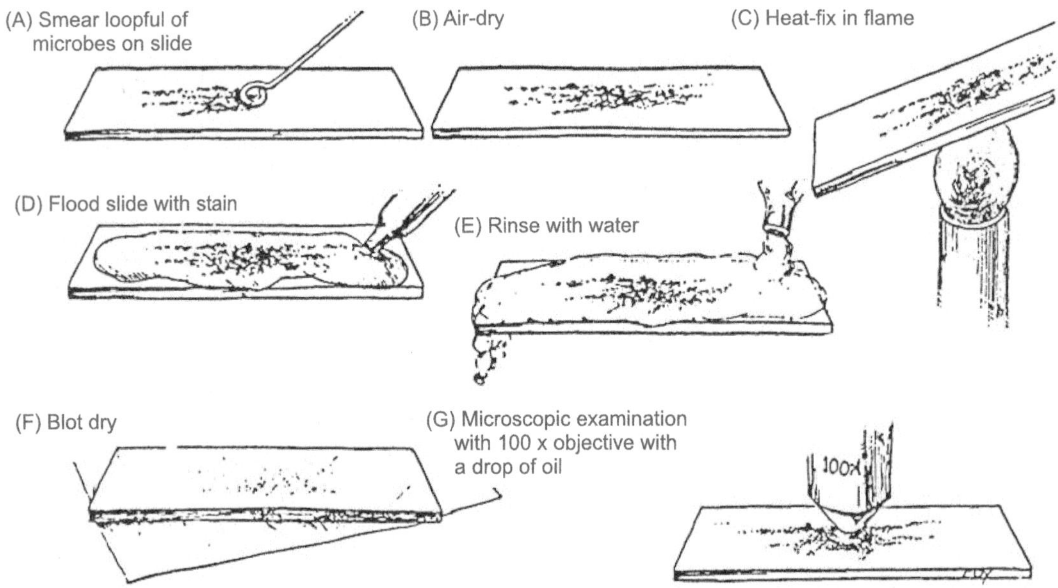

Fig. 4.9. Bacterial staining techniques. (A) With a flamed loop, smear a loopful of bacteria suspended in broth or water on a slide. (B) Allow slide to air-dry. (C) Pass the slide through the flame to heat-fix the bacteria to the slide. (D) Flood the slide with the stain. (E) Rinse with water. (F) Blot dry with bibulous paper or paper towel. (G) Examine the slide with the 100X microscope objective using a drop of immersion oil directly on the smear.

(Burton, p. 45)

Dry, dusty spores, such as those of aspergilli are difficult to moisten and are likely to carry bubbles of air into the mount. This difficulty can be overcome by touching the drop of lactophenol, after the fungus has been put with the tip of a glass rod moistened, but not dripping, with alcohol.

Fungal Stains

Congo Red :

Congo red (80% dry content)	2 g
Distilled water	100 ml

Cotton Blue Lactophenol :

Cotton blue (1% aqueous solution)	5 ml
Phenol Crystals	20 g
Lactic acid	20 g
Glycerine	40 g

SUMMARY

Fungi secure food and energy from the substrates upon which they live in nature. In order to culture fungi in the laboratory, it is necessary to furnish in the medium those essential elements and compounds they require for the synthesis of their cell constituents and for the operation of their life processes. The synthetic abilities of fungi differ. Some fungi are unable to synthesize certain key compounds that they require and must obtain them from the medium upon which they grow. All the fungi require much the same essential elements but differ widely in their ability to utilize compounds in which these elements occur. There is no universal natural substrate or artificial medium upon which all fungi will grow.

On the basis of composition there are three general types of media : natural media, which are composed entirely of natural products; semisynthetic media, which are composed in part of natural substances; and synthetic media, which are of known composition. Natural media are most useful for routine work, while synthetic media and, to a limited extent, semisynthetic media are used to investigate the nutritional requirements of the fungi. Media differ only with respect to constituents and concentrations.

The compounds and the amounts used in preparing a medium must be specified exactly. Media should be designated by naming the carbon and nitrogen sources used, e.g., glucose–asparagine medium.

The selection of a suitable medium depends upon the fungus under study and the purpose of the experiment. Not all media are equally suitable for all fungi, nor is one medium suitable for a complete physiological study of the fungus.

REFERENCES

1. Atlas, R.A. 1986. *Basic and Practical Microbiology,* p. 110–123. Macmillan Publishing Company, New York.
2. Birge, E.A. 1992. *Modern Microbiology,* p. 48–51. Wm. C. Brown Publishers, Dubuque, U.S.A.

3. Brock, T.D. 1970. *Biology of Microorganisms,* p. 759–762; 777–778. Prentice - Hall, Inc., Englewood Cliffs, New Jersey.
4. Gwynne–Vaughan, H.C.I. and B. Barnes. 1965. *The Structure and Development of The Fungi,* p. 361–381. University Press, Cambridge.
5. Johansen, D.A. 1940. *Plant Microtechnique,* Tata McGraw - Hill Publishing Co., Ltd., New Delhi.
6. Lilly, V.G. and H.L. Barnett. 1951. *Physiology of The Fungi,* p. 8–23. McGraw - Hill Book Company, New York.
7. Prescott, G.W. 1969. *The Algae : A Review,* p. 375–381. Thomas Nelson and Sons Ltd., London.
8. Rangaswami, G. 1979. *Diseases of Crop Plants in India,* p. 55–60; 487–493. Prentice–Hall of India Private Ltd., New Delhi.
9. Wilson, C.L. and Walter E. Loomis. 1957. *Botany,* p. 333–345. Holt, Rinehart and Winston, Inc. New York.
10. Wolf, F.A. and F.T. Wolf 1969. *The Fungi,* Vol. I. p. 14–27. Hafner Publishing Company, New York.

Chapter 5

Mushroom Culture

INTRODUCTION

Mushrooms have fascinated man since time immemorial. Some of them are a good source of food. About 2000 species of edible mushrooms have so far been reported from all over the world but only a few species like *Agaricus campestris,* the common field mushroom; *A. bisporus,* a cultivated variety of Europe and North America; *Lentinus edodes,* the *Shii–take* of Japan; and *Volvariella volvacea,* the paddy straw mushroom of China and South-East Asia are being cultivated on an extensive scale. In Europe and North America, the cultivation of common field mushroom has become a large industry. In Great Britain and America large quantities of mushrooms are available for canning and food manufacture as well as for direct consumption. Several million pounds of the *Shii–take* are produced annually in Japan and China. In India, large scale mushroom cultivation is in progress but is confined to a few states like Himachal Pradesh, Tamil Nadu, and Uttar Pradesh.

What is a Mushroom ?

The Greek work *Mykes* means mushroom or fungus but in strict sense mushrooms are agaricus or fleshy species of a group of fungi. They may be edible or non-edible, poisonous or non-poisonous. The poisonous mushrooms are, however, popularly known as toadstools. They may be seen growing on the trunks of trees, on stumps or logs of wood, coming out of leaf moulds on the forest floor, or in the grass in yards, lawns, fields land pastures or along the road sides. The plant may look like round balls or shelves or brackets, as caps supported on slender stems or branched and coral like.

Parts of a Mushroom

The simple agaricus has an aerial umbrella-like fruiting body which is formed by the aerial growth of subterranean mass of hyphae. This is commonly known as sporophore or basidiocarp.

It grows in decaying organic matter such as old leaves, straw, and rotting materials. The fruiting body consists of a fleshy cap or *pileus* two to three inches in diameter and is supported on a stalk or *stipe* which comes out of the ground. On the underside of the cap are *gills* or lamellae. These are thin plates standing on edge and radiating out from the stalk, like the spokes of a wheel. On both the surfaces of gills develop numerous spores (called basidiospores). These spores fall on the substratum and germinate under suitable conditions to send out mycelia in all directions. The underground mycelium is perennial and its growth is centrifugal, *i.e.*, its hyphae run from centre towards the periphery as a result of which, it increases in diameter year after year. By the death and decay of the older portion, *i.e.*, the central part, the mycelium forms a ring like structure in the soil. This structure is called as *spawn.* On maturation the fructifications appear in a ring as tiny, white balls. Such circles of mushrooms are often called as "fairy rings". As the ball grows further, a stem (stipe) is produced and later the cap (pileus) begins to open up like an umbrella tearing away the delicate membrane or veil (velum). From the under surface of the pileus, delicate gills (lamellae) are developed which become dark coloured as spores appear on them. When they attain maturity, the cap of the mushroom begins to flatten out. The spores are eventually released to the ground in millions.

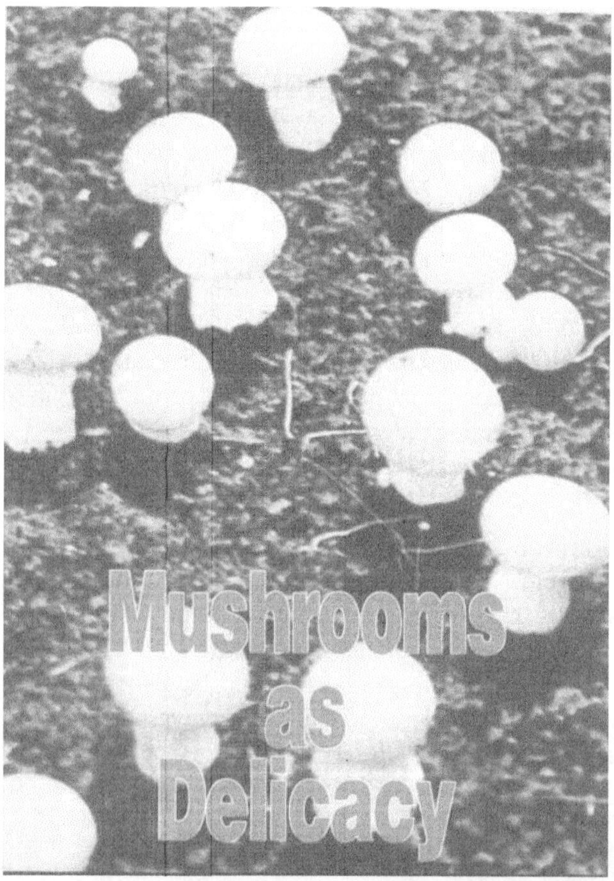

Fig. 5.1. Button mushrooms the most popular edibles

Mushrooms for Food

Until recently mushroom dishes have emerged as a delicacy on our dining tables in households as well as five-star hotels. This has resulted in their market demand during the last few years. White, round and fleshy mushrooms packed neatly in polythene bags can even be bought from local vegetable hawkers.

Fig. 5.2. Milky white mushrooms that thrive on organic waste

While purchasing make sure that they are firm and creamy white, not brown and droppy. If they have turned brownish, it means they are past their prime and the taste will not be the same. The white button mushrooms made a cautous entry on the Indian scene, somewhere in 1960s. Here was a total newcomer on the food front. Suppose it were poisonous ? But such fears have long since been allayed and this delectable item stands absolutely in the clear. In fact, it is now considered a delicacy on any menu, anywhere. The secret of their increasing popularity lies in having excellent nutritional qualities, novel taste, pleasant flavour and a meat-like texture. Surprisingly, this extraordinary gifted food item grows well on various waste materials and requires no land for cultivation.

History indicates that the great early civilisations of the Egyptians, Greeks, Romans, Chinese, and Mexicans prized mushrooms not only as a delicacy but for their medicinal or tonic attributes. The Ayurveda classics *Chark Samhita* and *Sushruta Samhita* also record the therapeutic properties of certain edible mushrooms.

Popular as Food

Why are mushrooms becoming popular ? Is it due to the changing food habits of modern people of the world under present circumstances, or have people developed some special attraction for them? Mushrooms are taken not only for their flavour and taste but also for their nutritive value. Like many other vegetables, mushrooms contain a high proportion of water apart from other constituents. They usually contain more protein than any other comparable fruits or vegetables (on a dry weight basis) and as such can claim to be a "protective food". Furthermore, mushrooms are a rich source of several vitamins, fibres, and important minerals.

Nutrition scientists attach high value to mushroom protein because it contains all the essential amino acids and is especially rich in lysine and leucine, which are lacking in most staple cereal foods. Thus, mushrooms can effectively supplement cereals in our diet. Recent nutritional assessments indicate that the overall nutritive value of high grade mushroom almost equals that of milk.

Fig. 5.3. Oyster mushrooms as nutritious as the button kind

In addition to good quality protein, mushrooms are the best plant sources of nicotinic acid and riboflavin, a good source of pantothenic acid, and a fair source of vitamins B, C, K, biotin, and thiamine. Folic acid and vitamin B_{12}, which are so much needed by pregnant and lactating women, are almost absent in vegetables, but are present in mushrooms. All these vitamins are well retained during the processes of cooking, canning, drying, and freezing. In some cases linoelic acid has been found to be the main fatty acid component. Moreover, mushrooms being completely devoid of starch, constitute an ideal dish for diabetic patients.

Among minerals, potassium, sodium and phosphorus are present in high quantities. Furthermore, as compared to other food substances, the potassium and sodium ratio in mushrooms is very high, which is desirable for patients of hypertension. Cholesterol, the dreaded sterol of the heart patients, is absent in mushrooms, and instead ergosterol is present which could be converted by the body to vitamin D. In addition, they have also been reported to contain specific blood cholesterol–reducing substances.

Mushrooms Can Keep You Slim

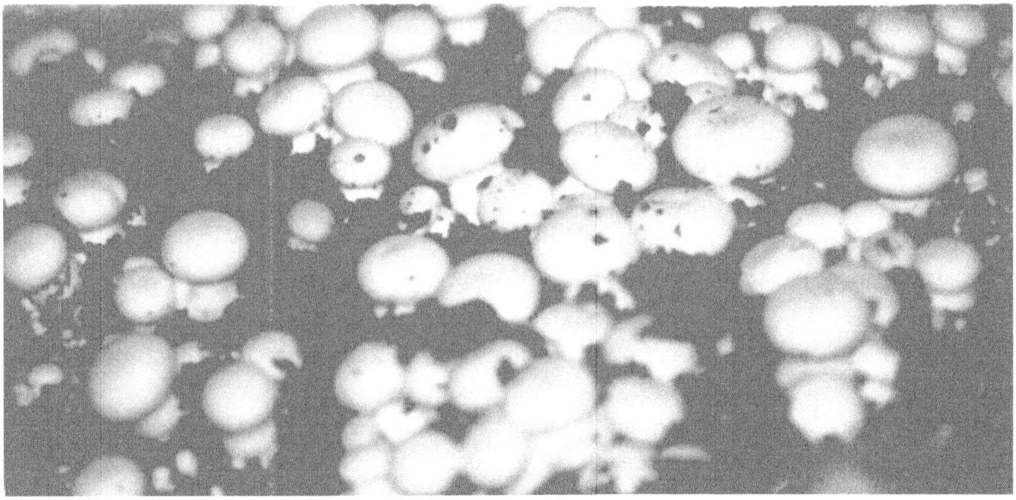

Fig. 5.4

Slimmer's Choice

Mushrooms shouldn't be eaten for taste alone. Nutrition experts say that mushroom dishes are highly suitable for obese persons and diabetics, because they provide less of calories, very little fat and almost no starch and sugars. This makes the mushrooms as a slimmer's delight. Hence, no wonder that chefs all over the world have developed several palatable recipes of mushrooms to be served hot or cold. Apart from routine curry vegetables, mushrooms can also be grated or sliced into a stuffing for sandwich, omellette, pizza, patty, hotdog, humburger, or even dosa, or may be used for garnishing rice, vegetables, dessert etc. The finely cut and aesthetically placed slices of mushroom can make a decorated salad salver even more attractive and nutritious too. Most of the five-star hotels provide mushroom soup and pickle as the latest appetisers on the menu.

Unbelievably, this nutritive and therapeutic marvel grows well on agricultural, industrial and forest wastes, such as cereal straws, coconut and coffee waste products, sugarcane bagasse, sawdust and cotton wastes, etc. These materials are used to make compost, which is the substrate on which mushrooms thrive. Usually the rural folks use wheat straw, dried leaves, weeds, dung, and waste from food and cottage industries to make the compost. But for maximum yield, it is recommended to add sufficient quantities of some chemical fertilisers such as urea, or other nitrogenous compounds, ammonium sulphate and gypsum to the mix. Some insecticides may also be added to avoid insect infestation during mushroom growth.

Mushrooms can be cultivated in compost-filled wooden shelves, trays or even in thick polythene bags, which can be stacked in a properly ventilated room or even verandah. Good quality mushroom "seeds" technically called spawns, are readily available in the market, packed in bottles or polythene bags. After spawning or "seeding", trays are left as such for about fifteen days and then it needs casing. The mushroom spawns are of the following two types :

(1) Blocks of richly manured soil containing the fungus mycelium and basidiospores in abundance,

(2) Pure culture of the fungus mcelium on some nutritive medium.

Casing is a process responsible for motivating mushrooms to enter the reproductive phase, in which "heads" or "buttons" appear on the white thread-like mycelium. During the process, compost in trays showing vegetative growth is covered by a fine layer of casing material (a mixture of fine dung powder and clayey loamy soil). After about eight to twelve days the first crop of mushrooms is ready for picking and the fruiting continues for about six to eight weeks. Mushroom heads or buttons are handpicked every week and a square meter area yields six to eight kilograms of mushrooms during the season.

Fresh mushrooms are very delicate and perishable, hence should be packed carefully and transported to nearby markets as early as possible. If grown scientifically, mushroom cultivators can earn a net profit of more than 100–200 per cent over and above the costs incurred. In addition, the material remaining after the mushroom crop has been harvested can often be used as a valuable additive to the soil, thus increasing its potential for the production of other agricultural and horticultural crops, or possibly being used directly as a source of animal feed.

Although nearly twenty types of edible mushrooms are cultivated commercially around the globe, cultivation of only three types suit the agroclimatic conditions of India. These are –

Button mushroom (*Agaricus bisporus*),

Paddy straw mushroom (*Volvariella volvacea*), and, Oyster mushroom (*Pleaurotus* sp.).

In the northern plains all these mushrooms can be grown sequentially throughout the year under normal weather conditions : mid-November to mid-March. Button mushroom February to mid-April and September to November; oyster mushroom mid-June to mid-September, and paddy straw mushroom mid-September. Due to popular demand, button mushroom is highest, but as far as nutrition is concerned all the three varieties exhibit more or less the same value.

The attractive enterprise of mushroom cultivation can be modestly started with fifty trays, but to gain maximum profit their number should be increased to hundred at the earliest. The National Bank for Agricultural and Rural Development (NABARD), some financial institutions, and several commercial banks provide long-term loans for mushroom cultivation to the extent of 90% of the total project cost. Apart from mushroom cultivation, banks also finance projects for setting up of spawn production, compost production, and mushroom canning units.

The mushroom cultivation has emerged as an enterprise which has great potential for economic growth of farmers and could be started with a very low capital base. Less time consumption, no drudgery indoor or nearby space availability are the additional advantages of mushroom cultivation. This activity can bring money in abundance when marketed to elite and city dwellers. If mushroom growers adopt modern post-harvest technologies, they can even earn valuable foreign exchange, as there is tremendous scope of mushroom export, especially to European countries.

Mushroom cultivation also offers additional avenues of employment for the unemployed and under-employed rural youths, women, etc. Hence, this income-generating activity fits very well in the present day scheme of rural development, which encompases poverty alleviation and employment generation as priorities. Furthermore, mushroom cultivation is an environmentally sound enterprise, because it converts the potential pollutants present in wastes to a nutritious food for human consumption. Thus, it needs a strong fillip from all sectors of society.

Mushrooms are best cooked with a dash of butter, some ground pepper, onions, and green chillies. Float them in heavily spiced gravies, and you lose the delicate flavours.

Flavour

The following compounds have been supposed to be associated with the flavour of mushrooms.

Lentinus Edoders

Guasnosine 5′ – monophosphate – lenthinonine,

Phallus Impudius

Methyl mercaptan hydrogen sulphide, and a number of aldehydes and acids,

Boletus Edulis

Seven compounds including 1–octen–3–1, 1–octen–3, Pyrazines and seven 2–formylpyroles,

Agaricus Bisporus

3–octason, 1–octen–3, benzaldehyde octanol, zocten–1.

Some Favourite Mushroom Recipes

Mushroom Soup

Ingredients –

1 tbsp flour
2 cups milk
5 cups water
1 tbsp butter
1 onion chopped
salt and pepper to taste.

200 g mushrooms, chopped. Fry the onions lightly in butter. Add mushrooms and fry lightly. Add flour and stir for a minute. Cool and add one cup of milk. Pass through a blender till smooth. Add another cup of water. Bring to boil and then simmer the mixture for 10 minutes, stirring occasionally. Add salt and pepper to taste.

Mushroom Pulao

Ingredients –

1 glass basmati rice
100 g mushrooms, sliced
1 onion, chopped
2 glass water
2 tbsp ghee
¼ tbsp cumin seeds
salt to taste.

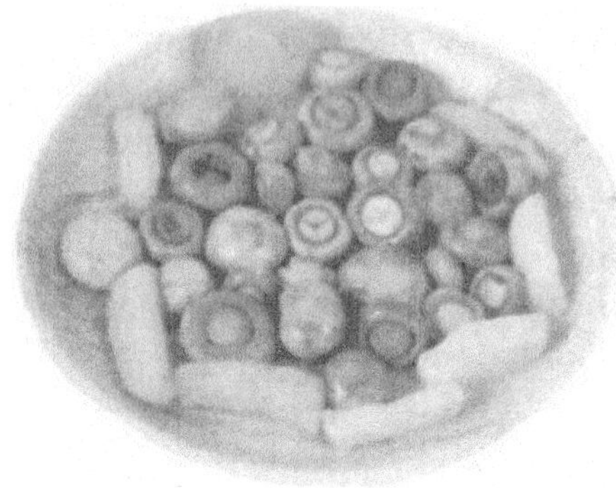

Fig. 5.5

Wash and soak rice for half an hour. Heat ghee in a heavy bottomed pan. Add cumin seeds and salt. Add onions and fry lightly. Now add the mushrooms – but take care that neither onions nor the mushrooms are browned. Add water and bring to boil. Now add the rice and cover tightly. Simmer on a low flame, till the water dries up.

Low Calorie Topping

Ingredients –

100 g mushrooms, sliced
400 ml skimmed milk
salt and pepper to taste.

Cook mushrooms and milk in a pan, stirring occasionally. Just before the liquid evaporates, add salt and pepper. Cook for a couple of minutes.

Mushroom Ketchup

Ingredients –

Fresh mushroom,	500 g
Salt,	12 g
Cardamom (ground),	2 g
Pepper (ground),	2 g
Dry ginger (ground),	2 g
Cloves,	1 g
Mace,	1 g
Cinnamon,	1 g
Vinegar,	500 g

Wipe the mushroom caps with a damp cloth. Keep them in a porcelain or an enamelled bowl for 13 hours after sprinkling with salt. Alternatively drop the mushrooms in vinegar for several days and blend the soaked or salted mushrooms to a fine slurry and add vinegar and spices in it. Heat the mass till it thickens to the consistency of ketchup. Small quantity of meat extract or monsodium glutamate can also be added to enhance the flavour. Fill the ketchup hot in bottles, seal and sterilize them in boiling water for 30 minutes. Cool the bottles slowly in water and stock them in a dry place.

Preservation of Mushrooms

The mushrooms can be preserved by canning or dehydration. The canned mushroom can be used without any further preparation in receipes after draining off the brine. The dehydrated product is to be rehydrated in boiling water before use.

Technical Know-how of Mushroom Cultivation

The technical know-how of mushroom cultivation and its further processing is available with the following :

(1) National centre for Mushroom Research and Training, Chambaghat, Solan (Himachal Pradesh).

(2) Dr. Y.S. Parmar University of Horticulture and Forestry, Solan.

(3) Division of Mycology, Indian Agricultural Research Institute, New Delhi.

TABLE 5.1. Mushroom Production and Consumption

Country	Production in tonnes	Consumption in tonnes
U.S.A.	127,000	163,000
France	114,000	64,000
United Kingdom	51,6000	55,500
Netherlands	40,000	11,000
Italy	38,000	39,000
Germany	29,200	125,500
Canada	19,000	28,000
Taiwan	60,000	9,000
South Korea	28,300	1,000
China	30,000	1,000

(Indian Mushroom Journal, April 1976)

TABLE 5.2. Nutritional Value of Edible Mushrooms and Egg

(amino acid in mg/1000 g of protein)

Amino acid	Volvariella duokasua	Pleurotus sp.	Termitomyces sp.	Lepiota	Agaricus bisporus	Hen's Egg
Leucine	3.5	4.4	6.5	8.7	6.1	8.8
Isoleucine	5.5	5.8	6.5	9.2	5.8	6.6
Valine	6.8	4.7	4.6	4.1	4.1	7.3
Tryptophan	1.1	0.9	0.9	0.9	0.1	1.6
Lysine	4.3	5.0	4.0	2.6	–	6.4
Histidine	2.1	2.1	5.0	2.6	–	–
Phenylalanine	4.9	2.0	3.2	2.6	–	5.8
Threonine	4.2	4.2	3.3	3.2	–	5.1
Arginine	4.1	6.7	8.0	6.5	2.2	–
Methionine	0.9	1.3	0.7	1.0	1.9	3.1

(Kannaiyan and Ramasamy, p. 40)

TABLE 5.3. Composition of Cultivated Mushrooms and Some Common Vegetables

Name	Calories %	Moisture %	Protein %	Fat %	Carbohydrate %	Protein percent (Dry wt. basis)
Lima beans	128	66.5	7.5	0.8	23.5	22.2
Green peas	98	74.3	6.7	0.4	17.7	26.1
Green beans	35	88.9	2.4	0.2	7.7	21.6
Cauliflower	25	91.7	2.4	0.2	4.9	28.8
Potato	83	73.8	2.0	0.1	19.1	7.6
Beetroot	42	87.6	1.6	0.1	9.6	12.9
Cabbage	24	92.4	1.4	0.2	5.3	18.4
Celery	18	93.7	1.3	0.2	3.7	20.6
Brinjal	24	92.7	1.1	0.2	5.5	15.1
Mushroom	16	91.1	2.4	0.3	4.0	26.9

(Kanaiyan and Ramasamy p. 4)

TABLE 5.4. Analysis of Edible Mushrooms (Fresh wt. Basis Percent)

Mushrooms	Moisture	Ash	Protein	Fat	Crude fibre
Agaricus bisporus	89.5	1.26	3.94	0.19	1.09
Pleurotus sp.	90.9	0.97	2.78	0.65	1.08
Volvariella diplasia	90.4	1.10	3.90	0.25	1.67
Volvariella volvacea	88.4	1.46	4.99	0.74	1.38
Lepiota sp.	91.0	1.09	3.3	0.18	0.86
Termitomyces sp.	91.3	0.81	4.1	0.22	1.13
Pleurotus ostreatus	92.5	–	2.15	–	–

(Kannaiyan and Ramasamy p. 41)

TABLE 5.5. Ash Analysis of *Agaricus Campestris*

Calcium	0.0024%
Phosphorus	0.15%
Potassium	0.50%
Total iron	19.50 ppm
Available iron	5.95%
Copper	1.35%

(Kannaiyan and Ramasamy p. 41)

Fig. 5.6

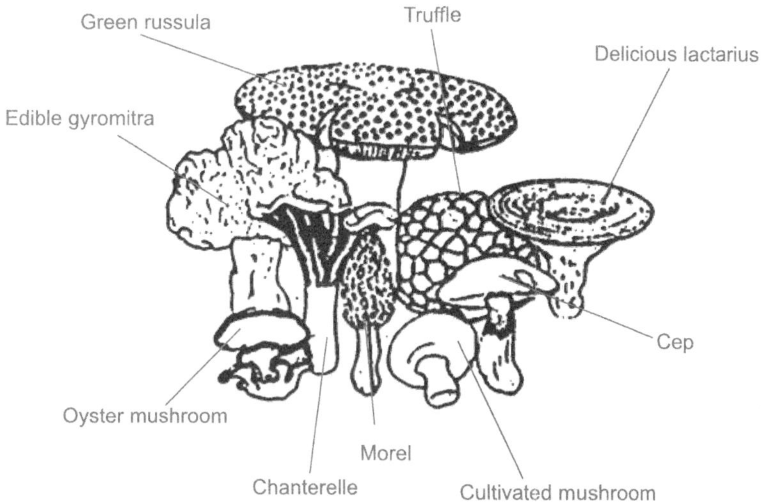

Fig. 5.7. Edible Mushrooms. Truffles are found in the wild, are a forest produce, and are not known in cultivated form. A type of fungus that fruits underground, generally under oak trees, it is quite difficult to locate. Truffles or 'guchchi' can be tracked down only by their pungent smell.

Vishal Agritech Launching 100% Export Oriented Mushroom's Unit

Vishal Agritech India Limited, an Indore based Company, is setting up a unit for the production and processing of white button mushrooms for the export market. The company has been set up as a joint venture with M/s Macon Agri Limited of UK who are also participating in the equity of the project as co-promoters. Macon has entered into an agreement towards 100 per cent buy-back of its entire production for five years and is also supplying the technology.

Macon Agri has over 30 years of experience in the mushroom industry and has successfully implemented numerous projects internationally.

A unique feature of this project that the project is based on the Irish growing pattern and the collaborators have assured a guaranteed yield. The project incorporates the innovative vapour absorption technology for air conditioning using steam instead of electricity maximum possible automation and computer aided environmental control techniques, and hence ensuring low operating costs.

The project which is being set up at cost of Rs. 16.05 crores will have an initial production capacity of 1200 mt. and has been appraised and funded by IDBI by way of a term loan of Rs. 650 lacs.

The project is fast coming up and almost 80 per cent of the construction has been completed. Imported machineries has already been started arriving at site and the erection work is in progress. The commercial production is scheduled to begin in April, 1996 and the company has projected that it will end the year 1996–97 with sales of Rs. 729.89 lacs and a net profit of Rs. 212.56 lacs. Over the next two years, the company expects its net profit to rise to Rs. 340.10 lacs. The EPS9 (Earning per share) is estimated to be around Rs 2.36 lacs for the year 1996-97 followed by Rs. 3.21 lacs and Rs. 3.78 lacs for the year 1997–98 and 1998–99.

The company is entering capital market on 22nd January 1996 for a public offer of 52,50,000 equity shares of Rs. 10/- each at part to part finance the cost of project.

The company has already made firm allotments to IDBI to the extent of Rs. 25.00 lacs alongwith other financial institutions like Unit Trust of India, Rs. 50.00 lacs, Madhya Pradesh Finance Corporation, Rs. 20.00 lacs and Madhya Pradesh Audhyogik Vikas Nigam (MPAVN), Rs. 15.00 lacs and NRI's Rs. 50.00 lacs.

Though the project is set up as an 100 per cent EOU, the company is entitled to sell 50 per cent of its produce in the domestic market under the present exim policy. The company besides exporting bulk of its production, also plans to market its branded mushrooms in the domestic market. Consumers from the neighbouring cities like Indore, Bhopal, and Baroda would be able to get Vishal's fresh mushrooms.

The world market for white button mushrooms, which is amongst the most favoured food in Europe and USA, is approximately 4 lacs mt and its growing rapidly. According to an APEDA study, against an export potential of 50,000 TPA, India Exports were a meger 3294.2 tonnes which indicates an enormous potential of growth for the industry.

Mushroom Cultivation

Mushroom cultivation is increasingly becoming popular it not only meets the dietry requirements but also adds to the income, especially of growers with insufficient land. It is considered to be a very rewarding and fascinating hobby for the retired persons as well as housewives who can grow mushrooms in small boxes or other containers while attending to household chores. Today, mushroom cultivation presents few difficulties provided the grower will follow simple rules of growing.

It is really amazing that a small quantity of spawn when planted in suitable growing medium can (within almost six weeks) grow into a highly profitable crop inside a room, where no other crop would grow. Moreover, mushrooms have more uses in modern culinary cuisine than any other food crop. Mushroom cultivation is carried out indoor in any room, shed, basement, garage, etc., which should be well ventilated. However, paddy straw mushrooms can be grown ouside in shady places also.

Of the many mushrooms only three kinds namely – button mushroom (*Agaricus bisporus*), straw mushroom (*Volvariella volvacea*), and oyster mushroom (*Pleurotus sajor – caju*) are suitable for cultivation in India.

Button mushroom is grown in winter. The most suitable temperature for the spread of the mycelium is 24–25°C while 16–18° is essential for the formation of fruit bodies. Higher temperature is harmful and low temperature retards the development of both mushroom mycelium and fruit bodies.

Paddy straw mushroom can be grown around 35°C. Temperature should not go below 30°C or above 40°C for more than 4–8hours during growing period. In northern India, it can be grown from April to September but the most suitable period is from middle of June to middle of September.

Oyster mushroom (commonly known as "dhingri") grows best between 22–28°C. It is grown in northern plains from October to March.

The methods of cultivation of these three mushrooms are given separately.

BUTTON MUSHROOM

Compost

The cultivated mushroom is grown on special composts, which are of the following two types, synthetic and natural. The compost should be prepared on well cleaned concrete or "pucca" floor, which should be at a higher level so that the run-off water does not collect near the heap. Composting is usually done in the open, but it has to be protected from rain by covering it with polythene sheet. It can also be done in a shed with open sides or a large room to shelter it from rain.

(a) **Synthetic Compost :** The following ingredients are required for 15–16 trays of size 100 cm × 50 cm × 15 cm.

Wheat straw (chopped 8–20 cm long)	250 kg
Wheat/rice bran	20 kg
Ammonium sulphate/Calcium ammonium nitrate	3 kg
Nitrate	3 kg
Urea	3 kg
Gypsum	20 kg

The straw is uniformly spread over the composting yard in a thin layer and wetted thoroughly by sprinkling water. All ingredients such as wheat bran, fertilizers, etc. except gypsum, are mixed thoroughly in the wetted straw, which is finally heaped into a pile. The pile, 1 m high, 1 m wide and length adjustable, can be made with hand or stack mould. The straw should be firmly but not compactly pressed into the mould.

It is essential to open the entire pile and remake it a number of times according to the following schedule :

Mixing of material and making pile	0 day
1st turning	4th day
2nd turning	8th day
3rd turning	12th day, add 10 kg gypsum
Final turning	20th day,

spray 10 ml malathion in 5 litre water (any other available pesticide like DDT, BHC, lindane can also be used).

At each turning, water should be sprinkled to make up the loss of water due to evaporation. If it is desired to add molasses, then 10 kg molasses diluted 20 times with water should be poured over the straw mixture during the first turning. 60 kg chicken manure, if available, can also be added at the time of start of pile.

(b) **Natural Compost**

Horse dung	1000 kg
Chopped wheat straw	300–350 kg
Gypsum	25 kg
Poultry manure	100–110 kg

(Poultry manure can be replaced by 3 kg urea).

It is prepared from pure horse dung (dung of other animals should not be admixed, which must be freshly collected and should not have been exposed to rain). Chopped wheat straw is mixed with horse dung, urea or poultry manure. The mixture is uniformly spread over the composting yard and water is sprinkled over it so that the straw becomes sufficiently wet. The manure is then heaped in a pile as for synthetic compost. After 3 days when the manure in the heap gets heated up due to fermentation and gives off an odour of ammonia, it is opened. The process is repeated 3 or 4 times after an interval of 3–4 days. 25 kg gypsum per tonne is added in two instalments at the 3rd and 4th turning. At the final turning, 10 ml malathion diluted in 5 litres of water is sprayed into the manure.

Filling

The compost when ready for filling and spawning has a dark brown colour and no trace of ammonia. There is no unpleasant odour but it smells like fresh hay. The pH is neutral or near neutral. The compost should not be too dry or too wet at the time of filling in the trays, which can be determined by the palm test. For this purpose, a small quantity of compost is taken into the hand and pressed lightly, if a few drops of water ooze out of the fingers then it is of right consistency. If relatively dry then the water should be made by sprinkling. If too wet, the excess water should be allowed to evaporate. The prepared compost is now filled in trays, which may be of any convenient size but depth should be 15–18 cm. A standard size of tray is 100 cm × 50 cm × 15 cm. The trays should be made of solft wood and provided with the pegs at the four corners so that they can be stacked one over the other leaving sufficient space (15 cm) between the two trays for various operations. The trays are completely filled with the compost, lightly compressed and the surface levelled.

Spawning

Spawning means sowing the beds with the mycelium (spawn) of the mushroom. Spawn can be obtained from any of the following source at a nominal cost –

(1) Mushroom Laboratory, Y.S. Parmar University, Chambaghat, Solan (H.P.)

(2) National Mushroom Research and Training Centre, Chambaghat, Solan and

(3) Division of Mycology and Plant Pathology, Indian Agricultural Research Institute, New Delhi.

The grain spawn in scattered on the surface of the tray bed which is covered with a thin layer of compost. Spawning can also be done by mixing the spawn with compost before filling it in trays. Five hundred gram of spawn is sufficient for five trays of standard size. After spawning, the compost surface is covered with old newspaper sheets, which are wetted by sprinkling water to provide humidity but no water is directly added to the compost during spawn running.

The trays after spawning are stacked vertically one over the other in 4–5 tiers. One meter clear space may be left in between the top tray and ceiling. There should be about 15–20 cm space between the two trays.

The room should be maintained around 25°C. The humidity should be built up by frequently watering the floor and walls. The room may be kept closed as no fresh air is needed during the spawn run. White cottony mycelium spreads and permeates through the compost. Eventually the compost surface gets covered with the mycelium. It takes 12–15 days for complete spawn run. Low temperature prolongs the spread of the mycelium.

Casing

After the spawn run is complete as is evident by white cottony growth, the surface of the compost is covered with 3 cm layer of casing soil. A suitable casing soil can be prepared by mixing equal parts of well rotten cow dung (finely crushed and coarsely sieved) and garden soil. The casing material shold possess high water holding capacity, good pore space and pH should not be lower than 7.4. The casing material is sterilized to kill insects,nematodes and molds. Sterilization can be ccomplished either by steaming or by treating with formalin solution. For one cubic metre of casing soil, half litre of formalin (40%) diluted with 10 litres of water is sufficient. The casing soil is spread over a plastic sheet and treated with formalin by sprinkling. The treated soil is piled up in a heap and covered with another plastic sheet for 48 hours. The soil is turned frequently for about a week to remove all traces of formalin which can be tested by smelling. After casing, the temperature of the room is maintained at 25°C for further three days, after which it must be lowered to below 18°C. At this stage lot of fresh air is needed and, therefore, the growing room should be ventilated by opening windows etc.

Cropping and Harvesting

The first flush of pin heads become visible 15–20 days after casing or 35–40 days after spawning. Small white buttons develop 5–6 days after pin head stage. The right stage of harvest is when the caps are still tight over the short stem. In case the buttons are allowed to mature further, the membrane below the cap will rupture and the cap will open up in umbrella-like shape. Such mushrooms are considered to be inferior. Harvesting is done by holding the cap with fore fingers slightly pressed against the soil and twisting it off. The soil particles and mycelial threads clinging to the base of the stalk are chopped off. Mushroom can also be harvested by cutting off with a sharp knife at soil level.

Yield

The average yield of 3–4 kg per tray is considered normal. However, if compost is carefully prepared, spawn is reliable and temperature is favourable, then a yield of 5–6 kg per tray is possible. partial or complete failure may also happen due to negligence.

Storage

The mushrooms are best consumed fresh. Storage in refrigerator for a few days is possible if they are placed between moist paper towel.

PADDY STRAW MUSHROOM

Paddy straw mushroom, also called Chinese Mushroom (*Volvariella* sp.) is grown in South–East Asia. This mushroom is dark in colour and is very delicious. It is usually grown in raised beds, which are laid in open, exposed or shady places over which temporary sheds are built to protect the beds from direct sunlight and rain. If the beds are made indoors in a well ventilated room which is not very dark, the production of mushrooms is more steady and reliable.

Spawn

Spawn of this mushroom is made on grains of cereals or millets and is called grain spawn. Sometimes it is also made on soaked chopped paddy straw. This spawn is called straw spawn. One bottle of spawn is sufficient for one bed.

Substrate or Bedding Material

This mushroom is usually grown in paddy straw. Thoroughly dried and long paddy straw is preferred. The straw is tied in bundles of about 8–10 cm in diameter. The bundles are then cut to a uniform length of about 70–80 cm, and are soaked by immersing them in a tank of water for about 12–16 hours. Later, excess water is allowed to drain off.

Procedure for Making the Bed

Make a 15–20 cm high foundation of soil or bricks. The size of the foundation is slightly larger than that of bed. The foundation must be strong and firm to hold the weight of the bed.

Make wooden bamboo frame of the size of the bed foundation and place it on top of the raised bed foundation.

Place four bundles of soaked straw side by side on the wooden/bamboo frame. Over these, place another set of four bundles similarly but the loose ends on the opposite sides. These 8 bundles constitute the first layer.

Scatter grain spawn about 8–12 cm from the edges of the first layer. If straw spawn is used, small pieces of the size of the thumb are planted about 4–6 cm deep and about 10–15 cm apart along the edges. Dust the spawn with powdered gram/arhar dal powder or rice/wheat bran.

Now, place a second layer of eight bundles across the first layer and spawn it as before.

Again place a third layer of straw bundles across the second layer and spawn it all over the surface.

Finally cover with a fourth layer of four straw bundles. Press it lightly.

Completely cover the bed with a transparent plastic sheet, taking care that the plastic sheet is not in contact with the bed.

Care of Bed

Remove the transparent plastic sheet after the mycelium has thoroughly permeated the straw. This would take about a week at 35°C. If the surface of the bed gets dry then it should be watered lightly by a sprayer at least once a day.

The Crop

Mushrooms begin to appear within 10–15 days after the beds are spawned and continue to do so for about a week or 10 days. The total yield of the bed is about 2–2.5 kg. The mushrooms should be picked when the volva (cup-like veil) just breaks to expose the mushroom inside. The mushrooms are very delicate and must be consumed fresh. If stored in fridge they can stand for 2–3 days. These mushrooms can easily be air dried in sun or shade.

OYSTER MUSHROOM

This mushroom is simple to grow and has excelled flavour and texture. It is very popular in many countries particularly in South-East Asia, where cultivation of button mushroom (*Agaricus bisporus*) is not possible under natural climatic conditions. Besides its delicious taste, it is known to be very nutritious and is recommended to control obesity and is reported to be beneficial for diabetics. Its low fat content makes it an ideal diet for the blood pressure patients.

Substrate

It can be cultivated on a wide range of cellulosic farm wastes or other materials. Cereal straws, banana pseudostems, waste paper, and cotton waste are particularly suitable. However, paddy straw is the most commonly used material.

Growing in Polythene Bags

- Paddy straw is chopped into small pieces 3–5 cm long. It is soaked in water for about 8 hours, after which water is squeezed out.
- About 200 g grain spawn (half litre bottle) is mixed thoroughly with about 5–6 kg of wet chopped straw (= 1½ kg dry straw).
- A polythene bag 45 cm long and 30 cm diameter is used. It is perforated with 2 mm diameter holes, about 4 cm apart, all over the surface. Spawned straw is filled about 2/3rd of the capacity of the bag and mouth tied.
- The bags containing spawned straw are palced in shelves in the growing room (R.H. 80–85% and room temperature 24–26°C).

Growing in Rectangular Blocks

- A wooden tray mould (50 cm × 33 cm × 15 cm) without bottom is required. A one m^2 piece of transparent polythene sheet is spread so that it forms the bottom of the tray mould and also lines the sides from inside. The loose edges hang out from the tray mould.
- Fill in the wetted chopped paddy straw (as described for bag method above) to make a 5 cm thick bottom layer. Scatter spawn uniformly. Lay another 5 cm thick layer over the top of bottom layer and spawn it in the same way. Finally, lay the third layer (final layer) over the second and repeat spawning. Spawn is covered with more wetted straw to bring it in level with the top of the mould. Compress it firmly by hands or a board. Two hundred grams (½ litre bottle) spawn is sufficient for two blocks.
- Fold the loose hanging edges of plastic sheet over the straw block, fasten with a string. Remove the rectangular block from the mold.

Spawn Run

- The spawned bags/blocks are placed on shelves in the growing room where R.H. is maintained at 80–85 percent and room temperature is about 24–26°C.
- Spawn run is complete in about 10–12 days and is indicated by white cottony mycelium which permeates throughout the straw. As a result, the straw becomes compact and does not split when handled. At this stage, polythene covering is removed by cutting it open in case of bags and by untying the polythene sheet in case of blocks. In case of bags, the straw gets compacted in the form of a cylinder.
- The cylinder/blocks are neatly arranged on shelves and gently watered at least twice a day.

The Crop

First mushrooms appear about 18–20 days after spawning. Two or three flushes appear at an interval of about a week. "Dhingri" should be harvested when the cap starts becoming folded. Harvesting can be done by cutting with a sharp knife or by twisting it off with fingers from the substrate. "Dhingri" is best consumed fresh. It can be dried in sun or in a mechanical drier, and stored in polythene bags.

Yield

About one kg mushroom can be obtained from 1½ kg dry straw (= 5–6 kg wet straw).

Diseases Caused by Mushroom

Toxins produced by some fungi are responsible for some serious cases of food poisoning (mushroom poisoning). Many of these mycotoxins are potent neurotoxins that can be absorbed through the gastrointestinal tract. The ingestion of poisonous mushrooms such as *Amanita phalloides*, is normally fatal. The toxins produced by *A. phalloides*, and other species of *Amanita*, produce symptoms of food poisoning 8–24 hours after their ingestion. Initial symptoms include vomitting and diarrhoea; later, degenerative changes occur in liver and kidney cells, and the patient often becomes comatose. Death may occur within a few days of ingesting as little as 5–10 mg of toxin.

OLDEST MUSHROOM FOSSIL FOUND

A 90–million-year-old mushroom has been found in America, reports ANI.

According to a report published here, the oldest known fossil mushroom was dug up in New Jersey.

Among the two fossils found, one was complete with its cap, gills and central stalk, report said.

The mushroom is one of the holobasidiomycetes, a group which includes puffballs, polypores, and stinkhorns.

"Their fossils are exceedingly rare" scientists have said.

MUSHROOMS BRING IN PROFIT ROUND THE YEAR

Fig. 5.8. Students at the Mushroom Grower's Training centre in Horst, Holland.

If mushrooms are devoured as a delicacy by connoiseurs of good food, they are also becoming an attractive business option for many enterprising farmers and businessmen. With mushroom growers the world over unable to meet the demand for this exotic vegetable, the Agricultural and Processed Food Products Export Development Authority (APEDA) in India, encourages mushroom exports to different parts of the globe.

"Mushrooms are anti-fat, anti-cholesterol, anti-hypertension and anti-blood pressure, and therefore, good for the heart," says S.S. Malik, a mushroom consultant. This delicious white fungus is rich in protein and contains vitamins in abundance.

As the Indian palate begins to relish mushrooms, an increasing number of entrepreneurs are cashing in on the booming mushroom business. The added attraction is that it promises quick returns to the grower.

Mushrooms grow best in cool temperatures, and till recently, were grown only seasonally between September and March. However, with the availability of indigenous environment controlled mushroom cultivation technology, they are now grown round the year in most parts of the country.

Mushroom cultivation consists of several stages including compost making, spawning, casing and harvesting (handpicking). While farm-fresh mushrooms are supplied to local markets, for the export market, they have to be processed and canned.

Low-cost farms : According to Sanjay Vohra of Acharya Narendra College, University of Delhi, who has trained scores of farmers and women entrepreneurs in mushroom growing as part of projects funded by the department of science and technology, FICCI ladies organisation and others, "seasonal mushroom cultivation does not require many scientific skills. Any person with low investment capital can set up a mushroom farm in a small room, especially if the person is an agriculturist. The raw material for compost — made of wheat or rice straw — for such a person is almost free and labour for picking mushrooms is cheap.

Short mushroom cultivation training programmes, lasting for a couple of days, according to Vohra, cannot really teach much. But they do introduce one to the subject. So, Vohra's advice is, "if you are interested, begin with a small farm and learn how to do the composting, spawning, casing, picking and packing, before you expand". With an experienced guide, you can develop the skills over a year's time, learning how to control or raise temperatures with coolers or heaters, in case you want to avoid investing in expensive technology.

The general opinion is that growing mushrooms is ideal for women entrepreneurs. They can easily grow 10 kg of mushrooms, a day, on a small farm indoors, provided they use high quality spawn, according to Malik. This can fetch them up to Rs. 300 a day, at the rate of Rs. 25 a kg whereas production cost with low-cost technology is roughly Rs. 6 per kg. "In an area of 60 × 30 sq. ft., you can grow mushrooms worth Rs. 1 lacs," adds Vohra.

Hi-tech projects : Those who wish to attempt year-round cultivation and exports must opt for hi-tech and quality farming. The slightest presence of any toxin or disease could make the foreign buyer consign your entire consignment to the seas. The American FDA standards are even more stringent than those of Europe :

While a few growers like Sahas Agro Pvt. Ltd. in UP. prefer to use indigenous machinery for an environmentally controlled growing unit, many mushroom growers are seeking foreign collaborations and even importing technology and plants to produce A-grade whole white button mushrooms for export.

Fig. 5.9. Mushrooms grow in an environmentally controlled hall.

At present, there are six or seven large-scale mushroom projects which have started exporting, though they are all still trying to perfect the art of composting and spawning.

Last year, India produced 30,000 tons of mushrooms. "But, it exported only 10,000 metric tons of the white button variety to Europe and America," says A.K. Gupta, general manager, APEDA.

He adds that Indian companies need foreign assistance and know-how if they are to become market leaders in mushrooms.

APEDA's list of incentives to mushroom growers includes 50 per cent financial assistance for feasibility studies with a ceiling of Rs. 1 lakh, 10 per cent for publicity for exporters, 25 per cent for purchase of transport units and quality control among other benefits.

If India continues to set up state-of-the-art environmentally controlled mushroom farms, it could soon become the largest exporter of quality mushrooms. "But it is not just a matter of getting the right technology," says Chris Bonnet, a mushroom consultant from Holland. "It is simple to make compost, grow mushrooms and sell. But it is not simple to make good compost, to grow mushrooms efficiently and to sell them at a good price.

The IPC Mushroom Growers, Training Institute at Horst in Holland offers a two-week practical training course at its modern 12-room farm. Further details can be had from the IPC Plant, Mushroom Growers' Training Centre, Westerholtstraat 2,5961 BJ Horst (Limburg); fax: (04709) 841600.

The training can be followed by a month or two of apprenticeship on a mushroom farm. The training, held twice a year, costs around 3,000 guilders with the apprenticeship costing an additional 500 guilders per week. "The training programme can be tailor-made for a particular country, if there are a sufficient number of students," says Bran Van Nieuwenhuijzen, director of the institute.

Dutch agro-horticultural companies like CMO Holland offer advanced mushroom technology for projects of at least half a crore. Such packages come complete with training in Dutch institutes, plus advice from consultants, till the project becomes self-sufficient. They ask for total commitment of the growers to mushrooms for horticulture is an industry that requires the grower's direct touch.

Before you settle for mushrooms, take into account that while it is a high-profit venture, it could prove to be a high-risk industry. Despite every precaution, crops can fail. Long power-cuts, not unusual to India, can be very hard on mushrooms. And one must know how to deal with diseases that can strike this fungus.

But if you are a true entrepreneur, it is not difficult to take care of such risks.

MUSHROOMS MEAN MONEY

The work 'mushroom' conjures up the vision of something that appears in plenty suddenly and disappears with equal suddenness. While this is quite true what is generally not known is that mushrooms are nutritious, quick growing and tasty food material, with medicinal qualities otherwise rare.

Fig. 5.10. Tray grown mushrooms
(Madhya Pradesh Chronicle, I. 9. 85)

Dhingri (plentorious), a mushroom variety grown in northern India, is used both as food and as medicine, and has a delicious flavour. It is rich in protein and vitamins and is helpful in treating diseases such as scurvy and gingivitis. With very low fat and carbohydrate content it is prescribed for treatment of diabetes and obesity. Besides protein, it contains fat, carbohydrates, vitamins and ash. They are rich in minerals such as nitrogen, phosphorous, potassium, calcium, magnesium, sodium, copper, zinc, manganese and iron. Amino acid contents are also relatively high.

Dhingri varieties grow well within a temperature range of 20°–30°C. This temperature range has to be maintained for at least 12 hours. In places 1300 m ASL mushrooms can be grown from March to October, in places 800; 1300 m from September to November and March to May and in low altitude areas (800, and below), from November to February.

For successful growing of mushrooms, a covered area, preferably a room is required. Collect wood trays (1mX 5mXX 18 cm), or a box, and paddy straw, and you are on the way to become a mushroom farmer. Spawn or mushroom seed, dal powder, single super phosphate are the other requirements. The trays are arranged in tiers, one above other, but clearing each other by at least 30 cms.

Straw, less than a year old and not drenched in rain or water is chopped into small bits, not exceeding 5 cms in length.

Mushroom seeds are available on payment from the offices of the Indian Council of Agricultural Research.

Chopped straw is filled in gunny bags, and immersed in water for upto 18 hours. They are taken out and placed on the ground to allow excess water to drain away.

Each tray will hold about 11 kg of sodden straw. It is mixed well with about 125 gram of dal powder, half a bottle of mushroom seeds and 125 gram of single super phosphate. One third of the mixture is put into the tray and a layer is patted into shape by hand. Then a second layer is laid, followed by a third. The tray is covered with 2 kgs of wet straw and the whole stuff is once again pressed down by hands. The tray should be covered with wet newspapers or gunny bags, supported by sticks placed crosswise across the tray. In about 10–20 days, the mushroom's mycelium (white cotton like strands) will appear. When the growth covers the entire surface of the tray, the newspaper/gunny bag cover can be removed. They may be wrapped around the trays so that humidity is maintained around 80 per cent. The trays should be sprayed with 0.1 per cent malathion and 0.2 per cent dithen Z–8, to keep away pests, from the third day of seeding. It should be continued 2–3 time a week. In about 15–18 days, pinheads appear. Some varieties take upto 30 days. These pinheads are the young fruits and pesticide spraying should be stopped when they appear. Within 3–4 days the pinheads develop into fully grown mushrooms.

The first flush of mushrooms are quite heavy, Subsequent flushes appear at 7–10 day intervals. Each tray will give at least three flushes, weighing around four kgs. Fruits are to be separated by twisting them clockwise. Then the stalk is cut off and your mushrooms are ready either for the kitchen or for the market or for the freezer.

The straw that remains in the tray is rich in nitrogen. It can be used as compost or can be mixed with fodder.

Among the tasty dishes made of mushrooms are mushroom pakodas, mushroom omelettes, mushroom muttar masala, mushroom soup and mushroom palak.

Grow mushrooms for wealth and eat them for health.

MUSHROOM AT THE TOP

For most of us the mention of mushrooms conjures up pictures of cans of salt water, in which are tightly-packed small buttons. Or at best, visions of snowy-white mushrooms placed in polythene bags and kept in the refrigerated compartment of an up-market food store.

Unfortunately, mushrooms are not an integral part of the Indian gastronomic tradition. The only mushroom to feature on medieval Indian menus was the gucchi (morrel), which was found in Kashmir and used in pulaos and other expensive Mughlai dishes.

In recent years, the cultivated mushroom (champignon de Paris to the French) has made an appearance on 'continental' vegetarian menus, either as mushrooms on toast or in baked dishes.

Fig. 5.11.
(Madhya Pradesh Chronicle, 24. 7. 96)

A few restaurants have tried to make up Indian recipes that use mushrooms but rarely have they got much further than dingri muttar, which is essentially the same recipe as muttar paneer with mushrooms standing in for the cottage cheese.

All of this is a pity because mushrooms are a great source of minerals and vitamins, are virtually non-fattening because of the high water content and offer endless possibilities to the imaginative cook.

There are broadly two kinds of mushrooms: the cultivated and the wild. Virtually every mushroom you get in India is cultivated. This makes—or should make—for a lower price because of economies of scale, but the truth is that India has the most expensive mushrooms in Asia.

Steer clear of the canned variety. They taste of nothing but brine. The fresh cultivated mushroom has enormous possibilities, provided you keep two basic ground rules in mind: it must be sauteed because it needs a little fat to bring out its flavour (butter or a good olive oil are perfect) and that any of the onion family makes an ideal match at the sautee stage.

If you keep these rules in mind, then it is difficult to go wrong, regardless of the cuisine: a mushroom sauce for pasta, masala mushrooms to eat with rotis or on toast, or mushrooms stir fried with Chinese vegetables or noodles in dark soya sauce.

Other mushrooms are more flavourful but less versatile. The oyster mushroom, a cheaper and lesser variety, is on sale in some parts of Calcutta and though you can substitute it for the champignon de Paris in most recipes, it is never quite as flavourful.

Also remember that chefs call the oyster mushroom 'the weeper' because it starts giving out water within three minutes of hitting the heat.

Show a Chinese chef a fresh shitake mushroom and he won't recognise it. Yet, rare is the Chinese restaurant that does not serve what are described as black mushrooms.

Unfortunately, these are sold dry and have to be re-constituted by soaking in warm water for half an hour or so. A good shitake mushroom should retain its flavour but the quality available in India is thin, and tastes like plastic.

If you are fortunate enough to find some thick, large, dehydrated shitakes, then re-constitute them after washing by soaking them in sherry or a mixture of sherry and port.

The mushrooms will take on an unusual flavour that you can further enhance by sauteeing them in butter with a dash of light soya sauce or by braising them with finely-chopped onion in oyster sauce.

There are many other mushrooms with much more flavour. Straw mushrooms are cheap and plentiful all over Asia. Open field mushrooms are much tastier than button mushrooms. And the Japanese have their own range of varieties.

Sadly, these are not available in India. Nor, of course, are the great mushrooms of Europe : Chanterrelles, girolles, porcini or the truffle.

But there is still a lot that you can do with what is available if you make the right effort.

REFERENCES

1. Atlas, R.A. 1986. *Basic and Practical Microbiology,* Macmillan Publishing Company, New York.
2. Brock, T.D. 1970. *Biology of Microorganisms,* Prentice-Hall, Inc., Englewood Cliffs, New Jersey.
3. Information Bulletin. 1991. *Technology for the cultivation of mushroom.* Division of Mycology and Plant Pathology, IARI. New Delhi.
4. Kannaiyan, S. and K. Ramasamy. 1980. *A Handbook of Edible Mushrooms,* Today and Tomorrow's Printers and Publishers, New Delhi.
5. Nath, P. 1995. *Mushrooms can keep you slim,* The Times of India, Saturday Times, April 8, p. 5.
6. Wistreich, G.A. and M.D. Lechtman. 1980. *Microbiology,* Macmillan Publishing Co., Inc. New York.

Chapter 6

MYCORRHIZAE

INTRODUCTION

The root hairs and other surfaces of the young root do not constitute the chief absorbing surfaces in all plants. Exceptions to this situation are found in a number of kinds of plants in which the young, active portions of the roots are invaded by specific soil fungi. The resulting association is intimate and complex, and the combination of root and fungus is termed as a *mycorrhiza* (literally, "fungus root").

WHAT IS A MYCORRHIZA ?

A mycorrhiza is mutualistic morphological and physiological association between a fungus and the root of some species of higher plant. These structures are formed as a result of the invasion of young roots by the hyphae of various species of fungi commonly found in forest soils.

It is likely that the great majority of land plants enter into this kind of relation with soil fungi. The mycorrhizal fungi are present in forest soil and become associated with seedlings that germinate there. Seedlings growing elsewhere frequently do not thrive because of the lack of mycorrhizae.

If soils low in available mineral nutrients, there is an increased absorption of nutrients by plants infected with mycorrhiza, especially in soils low in available minerals. This increase is usually attributed to increased soil exploration by an increased surface area of the root fungus association. Also it is possible that the fungus has the ability to dissolve and absorb certain minerals that the plant roots cannot readily obtain, due to an unfavourable soil pH (*e.g.*, low levels of available phosphate in very acidic soils). Higher plants that live naturally in association with mycorrhizal fungi are able to grow very well under laboratory conditions in which the fungus is absent, of course provided that inorganic nutrients are available in sufficient quantities.

OCCURRENCE OF MYCORRHIZAE

Mycorrhizae occur in many species of ferns, trees, shrubs, and herbs in all parts of the world; most species of woody plants appear to have them.

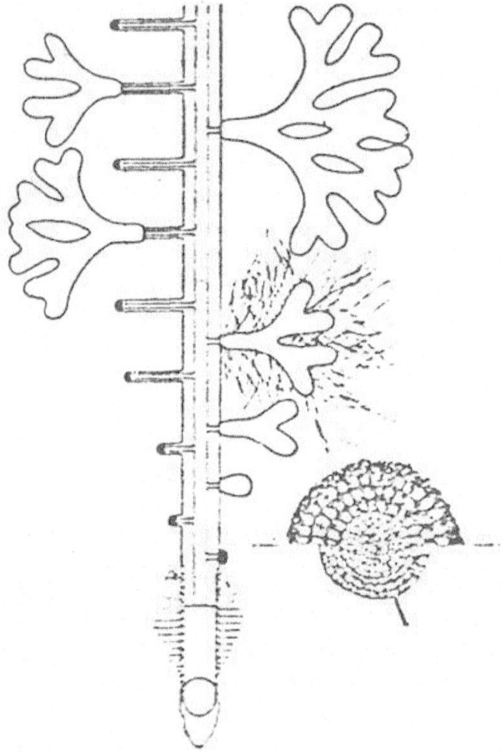

Fig. 6.1. Diagram of development of mycorrhizae on a pine root. The main axis is a long root bearing a root cap and root hairs on the unsuberized surface behind the meristematic region. Further back, mycorrhizal branches are developing from short roots. The upper of the cross section represents a mycorrhizal root; the lower part, an uninfected root.

(Kramer, p. 121)

In forest trees such as pine (*Pinus*), spruce (*Picea),* Iarch (*Larix*), oak (*Quercus*), beech (*Fagus*), poplar (*Populus*), and willow (*Salix*), the fungus is chiefly associated with the short side roots arising from a longer main root, although in some species the main roots are also invaded. The side roots are visibly swollen and otherwise modified by the effect of the fungus invasion. The swollen appearance results from changes in the cortical cells of the roots and also from the presence of felt like sheath, or mantle of fungus threads which closely surrounds the root. Threads arising from this sheath penetrate the epidermis and cortex, where they grow and branch between the cells. Root hairs are absent from these shorter side roots, which may constitute as much as 95% of the root tips of the tree. The susceptibility of fine roots to mycorrhizal fungi increases at first with their sugar content. Hence, all factors such as strong light and deficiency of phosphoric acid and nitrogen (inhibition of protein synthesis) which lead to an excess of soluble carbohydrates, favours the development of mycorrhiza.

TYPES OF MYCORRHIZAE

Two types of mycorrhizal roots are common: the *ectotrophic* and *endotrophic*. The line of demarcation between these types, however, is not sharp, and there are many mycorrhizae in which the fungal hyphae grow within the roots and on their surfaces.

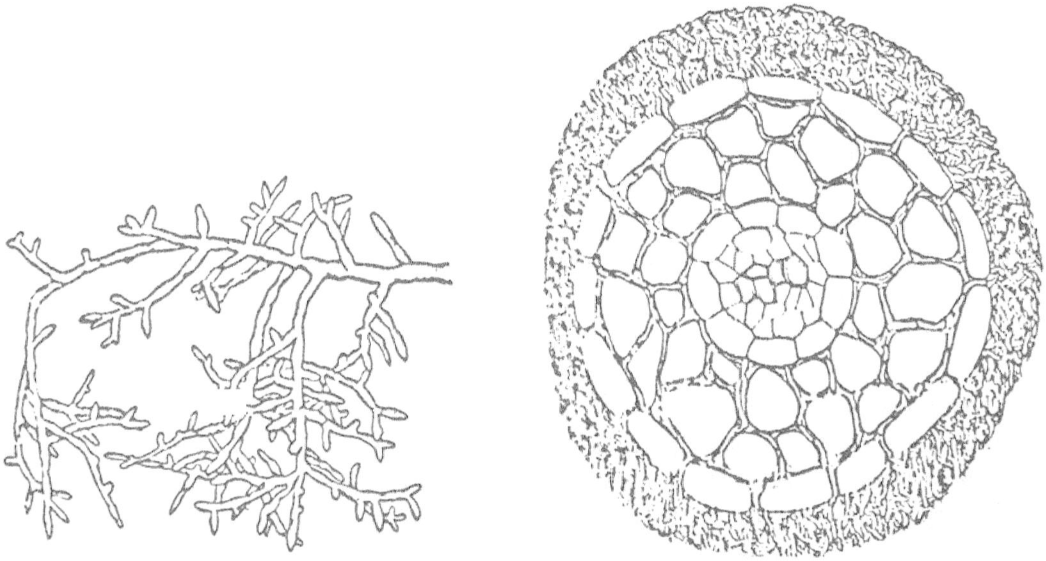

(a) (b)

Fig. 6.2. Mycorrhizae
(a) Swollen mycorrhizal rootlets of spruce (*Picea glauca*)
(b) Cross section of mycorrhizal rootlet, showing sheath of fungus threads around the root and threads between the cortical cells.
(Wilson & Loomis, p. 147)

Ectomycorrhiza

A majority of the deciduous or evergreen trees have ectomycorrhiza, in which the roots are entirely surrounded by the fungal hyphae. These fungi form a sheath around the root and penetrate the air spaces between the cells in the cortex, but do not enter cells. An extensive intercellular net is formed causing marked hypertrophy and extensive branching. They are found mainly in forest trees such as conifers, beech, oak and many others, and involve fungi of the Basidiomycota (like *Amanita, Boletus, Phallus, Tricholoma,* and *Scleroderma*). These fungi decompose the leaf litter and other forms of soil organic matter. Nutrients, like phosphorus, nitrogen, potassium, and calcium are easily absorbed by the fungal hyphae, and are then passed on to the root tissues, and hence to the rest of the plant. Their fruiting bodies (mushrooms) are commonly seen near the trees.

Endomycorrhiza

Endotrophic mycorrhizae occur in all other plants but affect the external appearance very little. They are found on many species of orchid, heath (Ericaceae), and gentian (Gentianaceae) families, and also on some trees, such as red maple (*Acer*), and walnut (*Juglans*). The fungi which associate in such mycorrhizae are apparently microscopic molds.

Like the ectotrophic types, the endotrophic mycorrhizae also form an intercellular network and extend into the soil, but they appear to penetrate cells (although in fact they do not break through the cell surface membranes of the root cells). *Liriodendron tulipifera* L. roots are example of the endotrophic type.

MORPHOLOGICAL NATURE OF MYCORRHIZAE

Mycorrhizal roots differ in certain structural features from uninfected roots; they are often short and thick, commonly lack root caps and root hairs, sometimes do not branch, and possess other morphological differences which distinguish them from normal roots. Roots which become infected with mycorrhizal fungi stop elongating and often branch extensively. In this association, fungal hyphae extend outwards from roots and act like extra long, nutrient absorbing hairs.

In mycorrhiza therefore, the parasite—host relationship is similar in character to that prevailing between leguminous plant and their nodule bacteria; this is true both of the mainly endotrophic mycorrhizae of orchids and of the ectotrophic mycorrhizae of conifer and broad-leaved trees.

DEVELOPMENT OF MYCORRHIZAE

Mycorrhizae are of great significance because they are probably the major route of entry of mineral elements into roots. The fungus receives organic nutrients, mainly carbohydrates and vitamins, from the plant and in return absorbs mineral salts (particularly phosphate, ammonium, potassium and nitrate) and water, which can pass to the plant root.

Generally only young roots are infected. Root hair production either ceases or is greatly reduced on infection. A network of hyphae spreads through the surrounding soil, covering a much larger surface area than the root could, even with root hairs. It has been suggested that plants of the same species, or even different species may have common interconnections with mycorrhizae.

The type of root system as well as the fungus determines the type of mycorrhizae. Trees such as yellow–poplar, maple, and members of the Ericaceae which have roots that can be differentiated only as fine or coarse produce endotrophic mycorrhizae when invaded by fungi. Trees of the Abietineae, Salicaceae, Betulaceae, Fagaceae and certain others produce roots of two types, long and short. Long roots grow relatively rapidly, produce root hairs which usually originate from a hypodermal layer, branch racemosely, are relatively long-lived, undergo secondary thickening, and seldom become mycorrhizal, possibly because they grow too rapidly to be affected by invading fungi. The short branch roots elongate slowly, bear few root hairs, do not have secondary growth, usually live one season or less, and are often invaded by fungi and become mycorrhizal. Mycorrhizal roots of this type are blunt and thickened. They often branch dichotomously and sometimes form coralloid clusters. The branching is caused by auxin produced by the fungus, but the development of mycorrhizal roots also seems to depend on an adequate supply of carbohydrates from the shoot.

The growth of mycorrhizae is also promoted by any factor that results in high carbohydrate levels in the roots, and moderate supplies of mineral elements, especially nitrogen and sulphur. High levels of these result in use of much of the carbohydrate in the synthesis of proteins and other substances. Fungal hyphae are intimately associated with the root and some extend out into the soil rather like an extended network of root hairs. The soil hyphae do indeed function like very long root hairs, absorbing mineral nutrients and passing some back to the root.

The ectomycorrhizas occur only in woody plants of cool climates, including trees such as birch, beech and pine. The tremendous effect of ectomycorrhizas on the growth and nutrient content of seedling of pine is shown in Table 6.1.

The growth and nutrient content of pine seedling (*Pinus elliotti*) with and without ectomycorrhizal infection given in Table 6.1.

TABLE 6.1

	Growth/g dry wt per plant	Macro nutrients/mg per plant		
		N	P	K
No mycorrhiza	0.26	2.4	0.22	2.27
+ ectomycorrhiza	2.81	38.9	3.90	35.6

(Ridge, p. 135)

The primary effect of the ectomycorrhizal fungus in this instance was on P uptake, phosphate being in limiting supply for growth; once the phosphate limitation was removed, growth and the uptake of N and K were greatly increased. When particular species of toadstools (fungal fruiting bodies) are found always associated with certain tree species, the fungi are nearly always ectomycorrhizal partners. The mycorrhizas are especially of the vesicular arbuscular (VAM) type (VA-mycorrhiza = the fungi form two characteristic types of structure, vesicles and arbuscules, inside the root but have little effect on the root's external appearance).

At least six other kinds of mycorrhizae occur but they are all internal types (endomycorrhizae) where fungal hyphae enter root cells. These can be formed in almost all higher plants, the Urticaceae and Cruciferae being two of the few exceptions. In general, the lower the level of soil nutrients (particularly phosphate), the greater the dependence of the plant on its fungal partner for obtaining nutrients, and thus for growth and survival. This is why the "cost" to host plants, in terms of supplying carbohydrates to support mycorrhizal fungi, is fully justified when soil nutrients are in short supply. In some ectomycorrhizae, 15% or more of host net production passes to the fungus. However, in nutrient—rich soils, the benefit to host plants is less certain and the "cost" of mycorrhizas may even reduce growth. With the emphasis in agriculture shifting towards "low input" systems, which means less use of fertilizer, ensuring that crops have the most effective mycorrhizal partners is receiving more attention, particularly in the topics.

SIGNIFICANCE OF MYCORRHIZAE

The physiological and biological significance of mycorrhizae is not thoroughly understood and therefore, is the cause of some disagreement among biologists in interpreting these structures.

Some botanists regard mycorrhizal fungi as parasites upon the roots which they inhabit and thus, regard mycorrhizae as purely pathological structures.

Other botanists believe that the relationship between these fungi and the roots which they inhabit is more complex than direct parasitism, and that, both fungi and roots benefit from their close association. According to the view, the fungi receive food from root tissues, and in turn facilitate certain physiological activities of roots, for example : (1) the ectotrophic fungi contribute substantially to the absorption of water and mineral ions by the host plant, particularly in species with limited number of root hairs, (2) mycorrhizal fungi promote the absorption of certain organic nitrogenous substances from the soil, (3) they carry on nitrogen fixation in much the same manner as nitrogen—fixing bacteria, or (4) the fungi may produce antibiotics that protect young roots from parasites.

The view that the fungi may have some beneficial effect upon roots is supported by a number of observations and experiments. In some species of plants, particularly certain species of trees, shrubs, and orchids, the mycorrhiza occurs in all individuals of the species. The invariability of this occurrence may be interpreted as indicating a condition of mutual benefit. In this type of symbiotic relationship, both the green plant and the fungus benefit from the association. It seems relatively certain that much of the water and mineral salts taken into the roots enter through the fungus threads which connect the sheath to the soil or directly through the sheath itself.

Controlled experiments have also indicated that some species of higher plants derive benefit from the mycorrhizal relationship; if seedlings of certain species of pine trees, for example, are grown in sterile soils, or in soils which do not contain suitable species of mycorrhizal fungi, the growth of the seedlings is slow and weak; if the soil is inoculated with these fungi and if the fungi then infect the roots, the growth of the seedlings quickly and conspicuously improves. Similar improvement of the growth of many other species of higher plants (many orchids, for example) has been demonstrated by similar experiments.

Mycorrhizae appear to be particularly useful in soils high in organic matter, as forest or bog soils. The fungi are thought to digest some of the organic matter and in this way make nitrogen and minerals available to the plant. The fungus, in turn, is believed to obtain from the root certain growth-promoting substances in which it may be deficient, such as hormones or vitamins, and also possibly some food materials in the form of simple sugars. Such relationships between the roots of green plants and fungi are more common than is generally believed.

In the roots of forest trees the parasitic fungus obtains carbohydrates, for which there is sharp competition in the soil, and perhaps also growth substances, so that in association with congenial species of trees it develops more vigorously, under certain conditions, than as a separate unit. On the other hand, the nitrogen balance of the tree may profit by the digestion of the hyphal coil. It is true that the mycorrhizal fungi (in contrast to many root nodule bacteria) do not assimilate atmospheric nitrogen and, therefore, they supply the host only with the nitrogen which they have obtained by decomposing humus.

The advantage to the host seems to be even greater in orchids than in forest trees. By digesting the fungal coils, orchids profit not only from the nitrogen assimilated by the fungus from the substrate and elaborated into protein compounds (so that, under certain conditions, the nitrogen content of infected orchid seedlings is three times as great as that of uninfected seedlings), but (especially in the heterotrophic species) also from the carbohydrates which the fungus decomposes in the humus and absorbs. In certain cases the fungus is also the purveyor (one that provides) of growth substances to the host plant. Thus, the Vandeae, a highly developed group of orchids with extremely degenerate embryos, have lost the capacity for the independent production of the growth substances essential to their development and procure them from their fungal partner; hence the latter is indispensible to these orchids when they are growing on a substrate deficient in growth substances. Knudson (1925) who made thorough investigation of the problem of orchid seed germination demonstrated that fungus infection of the seed or seedling was not necessary but harmful and that a high percentage of germination could be brought about by supplying the seeds with anyone of the sugars, glucose, fructose, or sucrose. It is, however, possible that the fungus or the sugar may also have supplied some favourable vitamin and that the absence of this vitamin limits the development of embryo.

The root hairs are most effective in the absorption of relatively immobile nutrients, such as phosphate, whose concentration near the epidermal surface is low. However, phosphate intake is influenced far more by the formation of this symbiotic association, *i.e.*, mycorrhiza.

CONCLUSIONS

Research in mycorrhizae will probably assume increasing importance in coming years, because the ever increasing cost of fertilizers will make it impossible to continue to use fertilizer chemicals in a wasteful fashion. Thus, it will become imperative to increase the efficiency of nutrient absorption by forest trees and crop plants with judicious management and selection of efficient mycorrhizal fungi.

As we learn more about mycorrhizae, it is likely that the knowledge will be applied with advantage to agriculture, forestry and land reclamation. For example, in tropical America, over 40% of soils are infertile and acidic, and inoculating soil with particular endomycorrhizae has improved growth for many crops, including tea and coffee. Mycorrhizal fungi increase plant uptake of various substances from soil, including Cu, and Zn. These two elements are essential micronutrients, so increased uptake is sometimes beneficial to the plant. But in higher concentrations they are toxic, so it is possible that in soil contaminated with large amounts of Zn, or Cu (or perhaps other heavy metals) mycorrhizal infection could increase the plant's uptake and therefore they harm to the plant.

REFERENCES

1. Curtis, O.F. and D.G. Clark, 1950. *An Introduction to Plant Physiology*, McGraw-Hill Book Company Inc. New York.
2. Fuller, H.J. and O. Tippo, 1961. *College Botany*, Oxford and IBH Publishing Co., Calcutta.
3. Greulach, V.A. 1973. *Plant Function and Structure*, Macmillan Publishing Co., Inc. New York.
4. Knudson, L. 1925. *Physiological study of the symbiotic germination of orchid seed.* Bot. Gaz. 79, 343–379.

5. Kramer, P.J. 1978. *Plant and Soil Water Relationships—A Modern Synthesis,* Tata McGraw-Hill Publishing Company Ltd., New Delhi.
6. Leopold, A.C. and P.E. Kriedemann, 1975. *Plant Growth and Development,* Tata McGraw-Hill Publishing Company Ltd. New Delhi.
7. Newman, E.I. 1993. *Applied Ecology,* Blackwell Scientific Publications, Oxford.
8. Noggle, G.R. and G.J. Fritz. 1991. *Introductory Plant Physiology,* Prentice-Hall of India Private Ltd., New Delhi.
9. Ridge, I. (ed.) 1991. *Plant Physiology,* Hodder and Soughton, Kent, England.
10. Sharma, O.P. 1990. *Text Book of Fungi,* Tata McGraw-Hill Publishing Company Ltd., New Delhi.
11. Soper, R. (ed.) 1990. *Biological Science,* 1&2. Cambridge University Press, Cambridge.
12. Wilson, C.L. and W.E. Loomis. 1957. *Botany,* Holt, Rinehart and Winston, Inc. New York.

Chapter 7

CHROMOSOMES AND DISEASE

INTRODUCTION

The congenital malformation called mongolism affects one in approximately 650. Ever since the condition was identified in the 1860's by the British neurologist John Langdon Down, concerned physicians and investigators had been seeking its cause. For many years, Mongolism was described as a heredity disease, the result of some unknown effects in the "germplasm". Investigators dissatisfied with such vague explanations have argued the opposite view that the disease is environmental, and have correlated its incidence with accidents due to the developing embryo during gestation, that is, the period between conception and birth.

The explanation of mongolism is now at hand. The disease is neither typically hereditary nor environmental, as these terms are commonly employed. It arises from a defect in the mechanism by which the heredity material is passed on from parent to offspring. This leaves certain questions unanswered: for example, whether the defect in the genetic mechanism is itself hereditary or environmental in origin. The explanation is nonetheless of great significance, for it is among the first findings to come from direct investigation of the genetic apparatus of the human cell. The techniques of cell genetics have mostly been restricted to more easily studied cells of lower animals and plants.

ADVANCES IN CELL GENETICS

During the present century, the science of cell genetics has made profound contributions to an understanding of the hereditary process. It was only in 1950's however that geneticist were able to establish the number of chromosomes in the human cell. These are the structure in the cell nucleus that encode the hereditary plan; the normal number in the human cell proved to be 46. Soon afterward came the discovery that mongolism is associated with the presence of 47 chromosomes. Similar gross abnormalities in the chromosome complement of the tissue cells have since been established in other diseases of man. To look through the microscope at the chromosomes of man is to see the very stuff that human life is made of.

The history of cell genetics can be said to date from 1903, when Walter S. Sutton, a graduate student at Columbia University in U.S.A., who later became a surgeon, pointed out in a classic paper the striking parallelism between the inheritance of genetic traits and the behaviour of chromosomes in cells. Subsequent investigations by Thomas Hunt Morgan and A.H. Sturtevant, also at Columbia, showed that the units of heredity, the "factors" of Gregor Mendel, must be arranged in linear order on the chromosomes.

These findings stimulated study of the chromosomes in various animals and plants. Investigators soon discovered that whereas the number of chromosomes varies in different organisms from 2 in some species of worm to 300 in some protozoa—the number in the cells of anyone species is constant. It was a relatively simple matter to count the chromosomes of organisms, notably the fruit fly, in which the chromosomes are few and large. In species in which they are small and numerous, however, the task was burdensome, and different investigators reported different results. This was the case with human chromosomes in particular, although the count of 48 obtained by T.S. Painter of the University of Texas in the early 1920's was accepted in the literature.

Meanwhile geneticists working with more favourable material were able to establish in the behaviour of the chromosomes the mechanisms that account for the segregation and independent assortment of traits first postulated by Mendel. Each cell of an organism, including the germ cells that give rise to sperm and eggs, bears two sets of chromosomes. The "homologous" chromosomes of each set (with the exception of the chromosomes that determine sex) are similar in appearance and carry genes affecting the same traits; one set is contributed by each parent. When the germ cells divide during meiosis, the chromosomes are replicated only once, before the first division. In the second division the two sets of chromosomes, pull apart or segregate to produce sperm or eggs (gametes) with only one set of chromosomes each. As a result, the gametes are haploid : they contain exactly half the number of chromosomes of the precursor germ cells. The subsequent union of the precursor germ cells. The subsequent union of the gametes produces a fertilized egg (zygote) with the full diploid number of chromosomes characteristic of the species, and the organism that then arises by mitotic division of the zygote exhibits a combination of traits inherited from its two parents.

CYTOLOGY AND CYTOGENETICS

For many years before 1956 the chromosome number in man was recorded as 48. Suitable material for human chromosome studies was difficult to obtain, and the older techniques were not satisfactory for critical chromosome counts. Human chromosomes are small, and in earlier preparations they overlapped on the microscope slide and were extremely difficult to distinguish. New methods of chemical treatment are now employed for fixing tissues with a minimum of distortion. Chemical treatment has also been developed for spreading chromosomes on the microscope slide, thus making it possible to identify each individual chromosome. Tissue culture methods have facilitated studies of cell division in living cells and have made metaphase stages available for chromosome counts. Recent developments in surgical treatment for cancer and other diseases have made more fresh human tissue available and modern techniques have provided better preparations for study. Testicular specimens have become particularly useful for chromosome investigations. Methods of freezing and fixing tissues have improved their value as cytological material.

In 1956, two cytologists, J.H. Tjio and A. Levan, working in Sweden, published a paper on the chromosome number in man, giving 46 as the $2n$ number. Their counts were made from tissue culture preparations of lung tissue representing four different human embryos. A pretreatment with certain solutions spread the chromosomes of dividing cells and made it possible to observe each chromosome separately. The investigators examined 265 dividing cells and concluded that the chromosome number in human embryo lung tissue was 46. Two more embryos were later included in the investigation and similar results were obtained. The lung tissue of all six Swedish fetuses had the same chromosome number. In the same year, two English investigators, C. E. Ford and J. L. Hammerton, reported 23 pairs of chromosomes in testicular preparations of each three adult Englishmen. In 1961, S. Makino and M. S. Sasaki studied 1422 cells from 54 different human embryos, ages 2 to 7 months, and found virtually all to have 46 chromosomes. Many specimens have now been obtained from widely separated parts of the world and the number 46 has become well established as the $2n$ number. No persistent difference in chromosome number has been detected among races of ethnic groups of mankind.

A group of cytogeneticists met at Denver, Colorado in 1960 and adopted a system for classifying and identifying human chromosomes. Chromosome length and the position of the centromere were the criteria for classification. The Denver classification, with refinements made at the London Conference (1963) and the Chicago Conference (1966), has become a standard for chromosome studies. Inspite of standardized preparations and numerous specimens for comparison, it has been impossible to identify consistently every chromosome pair. The 22 pairs of autosomes have, however, been divided successfully into 7 groups identified with letters A to G. Number 1 to 22 are associated with the autosomes in descending order by length. All autosomes can be placed satisfactorily within a group but the numbering within the groups is more or less tentative. The X chromosome is difficult to distinguish from members of the C group and the Y chromosome shows considerable variability in different preparations.

Equipped with newer techniques investigators proceeded to pair the homologous chromosomes of the human tissue cell, primarily on the basis of their over-all length and the position of the centromers : the site at which the twin strands of a replicated chromosome are held together at this stage of mitosis. They labeled the sex chromosomes X (female) and Y (male), and numbered the remaining somatic pairs from 1 to 22 in order of decreasing size.

As the first step toward relating the now identified human chromosomes to their genetic functions, investigators began to look abnormalities in chromosomal configuration in individuals with congenital disorders.

TYPES OF CHROMOSOME CHANGES

Cells of the same organism characteristically have the same number of chromosomes, with the exception of gemetes, which (in diploid organisms) have half as many chromosomes as somatic cells. Moreover, the organization and the number of genes on the chromosomes of an organism are the same from cell to cell. These characteristics of chromsome number and gene organization typically hold for all members of the same species. Exceptions to these normal conditions also occur and are known as *chromosome aberrations.* Chromosome aberrations may be defined as variations from the wild-type condition in either chromosome number or chromosome structure. Chromosome aberrations can arise spontaneously or can be induced experimentally by chemical or radiation mutagens, and in many instances they can be detected cytologically during mitosis and meiosis.

VARIATIONS IN CHROMOSOME NUMBER

Most eukaryotic organisms are diploid, meaning that they have two sets of chromosomes in their somatic cells ; the gametes of these organisms are haploid since they have only one set of chromosomes.

Variations may occur from the wild-type state in the total amount of genetic material in a cell. Such changes may occur spontaneously or may be experimentally induced. Variations in parts of chromosomes, give rise to aneuploidy and polyploidy.

CHANGES IN ONE OR A FEW CHROMOSOMES

Aneuploidy

Aneuploidy describes the abnormal case in which one to a few whole chromosomes are lost from or added to the normal set of chromosomes. In the case of diploid organisms, aneuploid variations take four main forms :

1. *Nullisomy* (a nullisomic cell) involves a loss of homologous chromosome pair, that is, the cell is $2n$-2.
2. *Monosomy* (a monosomic cell) involves a loss of single chromosome; that is, the cell is $2n$-1.
3. *Trisomy* (a trisomic cell) involves a single extra chromosome, that is, the cell has three copies of one chromosome type and two copies of every other chromosome type. A trisomic cell is $2n + 1$.
4. *Tetrasomy* (a tetrasomic cell) involves an extra chromosome pair, resulting in the presence of four copies of one chromosome type and two copies of every other chromosome type. A tetrasomic cell is $2n + 2$.

All forms of aneuploidy have serious consequences in meiosis and in most cases are lethal.

CHANGES IN COMPLETE SETS OF CHROMOSOMES

Monoploidy and *polyploidy* involve variations from the normal state in the number of complete sets of chromosomes. Considering an organism that is usually diploid, a monoploid individual has only one set of chromosomes.

Monoploidy occurs only rarely. Owing to the presence of recessive lethal mutations in the chromosomes of many diploid eukaryotic organisms, which are usually counteracted by dominant alleles in heterozygous individuals, many monoploids probably do not survive and thus are undetected.

Polyploidy describes the condition of a cell or organism that has more than its normal number of sets of chromosomes. Animals appear to have evolved to the point that the consequences of more than two sets of chromosomes in a diploid organism are often severe or lethal. The molecular basis for the severe or lethal effects is not well understood. Thus the evidence for polyploidy in animals often comes from studies of spontaneously aborted fetuses, so we know that polyploidy is one of the major bases for early, spontaneous abortions of human or other animal fetuses.

NONDISJUNCTION

The mechanism by which an individual's cells might come to have an extra chromosome was already familiar to cell biologists. Working with fruit flies in 1913, Calvin B. Bridges of Columbia University had demonstrated a failure, called nondisjunction, in the normal process of segregation at meiosis. In addition to three pairs of somatic chromosomes, the female fruit fly has a pair of X chromosomes (the male has an X and a Y chromosome). Normal meiosis in the female produces an ovum with a single X chromosome ; normal meiosis in the male, a sperm with an X or a Y chromosome. On rare occasions, Bridges found, the two X chromosomes of a female germ cell fail to "disjoin" in the first stage of meiosis. One of the ova produced by the next division accordingly receives both X chromosomes. Subsequent fertilization by a normal Y-bearing sperm gives rise to an exceptional fly female in appearance but with a chromosome complement of XXY.

Apparently, a similar accident during meiosis in the human germ cell can give rise to a mongoloid infant. In this case, however, the nondisjunction involves not the X chromosomes but a small somatic chromosome : Number 21. The ovum with two 21's is fertilized by a sperm also bearing a 21 ; the result is a zygote with three, instead of the normal pair, of these chromosomes. This is called "trisomy" of Chromosome 21.

The questions remains : What is it that causes this abnormality in the mechanism of heredity? It has long been known that the risk of bearing a mongoloid child rises steeply with the age of the mother : the mean age of mothers of affected children is 37 ; at the age of 45 the risk is nearly 3 per cent (Down named the disease for the supposedly oriental cast of the facial features associated with it and had been suggested that it might better be known as Down's syndrome). This suggests that the abnormal meiosis is triggered by sometime-linked mechanism, perhaps the overaccumulation of some metabolic product in the ovum of the older woman or even the activation of a "latent" virus. Another possibility is that the afflicted mothers bear a gene that increases the frequency of nondisjunction ; such a gene is well known in the fruit fly.

DOWN'S SYNDROME

A syndrome associated with trisomy of an autosomal chromosome was formerly known as "mongolism" but is now designated Down's syndrome. It was named after Langdon Down who first described the symptoms in 1866. The individuals he studied, and those with the same syndrome who have since been observed, were short in stature (about four feet tall) ; they had characteristic slanting eyes (thus the earlier name "mongolian"), stubby hands (particularly the fifth digit), broad, short skull, large tongues, with a characteristic furrowing, and general loose jointedness, particularly in the ankles. They were characterized as low in mentality.

A small chromosome presumably No. 21, is added to the normal complement in all cases where Down's syndrome is recognized. Chromosome No. 21 has been difficult to distinguish from No. 22 and it is not known which of these two is involved. On the basis of size, No. 21 is presumed to be associated with Down's syndrome.

Down's syndrome occurs once in about 700 births among European people, older mothers have a higher incidence than younger mothers. About 40% of Down's syndrome victims are born to women over 40 years of age, whereas this group of mothers produces only about 4% of all babies born in the general population. The mean maternal age for all babies born in the population is estimated to be 28.5 years compared with a mean age of 36.6 years for mothers of babies with Down's syndrome. The greater incidence among older mothers may be associated with the long delay in completion of meiosis. Potential eggs in the ovary begin the meiotic process before birth and proceed to primary oocyte stage. They then remain dormant until shortly before ovulation occurs. Cells of a woman 40 years of age have thus remained in an inactive stage for about twice as long as those of a woman 20 years old. This long period of quiescence may explain atleast partially why nondisjunction occurs more frequently in older than younger mothers.

Some patients of Down's syndrome have a total of only 46 instead of 47 chromsomes, but in such cases a translocation has been found to join chromosome No. 21 with another chromosome in the same complement.

KLINEFELTER'S SYNDROME

The very same anomaly that Bridges found in the chromosomes of the fruit fly actually showed up in another constitutional disorder of human beings that was linked to an abnormality of chromosome complement. This was Klinefelter's syndrome, named for Harry Fitch Klinefelter, the U.S. physician who described it. Individuals with this syndrome are phenotypically males but some tendency toward femaleness can also be recognized, particularly in secondary sex characteristics. It occurs once in about every 400 live male births and is characterized in the adult by tallness, enlargement of the breasts, small prostrate glands, testicular dysgenesis (defective development of testicles) and virtual nonproduction of sperm. The syndrome is sometimes also associated with severe mental retardation. Most of the patients show a chromosome complement of two X chromosomes and one Y chromosome : XXY. Bridge's XXY flies were female ; and XXY human is a sterile male.

An XXY zygote could result, of course, from nondisjunction during meiosis in either the father or the mother ; an abnormal ovum would supply two X chromosomes and the normal sperm a Y, or an abnormal sperm would contribute an X and a Y to the X of the ovum.

In one case it has been possible to establish in which parent the failure occurred. This victim of Klinefelter's syndrome was colour-blind, it was also determined that he had a colour-blind mother and a father with normal vision. The gene for colour-blindness is sex-linked (*i.e.*, it lies on the X chromosome) and recessive (*i.e.*, its effect is negated by another X chromosome with normal gene). The colour-blindness gene must therefore have been carried by both of the mother's X chromosomes, it could not have been carried by the father's X chromosomes or the father would have been colour-blind. It followed that the patient inherited both X chromosomes from his mother. In other instances, however, it is likely that the patient inherits an X chromosome as well as a Y from his father as the result of nondisjunction in meiosis of the sperm, and only the normal one – X complement from his mother. Additional cases have been identified, with XXXY and XXXXY cells. The genes on the little Y chromosome should not be underestimated ; even when it is outnumbered four to one, the patient still has masculine characteristics.

If nondisjunction occurs in the mother, the XX ovum need not, of course, be fertilized by a Y-bearing sperm. What would be the consequence of fertilization by an X-bearing sperm, yielding an XXX zygote? When this occurs in fruit flies, the resulting "super females" appear remarkably normal. When it happens in the human species, some of the XXX females are normal and fertile but many are mentally defective and infertile.

The few individuals that have been studied with tetrasomic X chromosomes (48, XXXX) are all mentally defective females. In the general population, a trisomic X chromosome occurs in the proportion of about one per 1000. Individuals with the "triple X syndrome" are comparable in some ways to *Drosophila* metafemales (XXX). In *Drosophila*, however, such individuals are strikingly abnormal and sterile, whereas human XXX individual, are visibly indistinguishable from individuals with the normal XX arrangement.

The best known symptoms associated with this syndrome are abnormalities associated with functional processes such as menstruation.

TURNER'S SYNDROME

If nondisjunction can result in ova with two X chromosomes, it can also produce ova with no X chromosome at all. When such a gamete is fertilized by an X-bearing sperm, the zygote will have only one sex chromosome – it is called XO. Such individuals have been found. Female in appearance, they lack many secondary sexual characteristics ; they may also display congenital defects such as usually short stature, a "webbed" neck, malformation of the aorta, deafness and mental deficiency. This syndrome was described by Henry Hubert Turner of the University of Oklahoma School of Medicine in 1938, but it was not until 1959 that Turner's syndrome was found to be the consequence of an XO chromosome constitution.

A final possibility would be fertilization of an ovum with no X chromosome by a Y-bearing sperm. This would produce a YO zygote, but such a chromosome complement is apparently incompatible with life—atleast no such individuals have been found.

CHROMOSOMAL ABERRATIONS

Aberrations can occur in the structure of the chromosomes as well as in their number. The number of such variations is essentially infinite. Two principal types are *deletions* and *translocations*.

Deletions

In deletions a piece of a chromosome breaks off and is lost. Such abnormalities, particularly if they involve sizeable amounts of the genetic material, could be lethal in many cases, and large deletions are in fact not commonly found.

Translocations

In translocation, breakage is followed by relocation of the fragments. A reciprocal translocation involves breaks in two nonhomologous chromosomes and thereafter a mutual exchange of the fragments. Depending on how the chromosomes behave during meiosis, some gametes will be formed with a normal, although displaced, gene content – in other words, a balanced translocation. But translocation may also give rise to gametes that have duplications in others. Some offspring of gametes that carry translocations may therefore be normal and some may not. When the germ cells of even the apparently normal person who bears a balanced translocation later undergo meiosis, an unbalanced gamete may result. In human pedigrees a translocation often manifests, itself when a disease is transmitted by unaffected individuals whose collateral relatives show the disease.

The first recognizable translocation was found in 1959 by Jérôme Lejeune of the University of Paris in a mentally deficient boy with congenital abnormalities of the backbone. The boy's cells had only 45 chromosomes, including one that was longer than its homologue. Lejeune suggested that part of the missing chromosome had been relocated, thus explaining the extra length of the abnormal chromosome ; the remaining fragment had been eliminated. Some patients suffering from Down's syndrome and having a normal number of chromosomes may be explained by translocation. In these cases the extra chromosomal material, usually present as an extra "free" chromosome (trisomy of chromosome Number 21), is translocated to another chromosome. The result is Down's syndrome with only 46 chromosomes. The disease would seem, therefore, to be related not so much to the number of chromosomes in the cell as to the extra chromosomal material.

THE PHILADELPHIA CHROMOSOME

A special case in which an extra chromosome or more, probably a chromosome fragment of No. 21, is associated with a chromosome fragment of No. 21, is associated with a specific disease. This is the Philadelphia chromosome which is found in leucocytes of patients with chronic granulocytic leukemia. It is apparently a deleted part of chromosome No. 21 and serves as an identification marker in the bone marrow cells of people who have this particular kind of leukemia.

CRI-DU-CHAT SYNDROME

A deletion of part of the short arm of chromosome No. 5 is associated with this syndrome which is characterized by microcephaly, a small head and mental disorder. Children with this abnormality have a cat-like cry.

CHROMOSOMAL MOSAICS

Chromosomal mosaics have been detected as in other animals. These are individuals who have atleast two cell lines with different karyotypes derived from the same zygote. Mosaics originate from a chromosome irregularity after fertilization. A chromosomae may lag in a mitotic division and become incorporated in the same nucleus with the chromosome from which it divided. One daughter cell would thus receive one too many and the other would be one deficient. Each cell would give rise to a cell line with the irregular chromosome number. Proportions of cells representing the different cell lines would vary in different tissues, making the extent of the mosaicism and the effect on the organism difficult to predict.

Many sex chromosome mosaics have been detected in human beings. Each has at least two cell lines with different sex chromosome arrangements. The main phenotypic characteristic is extreme variability. Some sex chromosome mosaics that have been reported are : XO/XX, XO/XY, XX/XY, XXY/XX, XX/XXX, XXX/XO, XXXX/XXXXY, and several other combinations reflecting two or three cell lines. Mild to severe phenotypic symptoms have been associated with these chromosome mosaics.

OTHER GENETIC DEFECTS IN HUMAN DEVELOPMENT

Much of the information about the regulation of gene expression during development has come from studies of nonhuman organisms. The reason for the relative dearth of information about human development is that studies of early human embryos is both technically and legally difficult. Nevertheless, several human developmental mutants have been identified. These mutants presumably result from defects in the regulation of expression of genes that are crucial for the developmental processes affected or from defects in structural genes whose products are needed for the developmental defects.

(1) *Achondroplasia* : Phenotypically, people with achondroplasia have a normal-sized trunk, enlarged head, and unusually short long-bones in the arms and legs. As a result, adults with achondroplasia average about 4 ft in height. Often other skeletal abnormalities can be seen upon X-ray examination.

The disorder is inherited as an autosomal dominant mutation. Most probably the mutation involved is in a gene that functions in bone formation.

(2) *Nail-patella Syndrome* : The phenotypes of an individual with nail-patella syndrome include deformed or absent fingernails toenails and a patella (knee-cap) that is abnormally small or absent. This syndrome is also the result of an autosomal dominant mutation.

(3) *Cleft Lip and Cleft Palate* : Cleft lip and cleft palate are also the result of gene mutations. In embryonic development the palate (roof of the mouth) halves do not fuse since they are not in the correct position at the appropriate time. Unlike the achondroplasia and nail-patella syndrome, most forms of cleft lip and palate are not the result of a single gene mutation. Rather, there appear to be a number of rare mutations that, as part of the phenotypic abnormalities they produce, include cleft lip, cleft palate or both.

One additional chromosome in the D group is associated with a cleft lip.

More complex arrangements of the sex chromosomes (such as 48, XXXY; 48, XXYY ; 49, XXXXY and 47, XYY, males and 48, XXXX and 49, XXXX females) are on record. Irregular numerical chromosome arrangements cited thus far have been associated with structural and mental abnormalities in the patients concerned.

TABLE 7.1. Human Chromosomes Grouped According to Size and Centromere

Location

Chromosome Group	Size	Centromere Location
1–3	Large	Approximately median
4–5	Large	Submedian
6–12 and X	Medium	Submedian
13–15	Medium	Acrocentric
16–18	Short Medium	Median or submedian
19–20	Short	Approximately median
21–22 and Y	Very short	Acrocentric

(Strickberger, p. 468)

TABLE 7.2. Sexual Aneuploids in Man and their Sexual Phenotypes

Sex Chromosomes		Phenotypes
Female		
XO	(monosomic)	Turner's Syndrome
XX	(disomic)	Normal
XXX	(trisomic)	Superfemale (mental abnormalities)
XXXX	(tetrasomic)	Superfemale (mental abnormalities)
XXXXX	(pentasomic)	Superfemale (mental abnormalities)
Male		
XY	(disomic)	Normal
XYY	(trisomic)	Normal
XXY	(trisomic)	Klinefelter's syndrome
XXYY	(double trisomic)	Klinefelter's syndrome
XXXY	(tetrasomic)	Extreme Klinefelter's
XXXXY	(pentasomic)	Extreme Klinefelter's

(Strickberger, p. 471)

TABLE 7.3. Phenotypic Abnormalities Resulting from Non-disjunction of the Sex Chromosomes

Condition/ Genotype	Symptoms	Frequency in Western Populations
Klinefelter's Syndrome (XXY)	Male, possessing some female secondary sexual characteristics, sterile, testes very small, little facial hair, breasts may develop, usually low intelligence.	0.02%
Turner's Syndrome (XO)	Female, lacking normal secondary sexual characteristics and very short, nipples close together.	0.03%
XXX	Female, normal appearance, fertile, but mentally retarded.	0.12%
XYY	Male, tall, variable intelligences possess psychopathic traits or tendency for petty criminal acts.	0.1%

(Soper, p. 856)

TABLE 7.4. Relationship between Age of Mother and Risk of a Trisomy-21

Age of Mother	Risk of Trisomy – 21 in Child
< 29	1/3000
30–34	1/600
35–39	1/280
40–44	1/70
45–49	1/40
All mothers combined	1/665

(Russell, p. 553)

TABLE 7.5. Aneuploid Abnormalities in the Human Population

Chromosomes	Syndrome	Frequency at Birth
Autosomes		
Trisomic 21	Down	1/700
Trisomic 13	Pataue	1/5000
Trisomic 18	Edwards	1/10000

Sex Chromosomes,	**Females**		
XO,	monosomic	Turner	1/5000
XXX,	trisomic		
XXXX,	tetrasomic		1/700
XXXXX,	pentasomic		
Sex Chromosomes,	**Males**		
XYY,	trisomic	Normal	1/1000
XXY,	trisomic		
XXYY,	tetrasomic	Klinefelter	1/500
XXXY,	tetrasomic		
XXXXY,	pentasomic		
XXXXY,	hexasomic		

(Russell, p. 556)

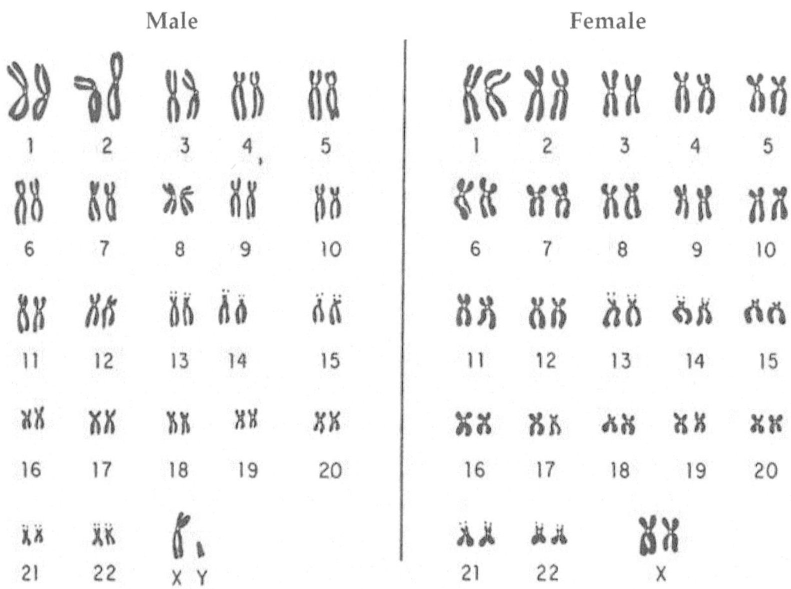

Fig. 7.1. Metaphase chromosomes of a mitosis in a human male and female arranged according to homologous pairs. (Strickberger, p. 469)

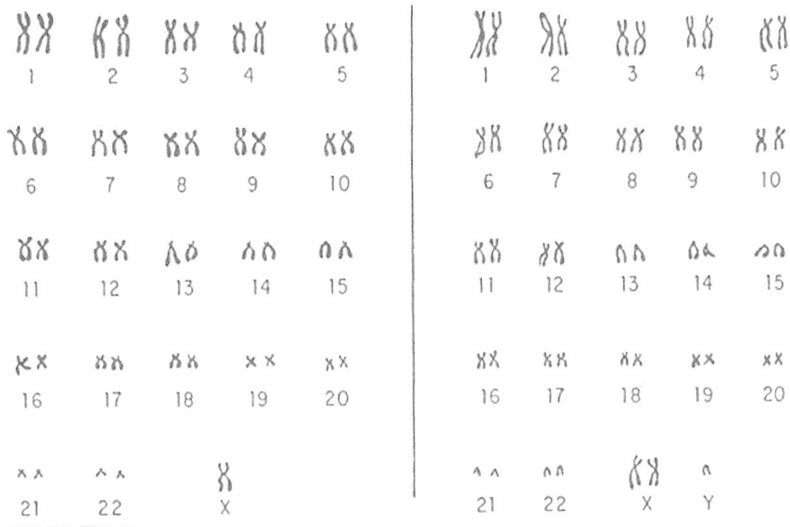

(a) (b)

Fig. 7.2. (a) Chromosome constitution (karyotype) of an individual with Turner's syndrome showing only one X chromosome without a Y. Such individuals are phenotypic females, but their ovaries do not develop properly. In addition, they are usually characterized by a webbed neck, short stature, and impaired intelligence.

(b) Karyotype of an individual with Klinefelter's syndrome bearing two X chromosomes with a Y. Such persons are phenotypic males but, as in Turner's syndrome, have underdeveloped gonads. People affected with Klinefelter's syndrome usually have long limbs, sparse body hair, and mental deficiency.

(Strickberger, p. 470)

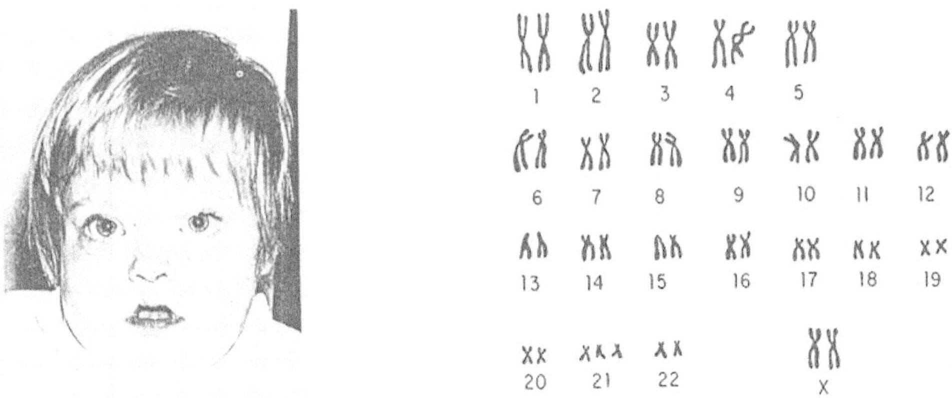

Fig. 7.3. Photograph and karyotype of a female with Down's syndrome showing three 21 chromosomes. (Actual differentiation between chromosomes 21 and 22 is difficult, and Lejeune has therefore proposed that the trisomic chrosome causing mongolism designated as 21, and the other simply 22) (Strickberger, p. 473)

EVERY 20 MINUTES, A CHILD IS BORN WITH DOWN SYNDROME

Every 20 minutes, a child is born in India with down syndrome—a disease characterised by mental and physical retardation—even though a blood test in the early stages of pregnancy can help detect the disease.

Lack of knowledge among couples about genetic counselling has resulted in the birth of tens of thousands of down syndrome-affected children in the country, say experts.

According to a report prepared by the genetic unit of the department of paediatrics in the All India Institute of Medical Sciences (AIIMS), all pregnant women should undergo a blood test between 15 to 18 weeks of pregnancy to assay maternal serum alpha feloprotein and human chorionic gonadotrophin as well as ultrasonography. These tests can help to detect down syndrome in the foetus.

If the foetus has down syndrome, abortion should be done within 20 weeks of pregnancy, according to experts.

Down syndrome is a condition with an extra chromosome in the body. Such children have 17 chromosomes in their cells in contrast to the normal 46 chromosomes. A child with this syndrome has eyes with an upward slant, flat face, small head and ears, weak muscles and short and broad hands with curved little fingers.

As the child grows, there is delay of mental and physical development, flattening of back of the head and furrowing of the tongue. In intancy there is a tendency to push out the tongue due to a small mouth.

Down syndrome occurs as a result of wrong division of chromosomes in the germ cells of the mother or the father. Normally cells in the ovaries in females have 46 chromosomes. Before conception, these cells divide into two daughter cells each containing 23 chromosomes. Thus, the sperm of the father containing 23 chromosomes combines with the mother's ovum which also contains 23 chromosomes. This results in the offspring having the normal number of 46 chromosomes.

In down syndrome, due to unequal division, one of the germ cells contains 24 chromosomes, while the other contains 22 chromosomes. When the germ cells having 24 chromosomes from one of the parents unites with its normal partner containing 23 chromosomes, the end result is a fertilised ovum containing 47 chromosomes.

The defect of cell division occurs more often in the mother, especially when the maternal age at conception is more than 35 years.

According to experts, analysis of data in India shows that the incidence of down syndrome is about one for every 1,000 births, But because of the large population and high birth rate, one infant with down syndrome is born every 20 minutes.

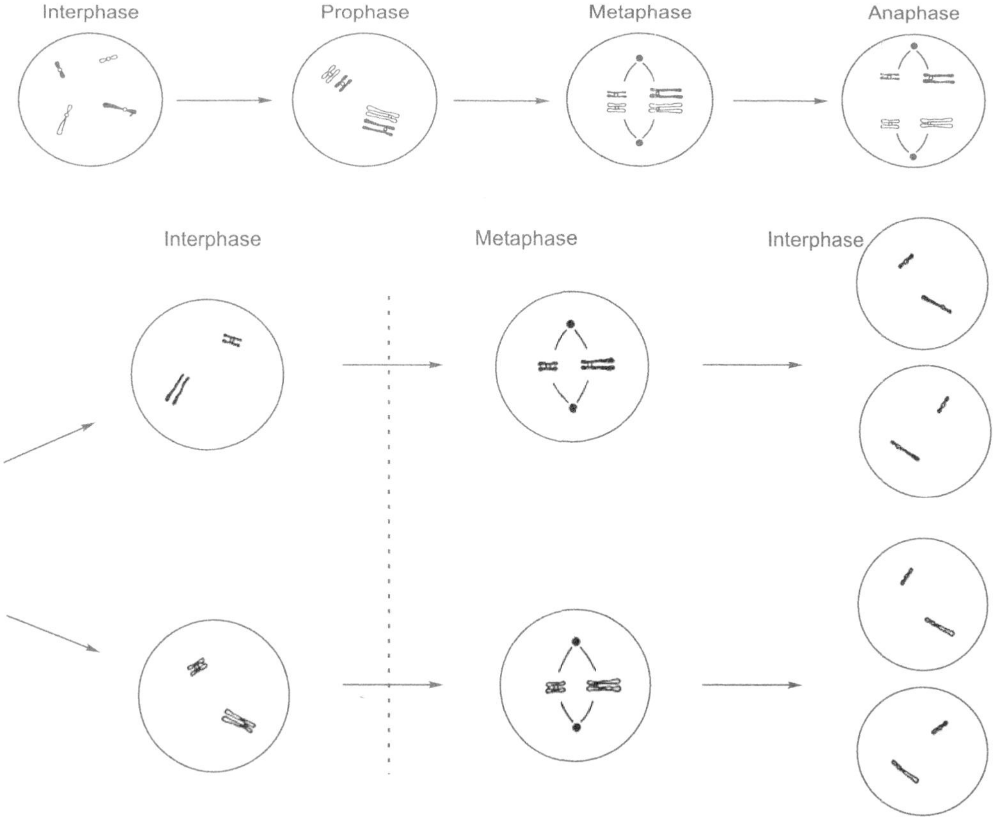

Fig. 7.4. Meiosis, the process by which germ cells divide to produce sperm and eggs with only half the full number of chromosomes, occurs in two stages diagramed here in simplified form. Before meiosis the germ cell has the full number of chromosomes (*left*). During prophase each chromosome, now replicated, pairs up with its homologue and assumes a visible two-stranded form. At metaphase the homologous pairs line up with each other and in anaphase one member of each pair, without splitting, goes to each pole. The new cell, shown at interphase, therefore has only one member of each pair. (The line up at metaphase can vary, so various combinations of maternal and paternal chromosomes can be found in the new cells). In the second stage of meiosis (*right of broken vertical line*) there is no further replication of the chromosome. The twinstranded chromosomes simply line up as in mitosis and divide. The four cells that result, precursors of the sperm and egg cells.

The report says that the state of children with down syndrome is much better than those with other forms of mental handicap. Though found to have a mental development, the former group is able to sit, walk and even speak although they being doing so slightly later than normal children.

It says about 40 per cent of the children with down syndrome have heart disease. They are also more prone to infections of the upper respiratory tract as well as pneumonia. Therefore these children should be kept away from those who have fever, cold or cough.

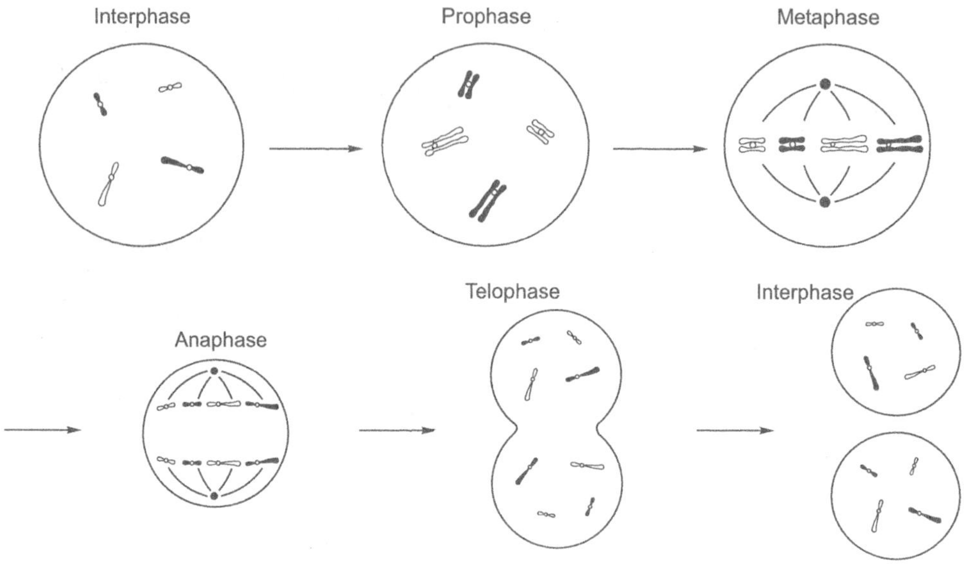

Fig. 7.5. Mitosis, the process by which body cells divide, is diagramed here for a hypothetical cell with four chromosomes : two homologous sets of two each. They are shown schematically at interphase (*left*), between cell divisions. By prophase each chromosome has replicated, forming two strands joined at the centromere. At metaphase the twin-stranded chromosomes line up at the equator of the cell. At anphase the strands split apart and move toward opposite poles. In telophase the cell splits to form two daughter cells. These, seen at interphase (*right*), each have four chromosomes, the same number as the original cell, and are exact genetic copies of the original.

(Scientific American)
November 1961, p 4 & 5

When a child with down syndrome is born, the risk of recurrence in subsequent pregnancies is only about one per cent. However, when the maternal age at conception is less than 30 years, about five per cent of cases with down syndrome has a different chromosomal abnormality called translocation.

In case were translocation is detected the risk of recurrence is higher, about five to ten per cent. In such cases geneticist should be consulted for precise advise. Geneticist will have to study the chromosomes of the parents before giving proper counselling.

Down syndrome can be prevented in high risk situations such as advanced maternal age by determining the chromosomes of the unborn baby. This is done by examining the amniotic fluid cells or by chorion villus biopsy.

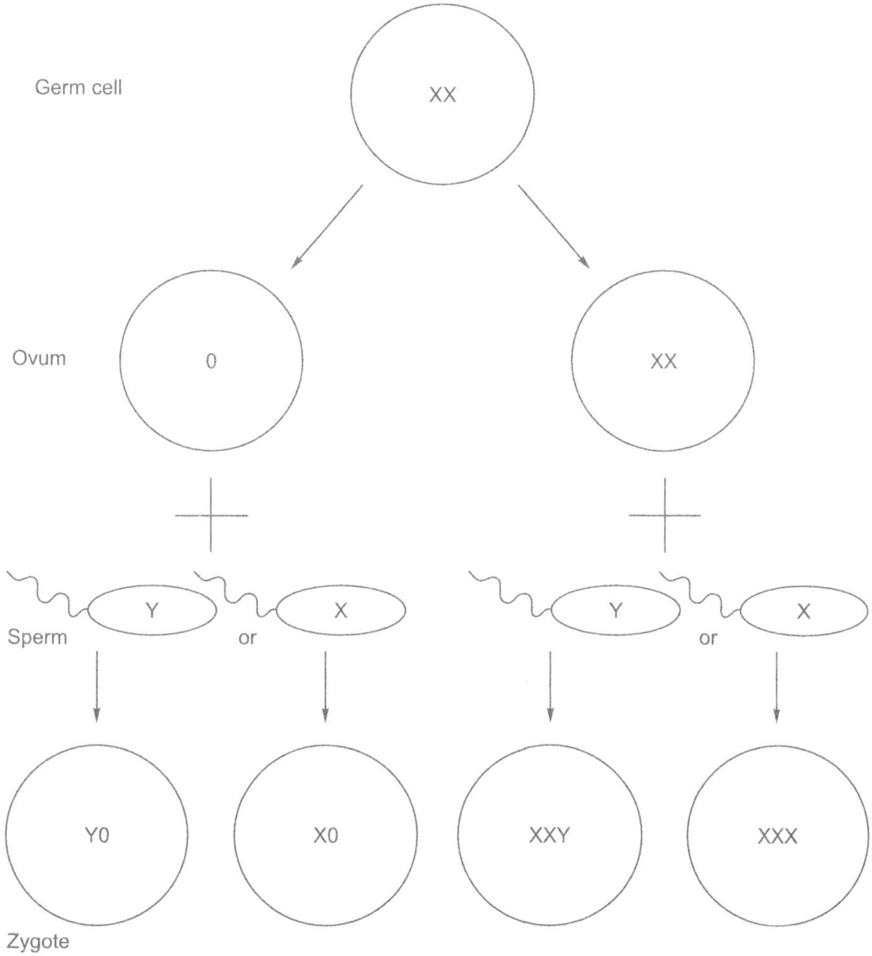

Fig. 7.6. NONDISJUNCTION IN MEIOSIS would explain several diseases. If two chromosomes end up in one ovum, fertilization by an X or a Y sperm will produce one of the abnormal zygotes shown. YO content apparently results in a nonviable zygote. XO is found in Turner's syndrome, XXY is characteristic of Klinefelter's syndrome and XXX is a "super female."

(Scientific American)

November 1961, p. 6

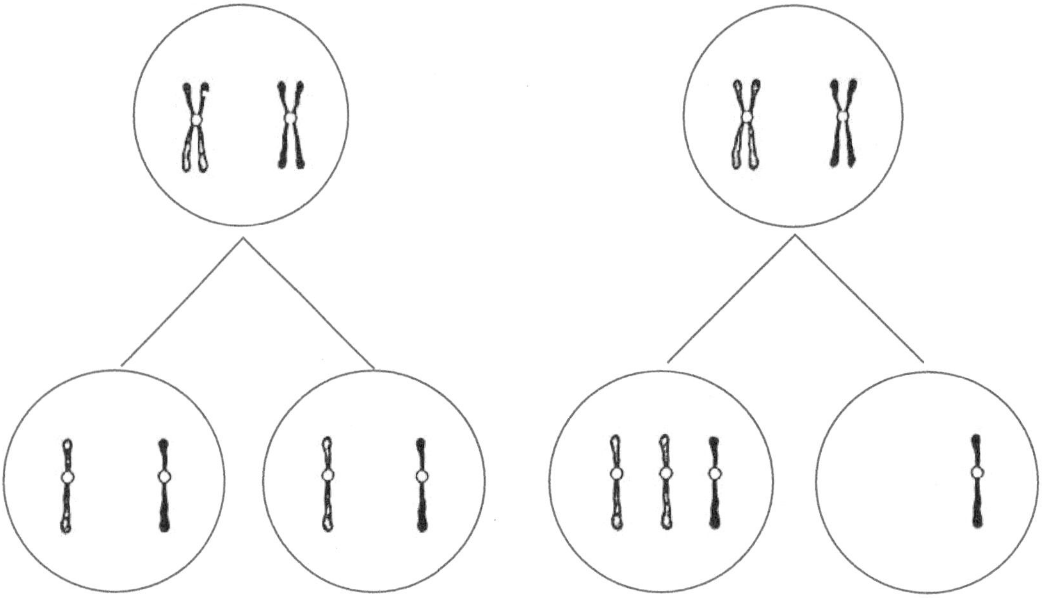

Fig. 7.7. NONDISJUNCTION is one means by which a cell can come to have the wrong number of chromosomes. Here it is diagramed in the case of one homologous pair. The chromosomes have replicated in preparation for division. Normally the two strands separate to opposite poles of the cell and one passes to each new cell (*left*). In nondisjunction (*right*) the strands of one chromosome fail to disjoin ; two abnormal cells result, one with three of the involved chromosomes (trisomy) and the other with one (monosomy). Through a similar process in meiosis either a sperm or an egg can be produced with too many or too few chromosomes.

(Scientific American)

November 1961, p. 6

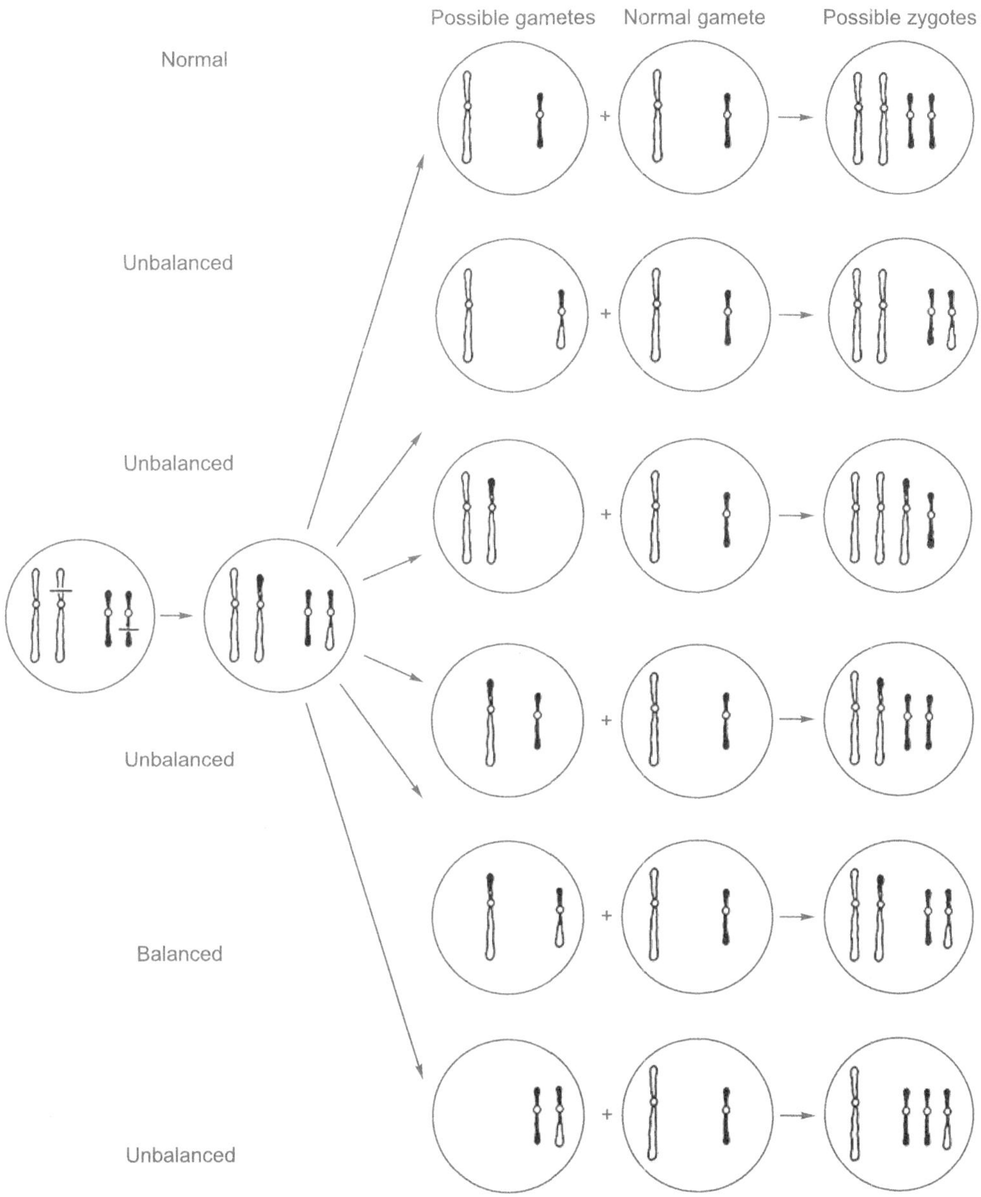

Fig. 7.8. IN TRANSLOCATION parts of two nonhomologous chromosomes of a germ cell break off (*left*) and are relocated (*second from left*). In meiosis abnormal pairing of these anomalous chromosomes is likely ; instead of a normal gamete (*top of first column*), meiosis could produce any of the other five gametes shown.

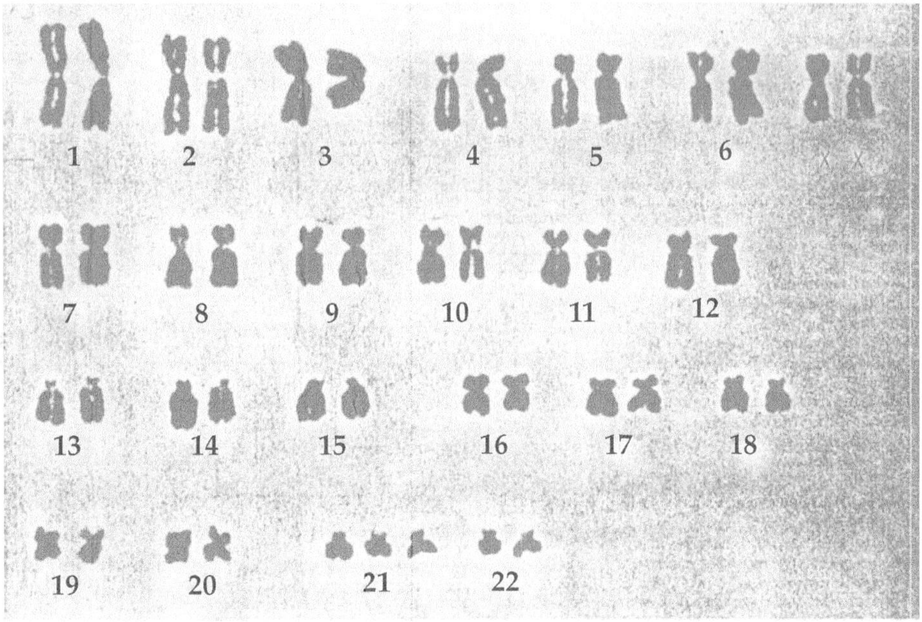

Fig. 7.9. TRISOMY OF CHROMOSOME 21 is revealed in this karyotype of a patient with Down's syndrome. During meiosis both Number 21's of the mother, instead of just one, went to the ovum. Fertilization added the father's homologous chromosome, resulting in trisomy.

(Scientific American) November 1961, p. 10

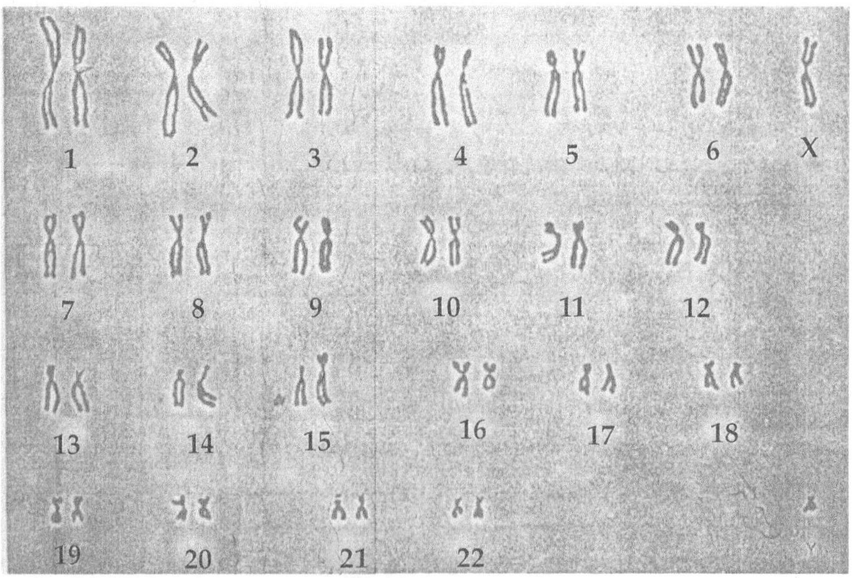

Fig. 7.10. "ESSENTIAL TRISOMY" OF 21, again resulting in Down's syndrome, occurs when extra Number 21 material is translocated instead of appearing in trisomic form as on the opposite page. In this case the extra material has been translocated to a Number 15.

(Scientific American) November 1961, p. 11

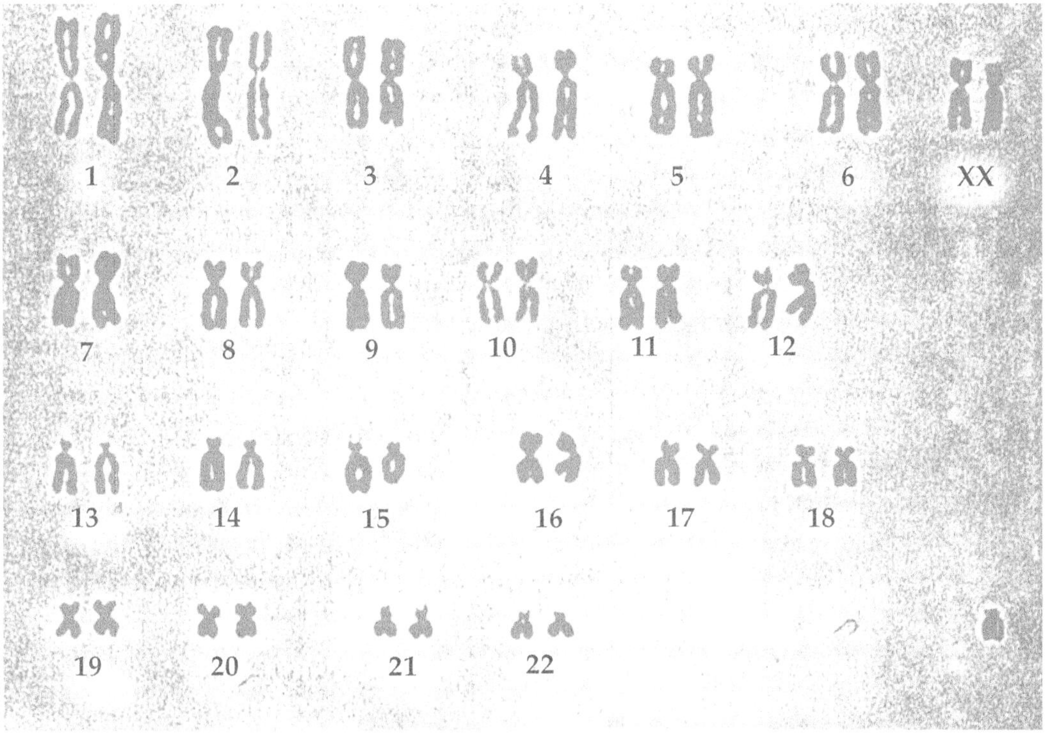

Fig. 7.11. KLINEFELTER'S SYNDROME is usually characterized by two X chromosomes and a Y, instead of XX (normal female) or XY (normal male). This can come as result of nondisjunction of sex chromosomes during gametogenesis in either the father or the mother.
(Scientific American) November 1961, p. 12

REFERENCES

1. Bearn, A.G. and James L. german III. 1961. *Chromosome and Disease*, reprint, Scientific American, Nov.
2. Edgar, S.R.B.O. 1965. *General Genetics*, Toppan Company Ltd., Japan.
3. Gardner, E.J. 1967. *Principles of Genetics*, Wiley Eastern, Japan.
4. Russell, P.J. 1990. *Genetics*, Harper Collins Publishers, U.S.A..
5. Sarin, C. 1985. *Genetics*, Tata McGraw-Hill Publishing Company Ltd., New Delhi.
6. Soper, R. (ed) 1990. *Biological Science*, 1 & 2. Cambridge University Press, Cambridge.
7. Strickberger, M.W. 1969. *Genetics*, The Macmillan Company, Canada.

Chapter **8**

ENZYMES

INTRODUCTION

An enzyme (from the Greek word *"enzymos"* meaning *"in yeast"*) is a biocatalyst. It is partly or entirely a protein that can tremendously increase the efficiency of a biochemical reaction without itself being used up. Although a biochemical reaction will proceed to completion in the absence of an enzyme, the process would be extremely slow – so slow in fact, as to make life as we know it impossible. Thus, it is safe to assume that most reactions occurring in the cells are catalyzed by enzymes. The term "catalyst" is used by the physical chemist to denote a substance which alters the rate of a reaction. Consequently, negative catalysis is possible. The word is used here in the popular sense, *i.e.,* a substance which accelerates a reaction.

NATURE OF ENZYMES

Enzymes, being organic catalysts, have many of the properties of inorganic catalysts and thus can be characterized as follows :

1. Enzymes are active in extremely small amounts-that is, in a biochemical reaction only a small amount of enzyme is necessary to convert a large amount of substrate to product. The two terms substrate and product signify the starting and ending material of a reaction respectively. The number of moles of substrate converted per minute by 1 mole of enzyme is called the turnover number of the enzyme. This number may vary from 100 to over 3,000,000.
2. True catalysts remain unaffected by the reaction they catalyze. Enzymes under stable conditions approach this property of the ideal catalyst very closely. Because of the protein nature of enzymes, however, their activities are confined to narrow ranges of temperature, pH, and so on.

3. Although an enzyme considerably hastens the completion of a reaction, it will not affect the equilibrium of that reaction. The reversible reactions usually present in the living system would proceed toward equilibrium at a very slow rate in the absence of enzymes. However, an enzyme will speed a reaction in either direction—that is, it will bring about the equilibrium of that reaction at a much faster rate.
4. Catalytic action is specific. Enzymes exhibit specificity for reactions they catalyze—that is, an enzyme that will catalyze one reaction may not catalyze another.

CATALYTIC ACTION

How does an enzyme influence the rate of a reaction ?

In ordinary chemical reactions, the beginning molecule (substrate) must be changed into one or more different unstable forms before it can be transformed to the product. We might refer to this unstable form as the *intermediate form* (transition state form). The intermediate form is usually at a higher energy level than is the starting molecule. Therefore, the probability is low that the intermediate form will occur unless additional energy is provided to the reaction. Thus, all reactions, even spontaneous ones, require "activation" of the starting molecule before conversion to product can takes place. Without a catalyst, most biological reactions would occur very slowly and be rate limited by the formation of an intermediate compound that requires a high energy of activation. One way to overcome the energy of activation barrier is to supply energy (heat). With an increase in temperature, a greater number of starting molecules will gain sufficient energy of activation to the intermediate form, and the intermediate form will convert spontaneously to the product. In enzymatically catalyzed reactions, the enzyme combines with the starting molecule in such a manner as to produce a change in the conformation of the substrate to that of an intermediate form. Because to form the enzyme – intermediate complex requires less energy than it does to activate molecules by heat, we can accurately say that the enzyme lowers the energy of activation required of the substrate and thus increases the rate of the unstable intermediate formed and hence the final product. If the energy of activation required for the formation of the enzyme-intermediate complex is low, many more starting molecules can participate in the reaction than would be the case if the enzyme were absent. For example, in the absence of the enzyme catalase, the energy of activation for the decomposition of hydrogen peroxide (H_2O_2) is 18,000 cal/mole ; but in the enzyme's presence, it, is 6,400 cal/mole. When the energy of activation of a reaction is lowered ; it is lowered for the backward as well as the forward reaction – in other words, an enzyme will speed a reaction to its equilibrium.

SPECIFICITY AND ENZYME—SUBSTRATE COMPLEX (MICHAELIS—MENTEN THEORY)

The enzymes (E) combine with their substrates (S) before yielding the products (P) of the reactions they catalyze. In other words, the enzyme and substrate form a complex before transformation of the substrate can occur.

$$E + S \rightleftharpoons E\,S \rightarrow E + P$$

Enzymes have active sites with which a substrate molecule can form an intimate association. If we visualize enzymes that have active sites and that are surrounded by numerous substrate molecules, which are very small by comparison, we can see that random collisions play an important part in enzyme-substrate reactions. Since the major portion of the enzyme molecule is devoid of active sites, we would expect many collisions between substrate and enzyme molecules before an active collision takes place. However, if enough substrate molecules are present, the active sites on an enzyme may become completely occupied, and at this point the rate of reaction will be at a maximum. The enzyme-substrate complex offers a good explanation of the specificity of enzymes. The active sites are specifically shaped within the numerous folds of the enzyme molecule. Only specifically shaped substrate molecules than, therefore, fit correctly into these active sites.

Indirect evidence supporting the validity of the enzyme – substrate complex theory can be found in a study of the action of inhibitors on enzyme activity. Structural analogs of a substrate molecule may, in some cases, occupy sites on an enzyme that are normally occupied by the substrate. The new complex formed is reversible and inactive with respect to the formation of products. In other words, these structural analogs compete with the normal substrate molecules for active sites on the enzyme. Substances that act in this manner are called *competitive inhibitors*, and their inhibition of enzyme activity is called *competitive inhibition*. Competitive inhibition can be overcome by increasing the concentration of substrate until all the active sites are occupied by substrate molecules.

One of the classic examples of competitive inhibition is the inhibition by malonic acid of the enzyme succinic dehydrogenase, which catalyzes the conversion of succinic acid to fumaric acid. The inhibitor, malonic acid, closely resembles the normal substrate, succinic acid, in chemical structure and as a result is able to occupy active sites normally occupied by succinic acid. Malonic acid is a competitive inhibitor, since inhibition can be overcome by increasing the concentration of succinic acid.

OCCURRENCE OF ENZYMES IN THE PLANT

Every living cell contains enzymes, usually hundreds of them, but they are not uniformly distributed throughout the cells of a plant. Certain enzymes notably those associated with respiration, are present in all living cells, but others may be restricted in their distribution to certain organs or tissues. Germinating seeds are the richest of all organs of higher plants in enzymes.

Ribosomes have been identified as cytoplasmic particles chiefly functioning in the process of protein synthesis. Thus, the enzymes catalyzing the formation of peptide chains should be located on the surface or in the very immediate vicinity of these particles.

Many of the enzymes of cellular metabolism are associated with the cytoplasmic particles of the cell. Perhaps the highest concentrations of enzymes may be found in the mitochondria and chloroplasts. All the enzymes necessary for the complete oxidation of pyruvate in the Krebs cycle to CO_2 and H_2O are present in the mitochondria. This would include the enzymes necessary for the passage of electrons to oxygen for the formation of H_2O. The passage of electrons form the intermediates of the Krebs cycle to oxygen occurs via the cytochrome or electron transport system and results in the formation of ATP.

The chloroplast is even more remarkable for the diverse array of enzymes that are within it. The enzymes necessary for the dark reactions of photosynthesis (assimilation of CO_2) are present in the matrix of the chloroplasts. Also, cytochrome enzymes have been located in this cell particle and, as in mitochondria, their activity leads to the production of ATP. In addition, the enzymes necessary for the synthesis of the pigments of the chloroplast (chlorophylls, carotenoids, etc.) are most likely present.

The enzyme deoxyribonuclease is located in the nucleus. This enzyme catalyzes the hydrolytic cleavage of deoxyribose nucleic acid (DNA). The ground phase of the cytoplasm (cytoplasm without formed particles), in contrast to the nucleus, abounds in enzymes. The enzymes of glycolysis and the hexose monophosphate shunt are located in the cytoplasm. Also various hydrolytic enzymes and phosphorylases are present.

In addition to enzymes associated with specific areas of the cell, there are enzymes that may be considered extracellular. Although extracellular enzymes are rare in higher plants, they are abundantly found in bacteria and fungi. These enzymes function in the extracellular digestion and transport of nutrients into the cell. For example, some bacteria utilize proteins and polysaccharides as nutrients. These molecules are very large and complex and could not possibly penetrate the cell membrane. However, the bacteria excrete enzymes that reduce these large molecules to smaller molecule, which are able then to penetrate the cell.

LYSOSOMES

Lysosomes are distinguished from other types of particles in the cytoplasm mostly on the basis of their biochemical properties, namely, their content of specific enzymes, which are primarily hydrolytic in function. About 36 hydrolytic enzymes have been identified in lysosomes, including those which digest proteins, nucleic acids, polysaccharides, lipids, organic linked sulphate, and organic linked phosphate. The lysosomes (also called "suicide bags") are ruptured after the death of a cell and are responsible for the process of self destruction (autolysis) of dead cells. Rupture of lysosomes in skin exposed for a long time to sunlight (ultraviolet light) leads to pathological changes following sunburn.

PROSTHETIC GROUPS, COFACTORS, AND COENZYMES

Many enzymes have, in addition to their protein structure, an attached nonprotein group. Proteins (in this case enzymes) with attached nonprotein groups are called *conjugated proteins.* Proteins or enzymes of this type may be thought of as consisting of two parts : an *apoenzyme,* composed only of amino acids (protein), and nonamino acid *prosthetic group* (the complete enzyme molecule is called *holoenzyme*). We can see a good example of this type of complex in enzymes that require a certain metal for activity. We refer to the metal as an *inorganic factor* (or activator). Separation of the apoenzyme from its metal component usually results in complete loss of activity, while restoration of the metal to the apoenzyme returns activity. Many of the enzymes associated with glycolysis require metal cofactors. Some metals known to be cofactors of enzyme systems are copper, iron, manganese, zinc, calcium, potassium and cobalt.

In contrast to the metal–requiring enzymes, some enzymes require an association with certain organic substances for activity. These organic prosthetic groups are sometimes an integral part of the enzyme and do not dissociate from the protein (apoenzyme) portion. Certain organic substances, however, dissociate readily and may act as prosthetic groups with many apoenzymes. These organic prosthetic groups are called *coenzymes.*

During enzymatic activity, a coenzyme generally acts as a donor or acceptor of atoms that are added to or removed from a substrate. Many coenzymes are very important in oxidation – reduction reactions. Further, they can be separated easily from the protein portion and when this occurs, the catalytic properties of the enzyme are greatly reduced. Some of the coenzymes that have been identified are nicotinamide adenine dinucleotide (NAD), nicotinamide adenine dinucleotide phosphate (NADP), coenzyme A (Co A), flavin mononucleotide (FMN), and flavin adenine dinucleotide (FAD). Most of the coenzymes are composed of vitamins, which are organic compounds that are synthesized in plants but not in mammals.

NOMENCLATURE AND CLASSIFICATION

Enzymes are generally named according to the substrate they complex with or the type of reaction they catalyze. The usual practice is to add the suffix – *ase* to the name of the substrate involved. Thus, the enzymes arginase and tyrosinase are so named because of their substrates arginine and tyrosine.

Enzymes may also be grouped under more general term describing a certain group of compounds with which they react. Thus, we have lipases, proteinases, carbohydrases, etc. Finally, enzymes may be named according to the type of reaction they catalyze. Some examples are hydrolases, oxidases, carboxylases, and phosphorylases.

For purposes of convenience, the following classification is followed.

1. Hydrolytic Enzymes

Catalyze the addition of water. *Hydrolysis* is the process by which a compound is split into fragments by the addition of water. The manner in which proteins, fats, and polysaccharides are hydrolyzed is of special importance because all these compounds are necessary for maintenance of life, yet they are too large to penetrate through the cell membrane. Those enzymes which help to hydrolyze or split proteins, carbohydrates, and fats into smaller residues are known as hydrolytic enzymes (*i.e., hydrolases*) or digestive enzymes, and the process of hydrolysis also spoken as digestion. Furthermore, by the activities of hydrolytic enzymes, stored foods can be mobilized when need arises.

The hydrolases are present inside all cells (in lysosomes) and make possible intracellular movements of proteins, lipids, and carbohydrates. Examples : Esterases, carbohydrases, proteases.

$$RCO - OR'\ HOH \rightleftharpoons RCOOH + R'OH$$

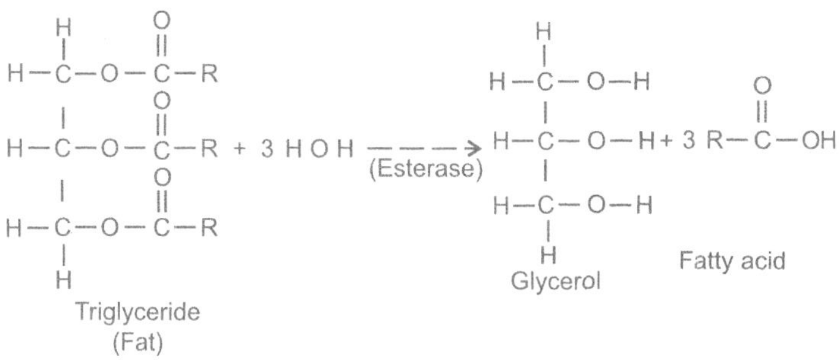

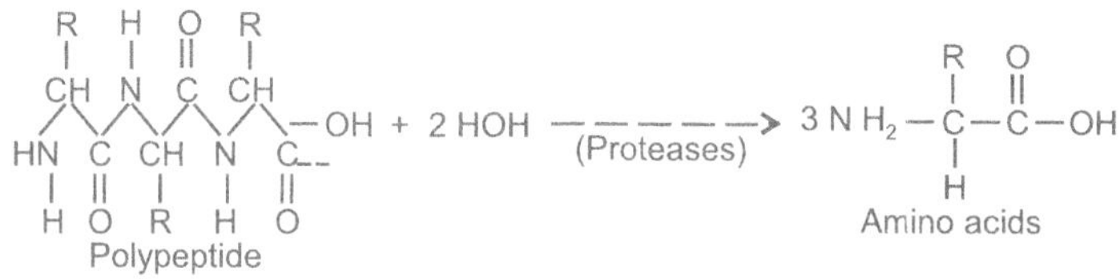

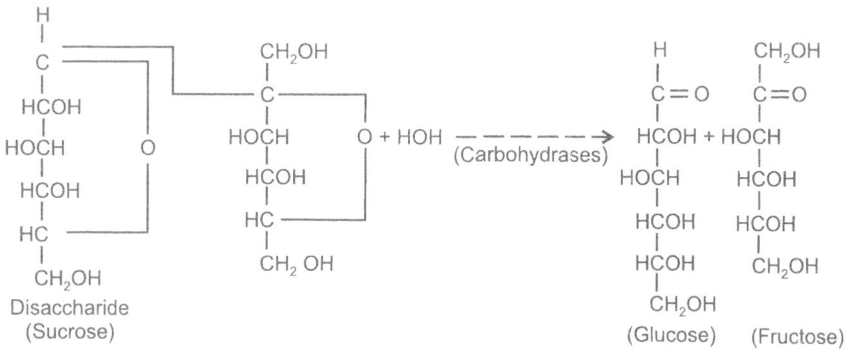

2. Oxidation–Reduction Enzymes

Catalyze the removal or addition of hydrogen, oxygen, or electrons from or to the substrate, which is oxidized or reduced in the process.

$$RH_2 + A \longrightarrow R + AH_2 \quad \text{(removal of hydrogen)}$$

$$RO + \frac{1}{2}O_2 \longrightarrow RO_2 \quad \text{(addition of oxygen)}$$

$$R^{2+} \longrightarrow R^{3+} + e^- \quad \text{(removal of electrons)}$$

These enzymes occupy a major position in cellular metabolism.

Examples : Dehydrogenases and oxidases (cytochrome oxidase, which transfers electrons, and coenzyme 1 – nicotinamide adenine dinucleotide or NAD, which transfers hydrogen ions and electrons. NAD is actually the coenzyme or prosthetic group of some of the dehydrogenases, (NADP is similar in structure except that it has an extra phosphoric acid group attached).

The reduced form of NAD may be designated $NADH_2$ or $NADH + H^+$ when it is desirable to show the release of an H^+ ion in the reduction. The oxidized form of NAD will be NAD^+. The same conventions are to be used with NADP.

3. Phosphorylases

Catalyze the reversible phosphorolytic cleavage of a specific bond on a substrate. The best known phosphorylases are those that catalyze the addition of the elements of phosphoric acid to starch and glycogen.

$$\text{Starch} + \text{phosphate} \rightleftharpoons \text{Glucose} - 1 - \text{phosphate}$$

The activity of these enzymes is somewhat analogous to that of the hydrolytic enzymes, except that the elements of phosphoric acid instead of water are added.

4. Transferases

Transferases catalyze the transfer of a group from a donor molecule to an acceptor molecule. This group of enzymes is very large and includes such enzymes as transglycosidases, transpeptidases, transaminases, transmethylases, and transacylases. Probably the best known example of a transferase is the enzyme glutamic – aspartic transaminase. This enzyme catalyzes the transfer of an amino group from glutamic acid to oxaloacetic acid to form aspartic acid.

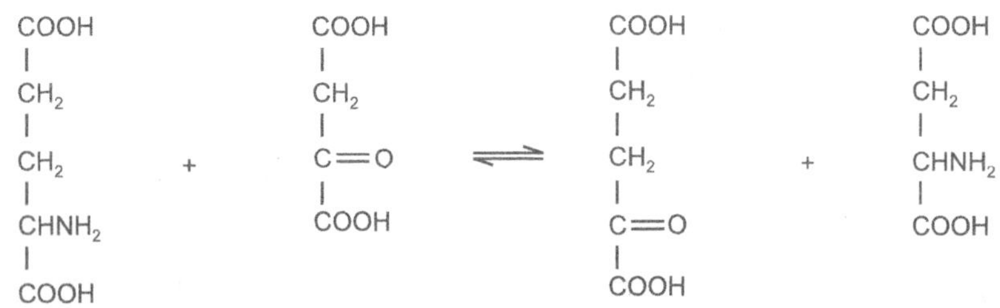

Glutamic acid   Oxaloacetic acid   α–Ketoglutaric acid   Aspartic acid

5. Carboxylases

The carboxylases catalyze the removal or addition of carbon dioxide. An example of an enzyme that removes CO_2 is glutamic decarboxylase. This enzyme catalyzes the removal of CO_2 from glutamic acid to yield γ-aminobutyric acid (γ–*Upsilon*, Greek alphabet)

$$\text{Glutamic acid} \rightleftharpoons \gamma\text{-aminobutyric acid} + CO_2$$

An example of an enzyme that catalyzes the addition of CO_2 is ribulose bisphosphate carboxylase. This enzyme is important in photosynthesis, during which it catalyzes the carboxylation of ribulose –1, 5 – bisphosphate to 3 – phosphoglyceric acid.

6. Isomerases

Catalyze internal changes in the molecules of compounds such as the interconversion of aldose and ketose sugars (isomerization reactions – different compounds with the same molecular formula are called isomers). For example, the interconversion of glucose – 6 – phosphate to fructose – 6 – phosphate is catalyzed by the enzyme phosphoglucoisomerase.

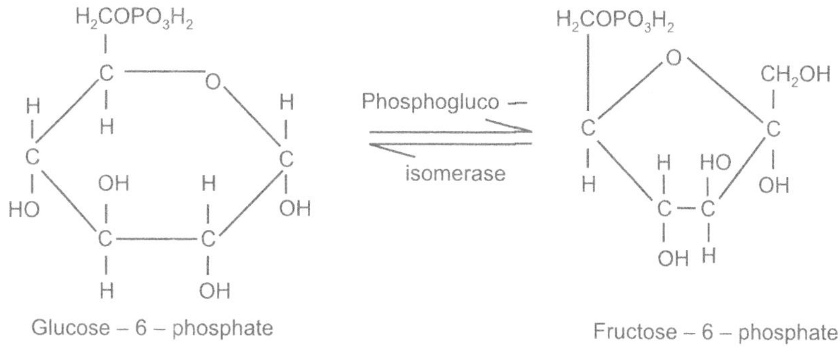

7. Epimerases

Epimerases are enzymes that catalyze the conversion of a sugar or a sugar derivative to its epimer. Epimers are molecules that differ only in the configuration of a single carbon atom. The conversion of a molecule to its epimer is called epimerization. An example of an epimerization is the reversible conversion of xylulose – 5 – phosphate to ribulose – 5 – phosphate.

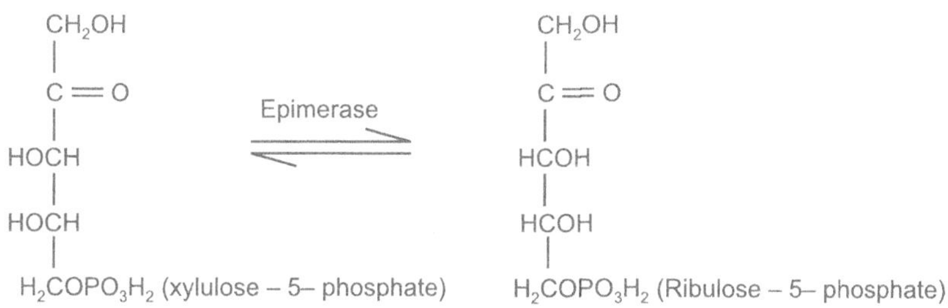

FACTORS AFFECTING ENZYME ACTIVITY

Like all chemical reactions, an enzyme – catalyzed reaction is susceptible to external conditions. Because of their protein nature, however, enzymes are unusually sensitive to the fluctuating influences of their immediate environment. Thus, the substrate concentration, enzyme concentration, temperature, and pH affect the rate of an enzyme – catalyzed reaction because they all affect the active site of the enzyme and the formation of the enzyme – substrate complex.

TABLE 8.1. Classification of Hydrolyzing Enzymes (Curtis & Clark, 1950)

Enzyme	Substrate	End Products
CARBOHYDRASES		
(*a*) Fructosidases :		
Sucrase, Invertase,	Sucrose, $C_{12}H_{22}O_{11}$	Fructose and glucose
Raffinase	Raffinose, $C_{18}H_{32}O_{16}$	Fructose and melibiose
Gentianase	Gentianose	Fructose and gentiobiose

(*b*) α-glucosidases :		
Maltase	Maltose, $C_{12}H_{22}O_{11}$	Glucose
Trehalase	Trehalose	Glucose
(*c*) β-glucosidases :		
Cellobiase	Cellobiose	Glucose
Gentiobiase	Gentiobiose	Glucose
Emulsin, amygdalase	β-glucosides (amygdalin, salicin, etc.)	Glucose and nonsugar
(*d*) β-galactosidases :		
Lactase	Lactose	Galactose and glucose
Melibiase	Melibiose	Galactose and glucose
Raffinase	Raffinose	Galactose and sucrose
Amylase	Starch $(C_6H_{10}O_5)_n$	Dextrins or maltose
Cellulase	Cellulose	Cellobiose
Hemicellulase or cytase	Hemicelluloses	Hexoses and pentoses
Inulase	Inulin	Fructose
EXTERASES		
Lipases (glyceridases)	Fats	Glycerol and fatty acids
Chlorophyllase	Chlorophyll *a*	Chlorophyllide *a* and phytol
Pectase	Pectin	Pectic acid and methyl alcohol
Phosphatases	Esters of phosphoric acid	Phosphoric acid and alcohol
Tannase	Tannin (digallic acid)	Gallic acid
ENZYMES HYDROLYZING NITROGEN COMPOUNDS		
(*a*) Proteases :		
Pepsin	Proteins	Proteoses and peptoses
Trypsin	Proteins	Polypeptide and amino acids
Papain	Proteins	Polypeptide and dipeptides
Bromelin	Proteins	Polypeptide and dipeptides
(*b*) Peptidases :	Peptides	Amino acids
(*c*) Desamidases :		
Urease	Urea	Ammonia and carbon dioxide
Asparaginase	Asparagine	Ammonia and aspartic acid
Arginase	Arginine	Urea and ornithine

STARCH

In plants starch synthesis occurs only within chloroplasts and leucoplasts, where the enzymes necessary for its synthesis are localized. The starch accumulates in the form of grains.

The starch making up a grain is usually of two different kinds : amylose and amylopectin. Amylose disperses in water much more readily than amylopectin and is sold commercially as "soluble starch". There is considerable variation with species in the number of glucose residues per molecule of either amylose or amylopectin ; the range is from several hundred to 1000 or more, so it is really more proper to consider the starches as a class of compounds rather than as a single compound.

Plants degrade starch primarily by hydrolysis (digestion). Most, if not all, species of plants produce two different enzymes that hydrolyze starch : *α-amylase* and *β-amylase*, the latter being particularly abundant in germinating seeds (*amylase* is often called *diastase*, but the latter term is much more loosely used to designate a mixture of enzymes, especially amylse, dextrinases, and maltase, which together comletely hydrolyze starch to glucose).

β-amylase hydrolyzes starch by splitting of one maltose molecule after another ; if the starch has an even number of glucose residues, it is eventually completely hydrolyzed to maltose. If it has an odd number of glucose residues, the final molecule is a maltotriose ($C_{18}H_{32}O_{16}$), which is further splitted into maltopentose and glucose.

$$\underset{\text{Amylose}}{(C_6H_{10}O_5)_{2n}} + \underset{\text{Water}}{nH_2O} \longrightarrow \underset{\text{Maltose}}{nC_{12}H_{22}O_{11}}$$

$$\underset{\text{Maltose}}{C_{12}H_{22}O_{11}} + \underset{\text{Water}}{H_2O} \longrightarrow \underset{\text{Glucose}}{2C_6H_{12}O_6}$$

$$\underset{\text{Maltotriose}}{C_{18}H_{32}O_{16}} + \underset{\text{Maltotriose}}{C_{18}H_{32}O_{16}} \longrightarrow \underset{\text{Maltopentose}}{C_{30}H_{52}O_{26}} + \underset{\text{Glucose}}{C_6H_{12}O_6}$$

α-amylase hydrolyzes starch, but the products of hydrolysis are dextrins rather than maltose (*dextrin* or starch gum is a polymer of glucose obtained by the hydrolysis of starch, being an intermediate during the conversion of the latter into monosaccharides). Initially at least the dextrins may be of rather high molecular weight with numerous glucose residues, but eventually much of the starch is hydrolyzed to dextrin with six or twelve glucose residues. The various dextrins may be subsequently hydrolyzed to maltose by β-amylase.

REFERENCE

1. O.F. Curtis and D.G. Clark 1950. *An Introduction to Plant Physiology*. McGraw-Hill Book Co., New York.

Chapter 9

Proteins and Nucleic Acids

INTRODUCTION

The chemistry of nucleic acids, as expressed through the proteins, regulates the intricate biochemical properties of life and the dynamics of evolution. A most significant influence of proteins resides in the fact that many are functionally active as enzymes. The enzymes are vital for the rapid rate of biochemical reactions. Although many biochemical reactions will proceed to completion in the absence of enzymes, these reactions are extremely slow. Indeed, we could go so far as to say that enzymes and life are inseparable.

Two other important functions of proteins are as major natural *hydrogen ion buffers* and *structural components* of cells. Certainly many important characteristics of proteins led scientists to the chemistry of regulators of cellular information, the nucleic acids.

AMINO ACIDS AND AMIDES

The acid hydrolysis of protein molecule reveals that it is composed of smaller repeating units, the amino acids. With the exception of two secondary amino acids, the amino acids found in protein have the following general structure.

$$\begin{array}{c} \mathrm{H} \\ | \\ \mathrm{R} - \mathrm{C} - \mathrm{COOH} \\ | \\ \mathrm{NH_2} \end{array}$$

This structure depicts a primary amino acid in which the amino group (—HN_2) is attached to the carbon (α–carbon) adjacent to the carboxyl group (—COOH). Individual differences between the primary amino acids are found in the R-group, which may be different from one amino acid to the next. For example, the amino acids glycine, valine, and leucine have R–groups that are quite different. The structures of these amino acids are given ahead, with the R–groups circled.

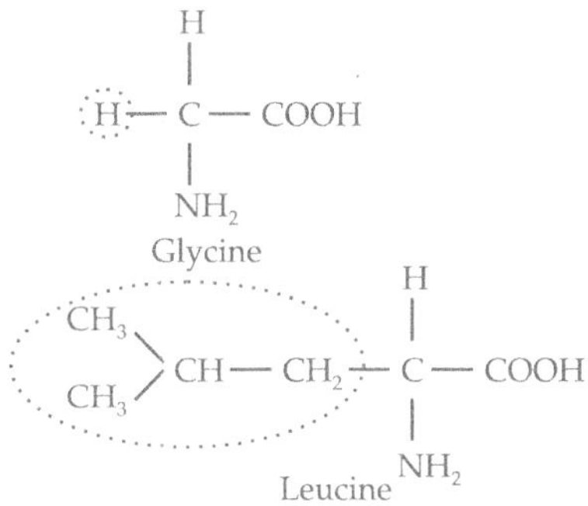

The amino acids found in plant protein are :

Glycine, Alanine, Valine, Leucine, Isoleucine, Serine, Threonine, Phenylalanine, Tryptophan, Tyrosine, Cysteine, Methionine, Proline, Hydroxylproline, Aspartic acid, Glutamic acid, Histidine, Arganine, and Lysine.

Amino Acids Found in Plant Proteins (Devlin & Witham 1986)

S.No.	Name	Formula
1.	Glycine - GLY	NH_2-CH_2-COOH
2.	Alanine - ALA	$CH_3-CH(NH_2)-COOH$
3.	Valine - VA	$(CH_3)_2>CH-CH(NH_2)-COOH$
4.	Leucine - LEU	$(CH_3)_2>CH-CH_2-CH(NH_2)-COOH$
5.	Isoleucine - ILEU	$CH_3-CH_2-CH(CH_3)-CH(NH_2)-COOH$
6.	Serine - SER	$CH_2(OH)-CH(NH_2)-COOH$
7.	Threonine - THR	$CH_3-CH(OH)-CH(NH_2)-COOH$

8.	Phenylalanine - PHE	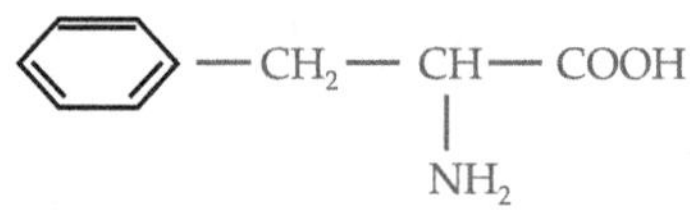
9.	Tyrosine - TYR	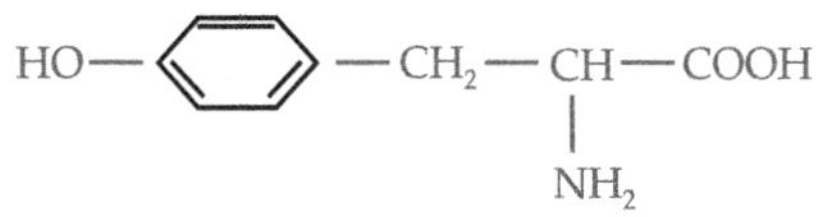
10.	Tryphtophan - TRY	$C_{11}H_{12}O_2N_2$
11.	Cysteine - CYSH	$CH - CH_2 - \underset{NH_2}{\underset{\vert}{CH}} - COOH$
12.	Methionine - MET	$CH_3-S-CH_2-CH_2-\underset{NH_2}{\underset{\vert}{CH}}-COOH$
13.	Proline - PRO	$C_5H_9O_2N$
14.	Hydroxyproline - HYP	$C_5H_9O_3N$
15.	Aspartic acid - ASP	$HOOC - CH_2-\underset{NH_2}{\underset{\vert}{CH}} - COOH$
16.	Glutamic acid - GLU	$HOOC - CH_2-CH_2-\underset{NH_2}{\underset{\vert}{CH}}-COOH$
17.	Histidine - HIS	$C_6H_9O_2N_3$
18.	Arginine - ARG	$H_2N-\underset{NH}{\underset{\Vert}{C}}-NH-CH_2-CH_2-CH_2-\underset{NH_2}{\underset{\vert}{CH}}-COOH$
19.	Lysine - LYS	$H_2N - CH_2 - CH_2 - CH_2 - CH_2 - \underset{NH_2}{\underset{\vert}{CH}} - COOH$

Depending on the pH of the solution, the alpha amino and alpha carboxyl functional groups of an amino acid may exist in one of the following forms :

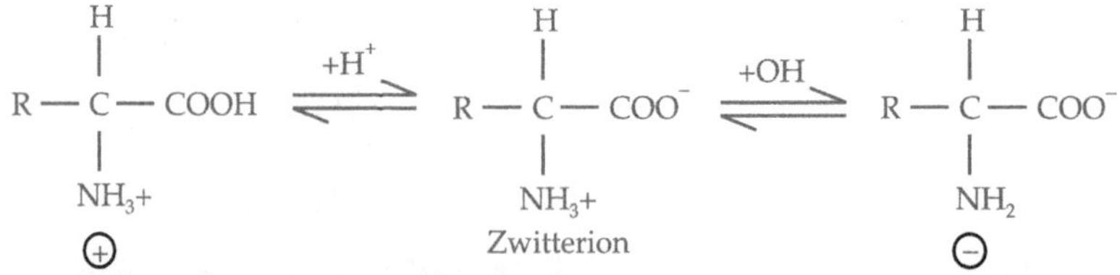

An amino acid might exist as a *Zwitterion,* or molecule consisting of both a positive and negative charge. In this form, an amino acid is dipolar and is considered to be *amphoteric, i.e.,* it can function as an acid or base. Further, the pH at which the *Zwitterion* form exists is usually referred to as the isoelectric point. In a solution very basic to the isoelectric point, an amino acid is an *anion* because of the predominating $-NH_2/-COO^-$ functional groups. Conversely, in solutions very acid (low pH) to its isoelectric point, an amino acid is a *cation* because of the predominating $-NH_3^+/COOH$ functional groups.

AMINO ACID SYNTHESIS

Amino acids are generally considered to be the initial products of nitrogen assimilation. Evidence obtained by following the assimilation of inorganic nutrients containing ^{15}N has shown that, in most cases, the initial recipients of the nitrogen of these compounds are the free α–keto acids in the cytoplasm. The α–keto acid is similar to the amino acid with the exception of an oxygen instead of an amino group attached to the α–carbon.

There are two ways by which nitrogen can be incorporated into α–keto acids.

Reductive Amination

Experiments with isotopically labeled nitrogen have shown that during the early stages of nitrogen assimilation, glutamate is by far the most–labeled compound. The reaction is reversible and proceeds as follows :

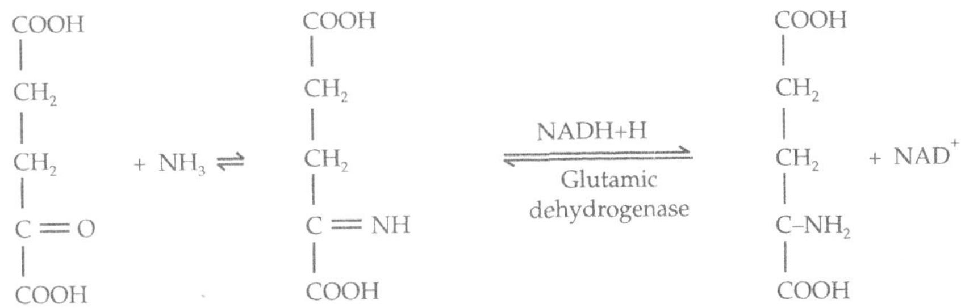

α-Ketoglutaric acid α-Iminogolutaric acid Glutamic acid

Because of the central importance of glutamic acid in the synthesis of other amino acids and because of the high proportion of glutamate formed in this manner by the plant, this reaction is of utmost importance to the nitrogen metabolism of the plant. The widespread occurrence of glutamic dehydrogenase in plants strongly supports the above statement.

Reductive amination as a means for the synthesis of amino acids other than glutamate is of limited importance. There is some indirect evidence for the direct amination of oxaloacetate and pyruvate to form aspartate and alanine, respectively.

Thus, there are four ways by which ammonia nitrogen is incorporated into organic compounds to form amino acids :

$$\alpha\text{–Ketoglutarate} + NH_3 \rightleftharpoons \text{Glutamate}$$

$$\text{Oxaloacetate} + NH_3 \rightleftharpoons \text{Aspartate}$$

$$\text{Fumarate} + NH_3 \rightleftharpoons \text{Aspartate}$$

$$\text{Pyruvate} + NH_3 \rightleftharpoons \text{Alanine}$$

Of these four pathways, the amination of α–ketoglutarate appears to be the major reaction in the assimilation of nitrogen by plants.

Transamination

A very important reaction in amino acid synthesis is *transamination*, which involves the transfer of an amino group of an amino acid to the carbonyl group of a keto acid. After inorganic nitrogen has gained entry primarily through the amination of α-ketoglutarate, the product, glutamate, is available for transamination reactions with keto acids to form the corresponding amino acids. The formation of 17 different amino acids takes place through transamination reactions with glutamate.

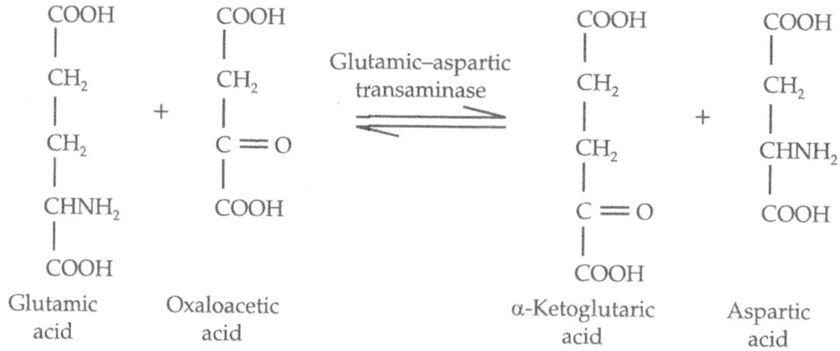

The enzymes catalyzing transamination reactions are called *transaminases*. Thus, the enzyme catalyzing the transfer of an amino group of glutamic acid (substrate) to the carbonyl group of oxaloacetate to form aspartate (product) is called *glutamic-aspartic-transaminase* (as shown above).

The amides *asparagine* and *glutamine* are found in relatively high quantities in many plants and appear to function in the transport and storage of nitrogen. In the synthesis of glutamine, the hydroxyl group of one of the carboxyl groups of glutamic acid is replaced by an $—NH_2$ group.

The enzyme catalyzing this reaction, *glutamine synthetase*, is activated by the metal cofactor ; Mg^{2+}. In addition, ATP is required, as shown below.

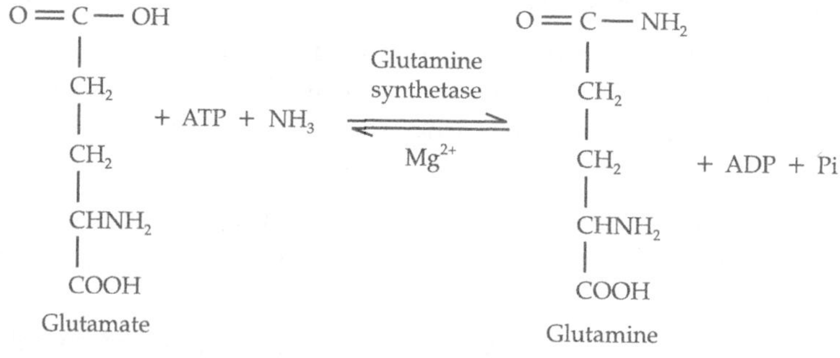

The synthesis of asparagine from aspartate is thought to takes place in the same manner and requires a metal activator and ATP. However, the enzyme *asparagine synthetase*, which would catalyze this reaction, has not as yet been isolated from plant tissues.

PROTEINS

Proteins are composed of repeating units, the amino acids. The amino acids are linked together by bonds connecting the carboxyl group of one amino acid with the amino group of another.

This type of linkage, which is repeated many times in the protein molecule, is called the *peptide bond.*

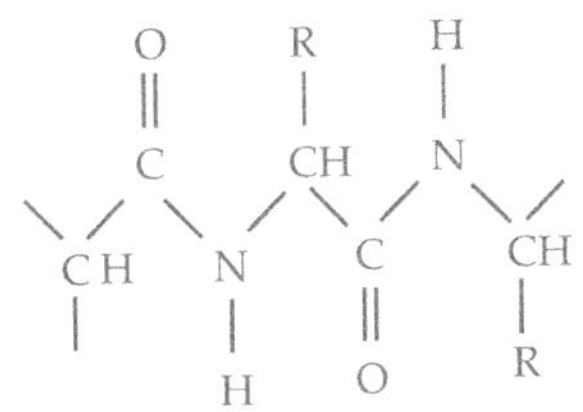

A compound composed of two amino acids linked together by two peptide bonds is called dipeptide-of three amino acids, a tripeptide, and so on. When a large number of amino acids are linked together in this fashion, the resulting compound is called a polypeptide. When we consider that a protein may be made up of twenty different amino acids, each one of which may be present many times and in different sequences, we get an idea of the complexity and size of the protein molecule. Proteins may range in size from the molecular weight of insulin, which is 6,000, to a molecular weight of several million.

PROTEIN STRUCTURE

Primary Structure

The biological properties of a protein molecule are related to its structure. The peptide bond and the definite sequence of amino acids give the protein to its primary structure. Since many proteins contain more than one polypeptide chain, a connection between them other than by a peptide bond is a necessary feature of the protein molecule. The *disulphide* (— S — S —) *bond* between two cysteine molecules is important in this respect.

Evidence from numerous investigations indicates that peptide and disulphide bonds are not the only links involved in protein structure. For example, dissociation of many proteins may occur under mild conditions that would not disturb the peptide or disulfide bond.

Secondary Structure

Polypeptide chains exhibit three major types of arrangement, or orientation : helical, pleated sheet, and random. These particular coil or spiral arrangements of the polypeptide chain constitute its secondary structure.

The *α–helix*, which is the most common helical arrangement, is maintained by an extensive assembly of hydrogen bonds throughout the chain. *Hydrogen bonds* are noncovalent links that occur as a result of the hydrogen atom sharing electrons with two oxygen atoms.

The *pleated sheet* orientation is formed when segments of the polypeptide chain are side by side and joined by hydrogen bonds in such a manner as to produce a zigzag appearance of the peptide backbone. Pleated sheets often take the appearance of chains running in opposite directions (*antiparallel*) or looped in such a manner that large portions of the chain run parallel (*parallel pleated sheets*).

With the so-called random arrangement, the polypeptide secondary structure may not appear to show geometrical order. This lack of geometry at the surface probably results from the folding of amino acid side chains rather than the folding of the peptide backbone.

Tertiary Structure

With complete folding or assumption of final form, by the chain, the secondary structure may take on its own specific final shape or pattern. This further folding of the various secondary structures and random segments is referred to as the tertiary structure of the protein. The tertiary structure is maintained primarily by hydrogen bonding.

The secondary and tertiary structures of the protein molecule seem to be intimately associated with the molecules biological function. In many instances when the secondary and tertiary structures are disrupted, certain specific functions (*e.g.*, enzyme activity) are irreversibly lost. This loss in activity may occur when proteins are exposed to relatively high temperatures, changes in pH, ultraviolet radiation, and so on. All of these conditions cause what is referred to as denaturation. Loss of many of the properties of the protein molecule such as solubility, specific activity, crystallizability – follows its denaturation. In many cases, these properties cannot be recovered on return to normal conditions.

Quaternary Structure

The quaternary structure of a protein molecule results form the association of two or more polypeptide chains. The final, three-dimensional shape (geometry) of the combined chain (each is referred to as a subunit) as formed into a complete molecule constitutes its quaternary structure. Hydrogen bonds are again involved in holding subunits together in the maintenance of the quaternary structure. However, another kind of important interaction that holds the quaternary structure together is the association of hydrophobic groups that combine and exclude water. Many proteins that contain multiple subunits seem to be constructed and held together by hydrophobic side chain interactions among subunits.

PROTEIN CLASSIFICATION

Because of the similar general structure found in the many different proteins, it is easy to separate them from other nitrogenous compounds into their own general group. However, because of this similarity, it is difficult to establish a classification for proteins themselves. Therefore, a classification has been attempted that is based in part on solubility properties and in part on known chemical and physical differences.

(A) **Simple Proteins :** Simple proteins are compounds that, on hydrolysis, yield only amino acids. Classification of the simple proteins is primarily based on solubility properties. The simple proteins can be divided into six major groups : albumins, globulins, glutelins, prolamines, histones, and protamines.

1. *Albumins :* The albumins are soluble in water and in dilute salt solutions. They can be coagulated by exposure to heat. The β–amylase of barley is a good example of an albumin.
2. *Globulins :* The globulins are either insoluble or sparingly soluble in water. They are soluble in dilute salt solutions. The globulins will also coagulate when exposed to heat. Many examples of globulins may be found in the storage proteins of seeds.
3. *Glutelins :* The glutelins are insoluble in neutral solutions but are soluble in weak acid or basic solutions. These proteins are chiefly found in the cereal grains. Glutenin is an example of a glutelin protein in wheat. Another example would be oryzenin from rice.

4. *Prolamines* : The prolamines are insoluble in water, but are soluble in 70 to 80 percent ethanol. They are not soluble in absolute ethanol. On hydrolysis, these proteins yield relatively large quantities of proline and ammonia – thus the term prolamine. Examples of plant prolamines are zein of maize, gliadin of wheat and rye, and hordein of barley.
5. *Histones* : The histones are rich in basic amino acids, such as arginine and lysine, and are soluble in water. These have been found in cell nuclei and may be associated with nucleic acids.
6. *Protamines* : The protamines, like the histones, are rich in basic amino acids and are soluble in water. Also, like the histones, they are found in the nucleus and are probably associated with the nucleic acids. The amino acids tyrosine and tryptophan are not found in these proteins. Also, the protamines contain no sulphur.

(B) **Conjugated Proteins :** The conjugated protein is associated with a nonamino acid component that may be referred to as a prosthetic group. These proteins are divided into five major groups: nucleoproteins, glycoproteins, lipoproteins, chromoproteins, and metalloproteins.

1. *Nucleoproteins* : On hydrolysis, the nucleoproteins yield a simple protein plus a nucleic acid.
2. *Glycoproteins* : As the name implies, glycoproteins are proteins containing small amounts of carbohydrates as prosthetic groups. Some of the proteins of the cell membrane may by glycoproteis.
3. *Lipoproteins* : The lipoproteins are generally insoluble in water and are common components of membranes.
4. *Chromoproteins* : The chromoproteins comprise a diverse group of compounds, which includes flavoproteins, biliproteins. (phycobilins and phytochrome), carotenoid proteins, chlorophyll proteins, and hemoglobins. The one property they all have in common is prosthetic group that is a pigment.
5. *Metalloproteins* : Many of the enzymes belong to the metalloprotein group. The enzymes require a metal as an activator.

NUCLEIC ACIDS

The nucleic acids (ribose nucleic acid – RNA and deoxyribose nucleic acid – DNA) are large polymeric molecule, composed of repeating units called nucleotides, which, in turn, are composed of three components : a purine or pyrimidine base, a pentose or deoxypentose sugar, and phosphoric acid. The nucleotides are bound together by sugar – phosphate linkages.

Separation of the nucleic acids into two large groups is determined by the sugar component present, a factor that has also influenced the naming of these groups. Thus, ribose nucleic acid (RNA) contains ribose, and deoxyribose nucleic acid (DNA) contains deoxyribose. The difference between these two sugars can be found in the second carbon.

Two types of nitrogenous bases occur in nucleic acids – the purines and pyrimidines. Adenine and guanine are the two purines commonly found, while thymine, cytosine, and uracil are the pyrimidines that are generally present. The pyrimidine thymine is found only in DNA, while the pyrimidine uracil is confined to the RNA molecule.

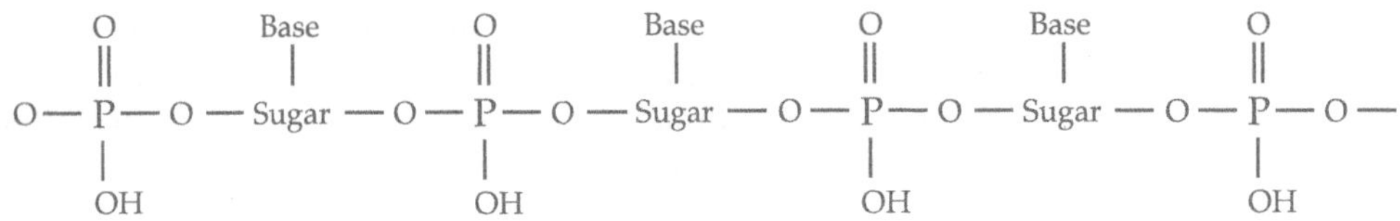

Sugar-phosphate linkages

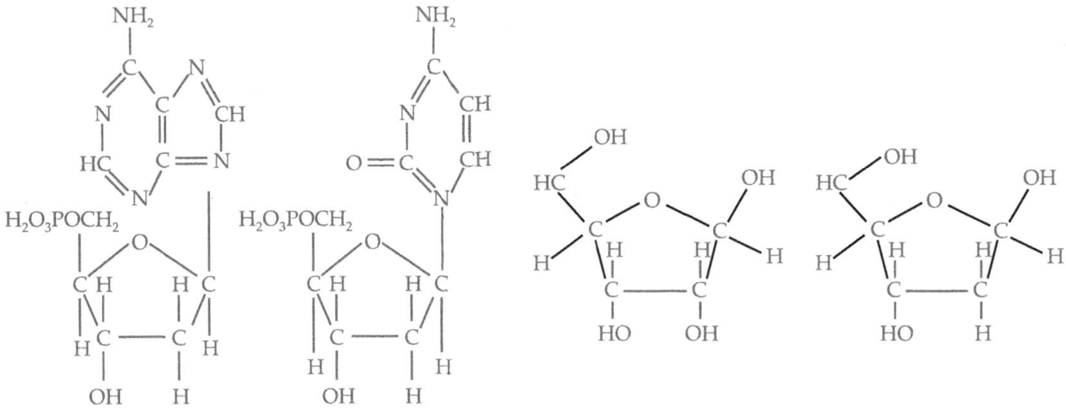

Purine nucleotide, Adenylic acid | Pyrimidine nucleotide Cytidylic acid | D-ribose | 2-deoxyribose

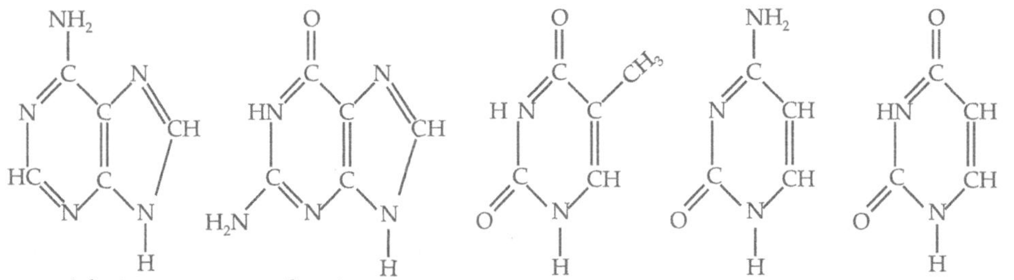

Adenine | Guanine | Thymine | Cytosine | Uracil

Sequence of DNA molecule showing sugar-phosphate linkages, purine nucleotide, adenylic acid, and pyrimidine nucleotide, cytidylic acid

DNA is found in chromosomes, in plastids, and in mitochondria. RNA is found in chromosomes, nucleoli, cytoplasm (as transfer RNA), ribosomes (ribosomal and messenger RNA), plastids, and mitochondria. The biological functions of DNA and RNA are the transmission of hereditary characteristics and biosynthesis of proteins. DNA is primarily associated with the transmission of genetic information, RNA is directly connected with protein synthesis.

The molecular structure and nucleotide sequence of the nucleic acids has been studied intensively. Evidence accumulated from X-ray diffraction studies indicates that the DNA molecule is a double helical structure ; the two strands are entwined and complementary. The two strands are linked together through hydrogen bonding between base pairs. Chemical analysis of the DNA molecule indicate that there is a 1 : 1 relationship between adenine and thymine and between guanine and cytosine. However, the ratio of adenine – thymine to guanine – cytosine can differ from one DNA molecule to another. The DNA molecule is a self – replicating molecule. Presumably, under the appropriate conditions and with the necessary enzymes, the two chains unwind from the double helix, draw from a base pool and duplicate each other.

RNA is also a helical structure (single strand) and is composed of a sequence of nucleotides spaced in much the same manner as those of DNA. Uracil replaces thymine in the RNA molecule. However, just as adenine pairs with thymine in DNA, so also does it pair with uracil in RNA.

Three types of RNA have been identified that vary in size and function. The largest RNA is found in ribosomes and is generally referred to as *ribosomal RNA* (rRNA). *Messenger RNA* (mRNA) is much smaller but is still of considerable size ; in electron micrographs mRNA can be identified as long fibrous molecules, with several attached ribosomes. The whole complex is called a *polysome* or *polyribosome.* Finally, small molecules of RNA, called *transfer* RNA (tRNA), have been identified and are associated with protein synthesis.

TRANSCRIPTION

DNA directs the synthesis of RNA essentially by acting as a template. In the process, DNA uncoils and the strands separate between the bases, thereby leaving single strands of DNA. At this point each strand consists of repeating nucleotides. If complementary nucleotides, which contain deoxyribose and which come from the cell nucleotide reservoir (referred to as a "pool"), join one of the open DNA strands by forming base pairs, another double – stranded DNA molecule results. However, if bases of ribose nucleotides complementary to the DNA strand are polymerized (catalyzed by RNA polymerase) of the template, a molecule of RNA is formed in which the sequence of nucleotides is complemetary to that of the DNA template strand. Since RNA is formed from the base sequence pattern of DNA, the process is sometimes referred to as *DNA – dependent* RNA *Synthesis.* Further, the DNA is said to be transcribed into a complementary RNA chain. Therefore, when we refer to the process from the stand point of informational transfer, we call it *transcription.* Once synthesized, a mRNA molecule may leave the nucleus via pores in the nuclear membrane and may become associated with ribosomes in the cytoplasm.

Although little is known of the process, rRNA is formed in the nucleolus before being released into the cytoplasm. Details of tRNA synthesis are still unknown.

The sequence of bases in the mRNA molecule is important since it represents a pattern that controls the synthesis of proteins. Plant proteins are composed of at least 20 different amino acids. A pattern of 1 base to 1 amino acid would account for only 4 different amino acids and combinations of 2 bases with 1 amino acid would account for only 16 different amino acids. However, with a *triplet* code, 64 combinations of bases are possible, which more than adequately account for the 20 different amino acids found in plants. The triplet groups of sequentially arranged bases on the mRNA molecule are called *codons,* and each codon represents the code for a specific amino acid. For example, the triplet sequence UUU (three uracil nucleotides) serves as the template for a molecule of the amino acid phenylalanine. The 64 possible codons and the amino acids they code for are given below. Note that three of the codons – UGA, UAA, and UAG – do not code for any amino acid. These are called nonsense triplets ; their function in protein synthesis may be to mark the ending of one protein and the beginning of another.

First Letter	Second Letter: U		C		A		G		Third Letter
U	UUU	Phenylanine	UCU	Serine	UAU	Tyrosine	UGU	Cysteine	U
	UUC		UCC		UAC		UGC		C
	UUA	Leucine	UCA		UAA		UGA		A
	UUG		UCG		UAG		UGG	Tryptophan	G
C	CUU	Leucuine	CCU	Proline	CAU	Histidine	CGU	Arginine	U
	CUC		CCC		CAC		CGC		C
	CUA		CCA		CAA	Glutamine	CGA		A
	CUG		CCG		CAG		CGG		G
A	AUU	Isoleucine	ACU	Threonine	AAU	Asparagine	AGU	Serine	U
	AUC		ACC		AAC		AGC		C
	AUA		ACA		AAA	Lysine	AGA	Arginine	A
	AUG	Methionine	ACG		AAG		AGG		G
G	GUU	Valine	GCU	Alanine	GAU	Aspartic acid	GGU	Glycine	U
	GUC		GCC		GAC		GGC		C
	GUA		GCA		GAA	Glutamic acid	GGA		A
	GUG		GCG		GAG		GGG		G

(Devlin & Witham, 1986)

Assignment of specific amino acids to 61 out of 64 possible codons. Remaining three codons are called nonsense codons, or triplets, because they do not code for any amino acid.

Adenine – Ribose – Phosphoric acid = Adenylic acid

Guanine – Ribose – Phosphoric acid = Guanylic acid

Cytosine – Ribose – Phosphoric acid = Cytidylic acid

Uracil – Ribose Phosphoric acid = Uridylic acid

Nucleotides of Ribonucleic acid

TRANSLATION (Protein Synthesis)

Activation of amino acids : Amino acid activation, the first step in protein synthesis, consists of the selection of specific amino acids from a heterogenous pool in the cytoplasm. The selection of amino acids is accomplished through the aid of highly specific enzymes, each amino acid having at least one activating enzyme. In the presence of ATP, the activating enzyme catalyzes the formation of an energy-rich, enzyme-bound amino acid, adenylate (E–AA–AMP) and the release of pyrophosphate.

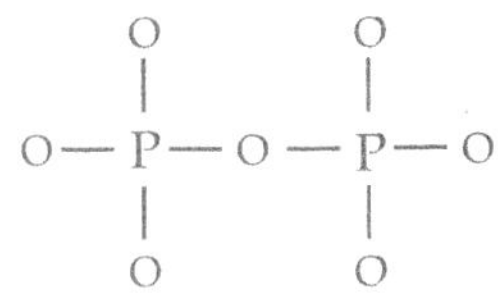

Amino acid – tRNA Complex : The activation of an amino acid is followed by its attachment to transfer RNA (tRNA), which is a relatively small molecule containing anywhere form 70 to 100 nucleotides. A specific tRNA molecule for each amino acid leads us to assume that the transfer of amino acids from the enzyme–bound active complex to tRNA is an additive rather than a competitive process. The point of attachment between tRNA and the activated amino acid is at the second or third carbon of the ribose sugar of a terminal adenylic acid.

Polypeptide Formation : Once formed, messenger RNA (mRNA) becomes associated with ribosomes in the cytoplasm to form a polysome. Amino acids are conveyed to the polysomes by tRNA, the amino acid being attached to one end of the tRNA molecule. At the other end of the tRNA molecule is a triplet of nucleotides, or an *anticodon*, which is the complement of the messenger codon for that amino acid. For example, for the messenger codon UUG, the tRNA anticodon would be AAC (the anticodon for leucine). The codon and anticodon lock rapidly into place by complementary attraction, as they do in other nucleic acid activities such as DNA replication and mRNA transcription. When the anticodons of a number of tRNA molecules are locked in place, the amino acids at their opposite ends are lined up in sequence for polypeptide formation. The ribosomes are thought to move along the mRNA molecule from one end to the other and join the amino acids in peptide linkages with the aid of specific enzymes. Thus, the role of tRNA appears to be to convey amino acids to the mRNA-ribosome complex (polysome) and to hold them in place according to the pattern dictated by RNA codons. The pattern is fixed, in turn, by the DNA on which it was transcribed. The actual joining of the amino acids is under the control of protein – synthesizing enzymes.

PROTEIN DEGRADATION

Protein metabolism in plants is in a continuous state of flux between synthesis and breakdown. Observers have found proteolytic enzymes, such as protease and peptidase, in various plant organs. This finding suggests that the activity of these enzymes may, in part, control protein degradation.

Investigators have studied protein degradation principally in germinating seeds and in detached leaves. During germination, a massive breakdown of storage protein occurs in the cotyledon or endosperm, parallel by a rapid synthesis of protein by the embryo. Observers have also found an accumulation of amino acids and amides in the embryo. Apparently, the physiological factors leading to germination set in motion the degradation of storage protein, the migration of the products of this degradation (amino acids) to the embryo, and the synthesis of new protein from these amino acids.

REFERENCE

1. Robert M. Devlin and F. H. Witham. 1986 *Plant Physiology*, C.B.S. Publishers & Distributors, Delhi.

Chapter 10

CARBOHYDRATES

INTRODUCTION

Carbohydrates are a group of organic compounds containing the elements carbon, hydrogen, and oxygen, generally in the ratio of 1 : 2 : 1, with the general formula $(CH_2O)_n$ (the n in the formula means that CH_2O is repeated a certain number of times ; if 6 times, a molecule of glucose –$C_6H_{12}O_6$ is formed.) One carbon bears a carbonyl group and the others hydroxyl groups. However, the definition of this group (as hydrates of carbon) has been broadened to include compounds containing nitrogen and sulphur, and compounds that do not conform to a strict 1 : 2 : 1 ratio of carbon, hydrogen, and oxygen. Carbohydrates, therefore, are considered to be *polyhydroxy—aldehydes* or *polyhydroxy-ketones,* and their derivatives.

The carbohydrates are derived more or less directly from carbon dioxide and water in photosynthesis. Sugar, starch, and cellulose are examples of carbohydrates that illustrate the importance of this class of compounds to life. Cellulose is the principal constituent of wood (and paper) ; starch is the principal constituent of the grains and other seeds, and the sugars, in addition to their widespread use as a foodstuff, are important functioning parts of all living organisms. To the plant, carbohydrates are important in several ways :

1. They represent a means for the storage of energy that is converted from light, in the process of photosynthesis—a function of utmost importance to animals as well as plants.
2. They are important constituents of the supporting tissues that enable the plant to achieve erect growth, in some cases to the height of 400 feet,
3. They provide the carbon skeletons for the organic compounds that make up the plant (*e.g.*, the synthesis of fats and proteins).

CLASSIFICATION

Carbohydrates may be roughly classified into three categories—the monosaccharides, the oligosaccharides, and the polysaccharides.

The *monosaccharides* are the group of carbohydrates from which no simpler carbohydrates can be produced by hydrolysis. They are the building units for the more complex oligosaccharides and polysaccharides. The *oligosaccharides* are also relatively simple ; they are made up of two or more monosaccharides held together by *glycosidic linkages.* The *polysaccharides,* in contrast, are complex molecules of high molecular weight composed of a large number of monosaccharides joined through glycosidic linkages (covalent bonds between sugars). They may therefore be regarded as condensation products of the simple sugars. Disaccharides yields two molecules of monosaccharides, three of trisaccharides and four of tetrasaccharide upon hydrolysis. All of the important compound sugars known to occur in plants are condensation products of hexose sugars. The hexose molecules formed on hydrolysis of compound sugars may be all of one kind, or of more than one kind, depending upon the specific oligosaccharide.

Monosaccharides

The simplest soluble carbohydrates are the 3-carbon compounds *glyceraldehyde* and *dihydroxyacetone.*

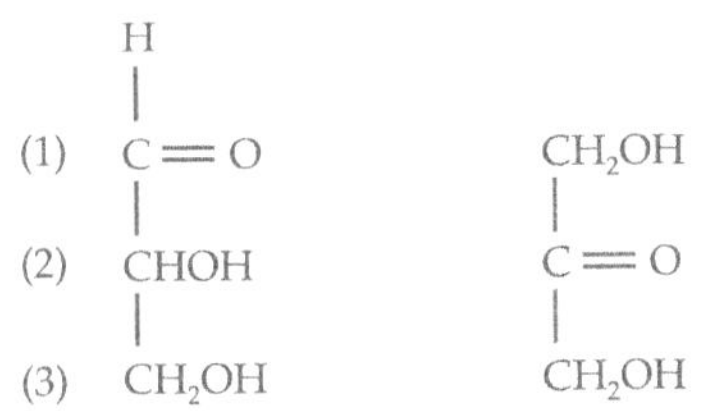

Glyceraldehyde Dihydroxyacetone

A consideration of the above compounds will help us with the general terminology used to describe sugars. For example, we classify the monosaccharides in accordance with the number of carbons present. Thus, we call glyceraldehyde and dihydroxyacetone as *trioses.* Note also that on these compounds one of the carbons bears a carbonyl oxygen, on the first carbon of glyceraldehyde to form an *aldehyde group* —(C═O) with H, and on the second carbon of dihydroxyacetone to form a *ketone group* (C═O). Thus, we can differentiate between the two trioses by calling glyceraldehyde *an aldose* and dihydroxyacetone *a ketose.* The aldehyde group and the ketone group are known as *reducing groups* because of their ready oxidation by certain compounds, which are themselves reduced in the reaction. Sugars possessing these groups are called *reducing sugars.*

Pentoses

The pentoses are five – carbon sugars, which are rarely found dissolved in the free state in the cell cytoplasm. However, they are quite abundantly found as constituents of some of the more complex carbohydrates of the plant. Thus, *D–xylose* and *L–arabinose* occur in plants as constituents of *xylans* and *arabans,* respectively – large polysaccharides with a structural function in the cell wall.

Classification of Principal Carbohydrates

- **(I) Monosaccharides or Simple Sugars**
  - 1. Bioses
  - 2. Trioses
  - 3. Tentroses
  - 4. Pentoses
    - Aldoses
      - 1. Xylose
      - 2. Arabinose
      - 3. Ribose
    - Ketoses
  - 5. Hexoses
    - Aldoses
      - 1. Glucose
      - 2. Mannose
      - 3. Galactose
    - Ketoses
      - 1. Fructose
      - 2. Sorbose
  - (D- and L-Forms)
  - 6. Heptoses
  - 7. Octoses
  - 8. Nonoses
  - 9. Decoses
- **(II) Oligosaccharides or Compound Sugars**
  - 1. Disaccharides
    - 1. Sucrose
    - 2. Maltoses
    - 3. Gentoiobiose
    - 4. Trehalose
    - 5. Melibiose
    - 6. Cellobiose
    - 7. Lactose
  - 2. Trisaccharides
    - 1. Raffinose
    - 2. Gentianose
    - 3. Melezitose
  - 3. Tetrasaccharides
    - Stachyose
- **(III) Polysaccharides**
  - 1. Pentosans
    - 1. Araban
    - 2. Xylan
  - 2. Hexosans
    - 1. Glucosans
      - 1. Starch
      - 2. Glycogen
      - 3. Lichenin
    - 2. Fructosans
      - Inulin
    - 3. Mannans
    - 4. Galactans
  - 3. Pectic compounds
    - Pectic acid
    - Pectin
    - Protopectin
  - 4. Gums ; 5. Mucilages
  - 6. Amines – hexosans
    - Chitin

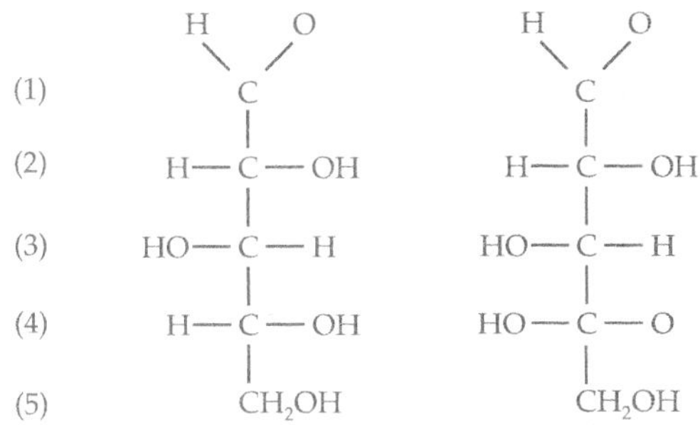

D-Xylose L-Arabinose

In addition to xylose and arabinose, the five – carbon sugars *D-ribose* and *2-deoxy-D-ribose* are also commonly found in plants as components of nucleic acids. Certain coenzymes important in hydrogen and group transfer reactions have *D*-ribose as a component of their structure. *D*-ribose and 2-deoxy-*D*-ribose differ only in the substituents about the second carbon. In place of hydroxyl group, 2-deoxy-ribose has a hydrogen atom (*deoxy* means "without oxygen"). These sugars are part of the nucleotide structure of the nucleic acids.

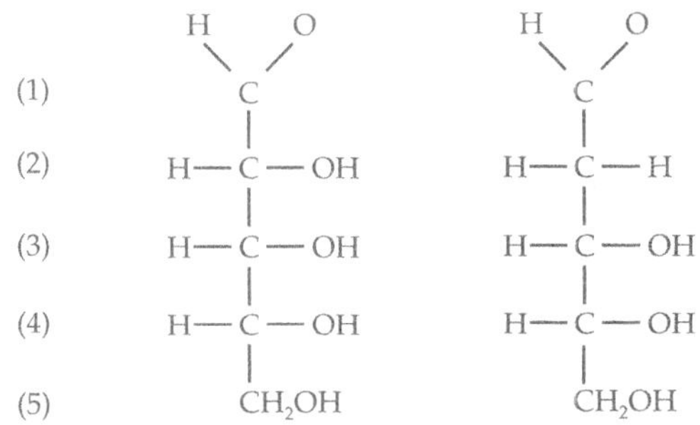

D-ribose 2-deoxy-*D*-ribose

Hexoses

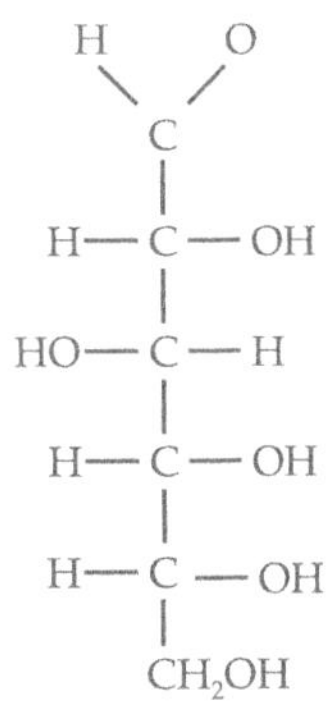

D-Glucose

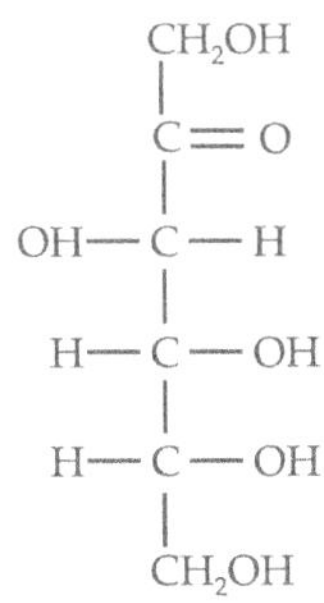

D-Fructose

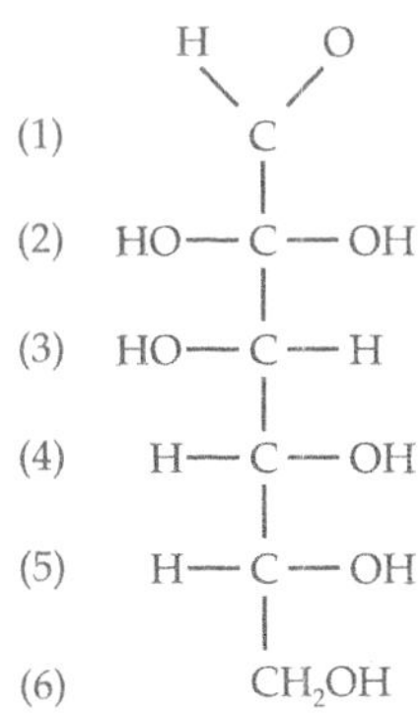

D-Mannose

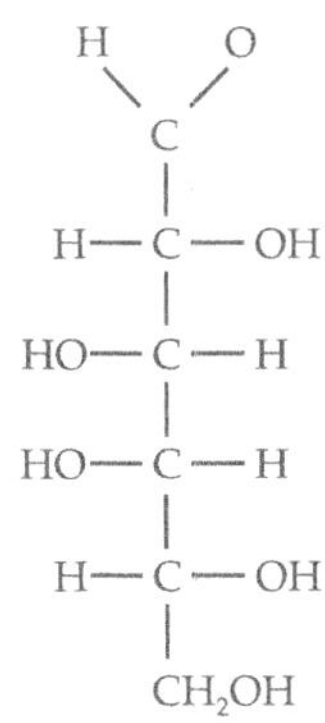

D-Galactose

The hexoses are six-carbon sugars. Four hexoses-*D*-glucose, *D*-fructose, *D*-mannose, and *D*-galactose are commonly found in most plants, either as a component of some more complex carbohydrates or free in the cell. Generally, glucose and fructose are the only hexoses found dissolved in the free form.

We can see immediately that there appearances to be very little difference in the structure of the hexoses. In the first three, the only differences are on the first and second carbons.

Fructose differs from both glucose and mannose in that it is a ketose and the other two are aldoses. Galactose differs from glucose only in the position of the hydroxyl group on carbon number 4.

General Properties of the Sugars

The mono, di, tri, and tetrasaccharides are collectively called the sugars. All of these compounds possess the property of sweetness and all of them are white, more or less crystalline compounds which are soluble in water.

1. Reducing and Nonreducing Sugars

All of the monosaccharides and some of the more complex sugars act as reducing agents. This action is made possible by the presence of an aldehyde or ketonic group in the sugar molecule. Those compound sugars in which the linking of the monosaccharides has occurred in such a manner that the aldehydic or ketonic groups have lost their usual reactivity, are nonreducing. Sugars are commonly classified on the basis as *reducing sugar* or *nonreducing sugars.* The reducing action of sugars is most commonly determined by means of Fehling's or Benedict's solution, in which, upon heating, a reducing sugar converts cupric hydroxide into cuprous oxide. The latter compound separates from the solution in the form of a reddish precipitate.

$$\text{Reducing sugar} + 2\,Cu^{++} \longrightarrow \text{Oxidized sugar} + Cu^{+}$$

$$2Cu^{+} + 2OH^{-} \longrightarrow \underset{\text{Yellow}}{2CuOH} + \underset{\text{Red}}{Cu_2O} + H_2O$$

2. Optical Activity

Most of the soluble carbohydrates are, like many other organic compounds, optically active in solution. The *optical activity* of a solution of carbohydrate refers to its property of rotating the plane of polarized light. Plane-polarized light may be obtained by passing ordinary light through certain minerals such as Iceland spar (a form of $CaCO_3$). Ordinary visible light is a from of electromagnetic radiation for which the vectors used to represent the changes in the electrical or magnetic fields are randomly oriented.

A properly prepared section of Iceland spar has the property of transmitting only these vectors oriented in one particular direction ; light of this type is called plane-polarized light.

(Vector=Any physical quantity that requires a direction to be stated in order to define it completely, for example, velocity).

The *specific rotary power* of a sugar is expressed in terms of the number of degrees of angular rotation of the plane of polarized sodium light caused by a solution made in the proportion of 1 gram of the sugar to 1 ml of solution, observed through a depth of 10 cm at a temperature of 20°C. Measurements of the rotary power of substances are made with an instrument known as a *polarimeter.* Compounds which rotate the plane of polarized light to the right (in clockwise direction) are called *dextrorotatory* (*D* or +); those which rotate it to the left (in a counterclockwise direction) are called *levorotatory* (*L* or –). In either case, the compounds are called optically active compounds.

3. Stereoisomerism

Differences between monosaccharides depend not only on variation in the number of carbon atoms present and the identity of the active group, but also on the presence of the active groups, but also on the presence of chiral of asymmetric carbon atoms (an asymmetric carbon atom is one in which each of the four valences is satisfied with a different kind of atom or group).

For example, carbon-2 (C-2) of glyceraldehyde, probably the simplest sugar has four different substituents. The two isomeric forms are possible since the groups can be arranged in two ways, each of which is a mirror image of the other. Such carbon atoms, as mentioned earlier, are said to be *chiral* or *asymmetric,* and the two forms of glyceraldehyde are referred to as *stereoisomers.*

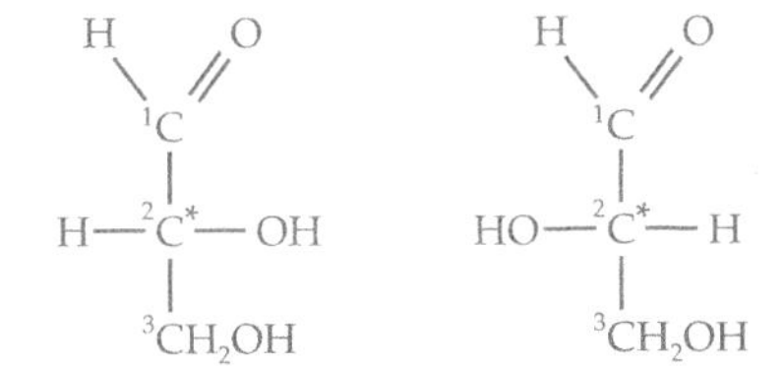

D-Glyceraldehyde *L*-Glyceraldehyde

(Each is the mirror image of the other)

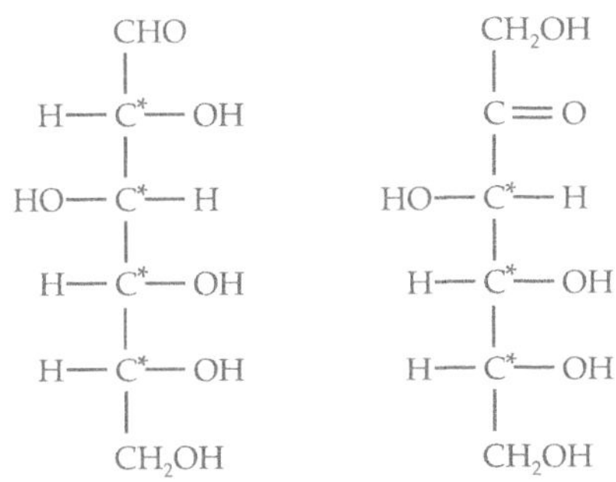

D-Glucose *D*-Fructose

*Chiral Groups

One form is termed the *D*-from since it rotates the plane-polarized light to the right, and the other is termed the *L*-form since it rotates plane -polarized to the left. By convention, the *D*-form is written with the hydroxyl group of C-2 at the right-hand side and the *L*- form with the hydroxyl group of C-2 at the left-hand side.

As the number of carbon atoms increases, so does the number of chiral carbons. In general, the number of possible stereoisomers is 2^n where n is the number of chiral carbons. Thus, in the aldohexoses there are 16 possible isomers of which eight are in the *L*-series and eight in the *D*-series. Some of these are illustrated below. Among these, some differ only in their configuration at a single chiral carbon. They are know as *epimers.* Thus, *D*-glucose is an epimer of *D*-mannose and *D*-galactose. However, there is no such relationship between *D*-mannose and *D*-galactose.

A monosaccharide is allocated the *D*- or *L*-series on the basis of the configuration of groups at the highest numbered chiral carbon, using glyceraldehyde as the standard. Thus, in the aldohexoses, where the highest numbered chiral carbon is five, those with the same configuration at C-5 as the chiral carbon of *L*-glyceraldehyde are allocated to the *L*-series. Similarly, those with the same configuration at C-5, as the chiral carbon of *D*-glyceraldehyde are allocated to the *D*-series.

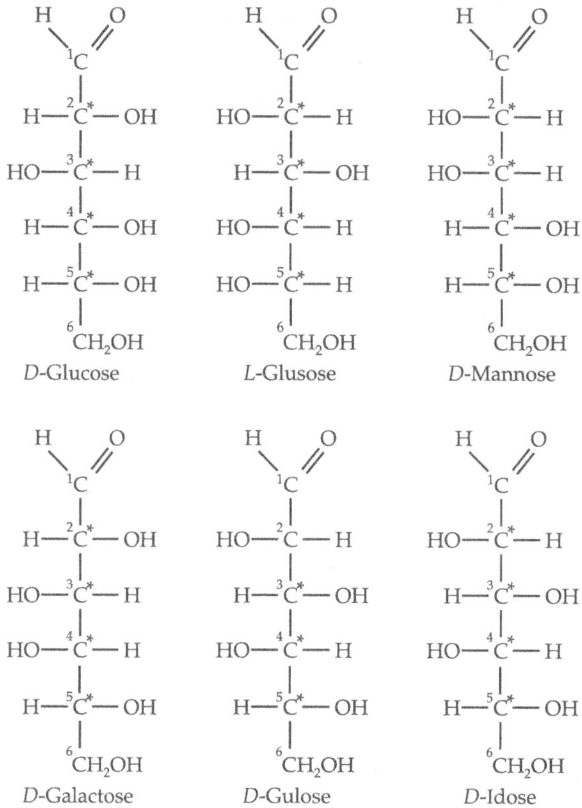

Some stereoisomers from the aldohexose series (* indicates chiral carbon). *D*-and *L*-Glucose are non-superimposable mirror images. All differ in the orientation of substituent groups about the chiral carbons.

Although the terms *L* and *D* when applied to glyceraldehyde refer to the direction of rolation of plane–polarized light, this does not necessarily apply to the higher monosaccharides since the direction of rotation depends on the individual contributions or more chiral bonds. Inspite of the similarity of structure and properties in the equivalent *L*-and *D*-forms, most naturally occurring monosaccharides belong to the *D*-series.

Configurational differences in molecular structure, such as those between *D*-and *L*-glucose, results not only in appreciable chemical differences between similar compounds, but also in marked differences in physical behaviour and in biological properties. *D*-glucose is readily utilized as a substrate for respiration by plant tissues, while *D*-galactose, *D*-mannose, and even *L*-glucose are utilized much more sluggishly or not at all.

Of the great number of possible stereoisomers of the hexose sugars, only small number are formed as a plant product. The hexose sugars–glucose, fructose, mannose, and galactose appear to be present in every living plant. Of these, only *D*-glucose and *D*-fructose primarily occur dissolved in molecular form in the cytoplasm or vacuole of the plant. *D*-mannose and *D*-galactose are found in the plant only in the form of derivatives and particularly as component sugars of polysaccharides of the cell wall. Still other hexose sugars occur scattered through the plant kingdom, generally bound with other materials in the form of rare or unusual compounds.

4. Cyclic Structure

The open chain formulation in which the monosaccharides have a free carbonyl group suggests that they should behave like normal aldehydes or ketones. However, the evidence indicates that these active groups are partially blocked. For example, they do not give reagent test for aldehydes. Furthermore, the aldehyde and ketone structures do not account for the change in optical rotation (mutarotation) characteristic of freshly prepared solutions of many monosaccharides.

It is now known that monosaccharides with five or more carbons usually exist as cyclic (ring) structures with an oxygen bridge between the carbonyl group and another carbon of the same molecule. In glucose, for example, a 6–membered ring may be formed by formation of an oxygen bridge between the terminal aldehydic carbon atom (C-1) and the fifth carbon atom (as might occur in solution). With the closing of this 6-membered, or pyranose ring, a new hydroxyl group is formed at the aldehydic end of the molecule (a 5 membered ring containing four carbons and an oxygen is called a furanose ring, and a 6–membered ring containing five carbons and an oxygen is called a pyranose ring). Since this new hydroxyl group may have either of two possible orientations with regard to the remainder of the molecule, there are two forms of the cyclic pyranose *D*-glucose forms which are known as α– and β– respectively (when they differ at C-I only, as in glucose, they are known as *anomers*). In the α- form the hydroxyl group on C-1 is *cis* to the hydroxyl on C-2 and in the β-form it is *trans* (*cis* = on the same side ; *trans* = on the opposite sides). This same hydroxyl group is involved in the formation of many of the derivatives of *D*-glucose, which may be derivatives either of α *D*-glucose, as starch, or derivatives of β-D-glucose, as cellulose.

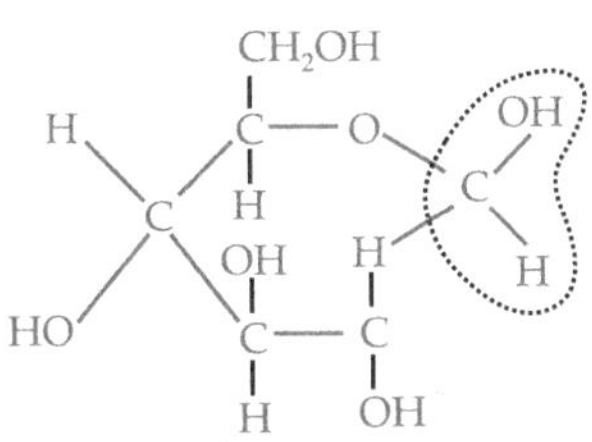

β-D-Glucopyranose

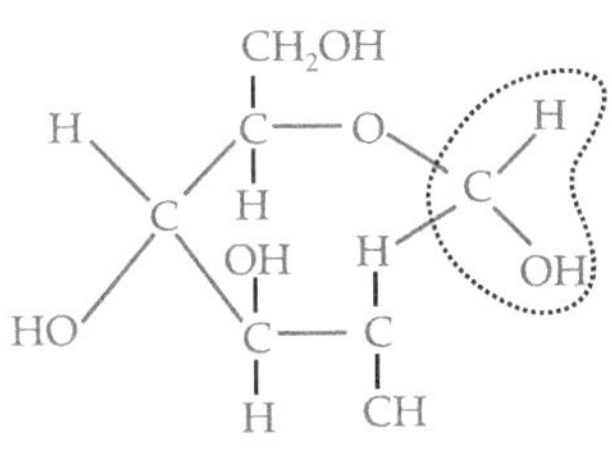

α-D-Glucopyranose

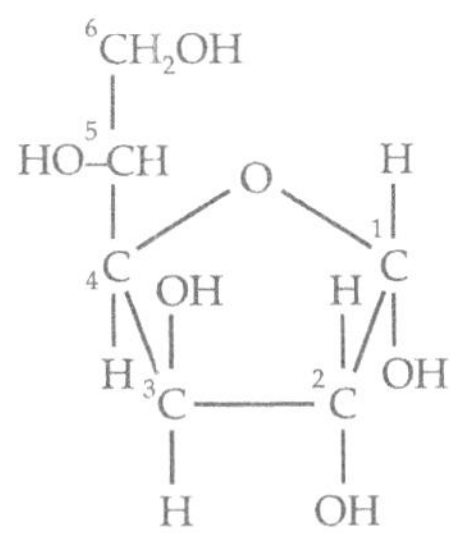

∝-D-Glucofuranose

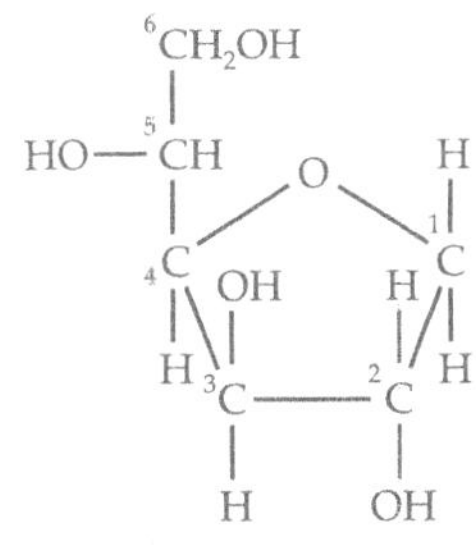

β-D-Glucofuranose

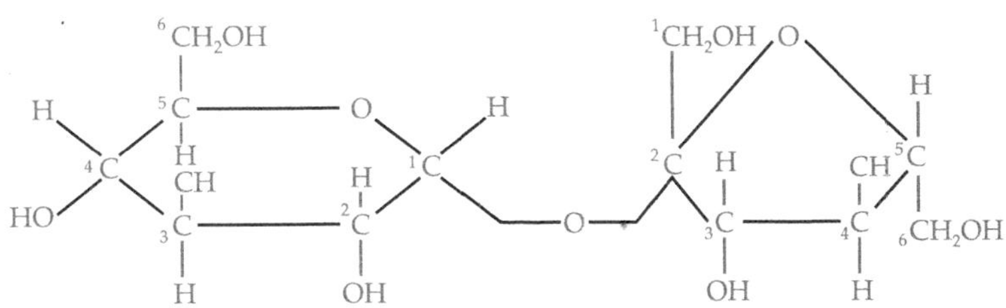

Sucrose

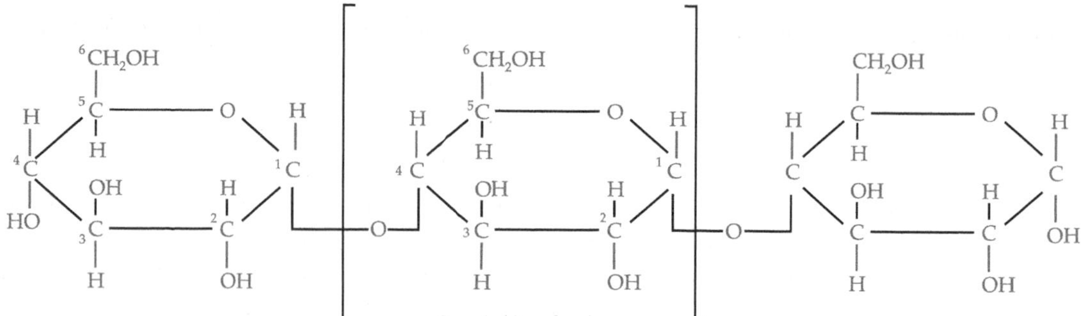

Starch (Amylase)

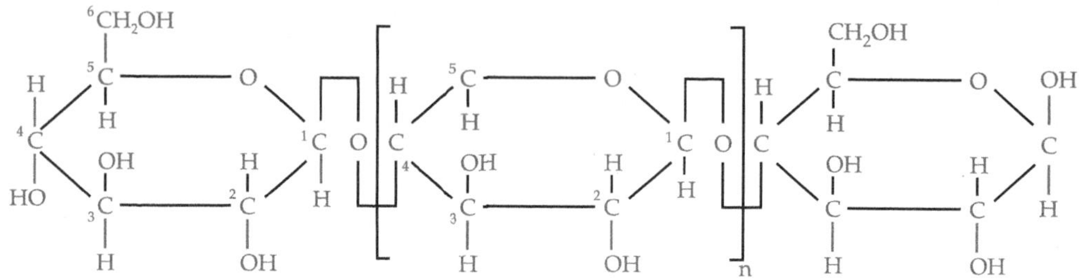

Cellulose

Monosaccharide Derivatives

A number of modifications of the simple monosaccharides exist. These include the sugar, phosphates, acids, alcohols, and glycerides as well as those substituted by amino and N-acetyl groupings.

Of major importance in plant metabolism are the phosphorylated derivatives. The first formed stable product of photosynthetic carbon dioxide fixation is 3-phosphoglyceric acid and, as in all living tissues, phosphorylated sugars are the key intermediates of carbohydrate metabolism.

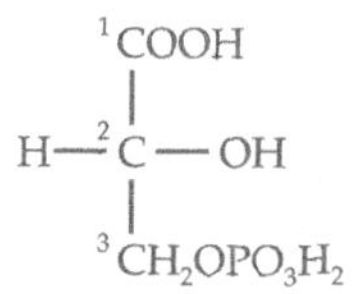

3-Phosphoglyceric acid

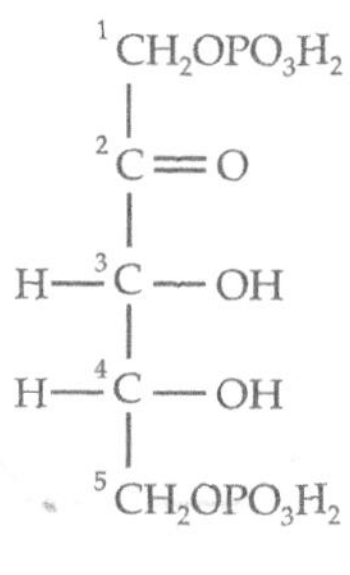

D-Ribulose–1,5 biophosphate

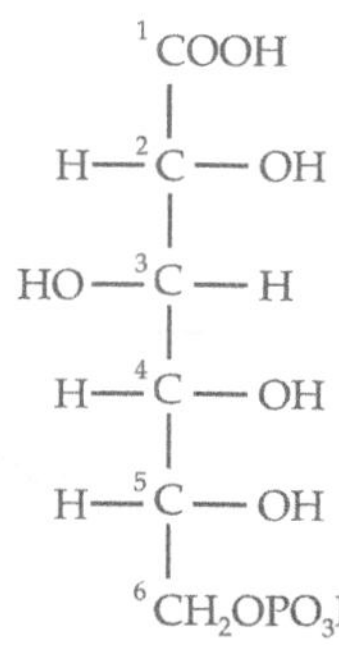

6-Phosphogluconic acid

Some Monosaccharide Derivatives

The Glycosidic Linkage

In living cells, free monosaccharides are a minor component of the total plant cell carbohydrate. They are generally found in combination with other monosaccharides as in oligo-or polysaccharides, or in combination with a non-carbohydrate component (aglycon) such as alcohol or ether. The two components are joined via the carbon through an oxygen bridge following the elimination of water. The bond joining the two components is referred to as a glycosidic linkage. Occasionally, the two components may be linked via nitrogen or sulphur atoms. Such compounds which are formed when sugars combine among themselves or react with non-carbohydrates with the formation of glycosidic linkages, are called *glycosides.* The non-sugar unit (moiety) is known as aglycone or genin. The most common glycosides are the O-glycosides where the sugar is linked through the oxygen atom to an alcohol, phenol, or acid. The sugar can also be linked directly to a carbon atom (C-glycoside), to a nitrogen atom (N-glycoside), or to a sulphur atom (S –glycoside). Aglycons can be phenols, terpenoids, steroids, or various nitrogen and sulphur containing compounds. D-glucose is the most common sugar to be found in glycosides.

Oligosaccharides

An oligosaccharide may be defined as a small polymer formed from n monosaccharide subunits joined by glycosidic linkages with the elimination of n–I molecules of water. The lower limit for n is 2, and the upper limit is generally around n = 20.

Oligosaccharides do not comprise a single class of compound and in many cases are simply short-chain intermediates in the synthesis and degradation of polysaccharides. They may be classified into di-, tri, tetra-, etc. saccharides depending on the number of monosaccharide subunits present.

The disaccharide sucrose is the most abundant oligosaccharide present in plants.

Sucrose is a condensation product of glucose and fructose-that is, in the formation of sucrose, glucose and fructose are linked together with the elimination of water. Thus, sucrose has the emprical formula $C_{12}H_{22}O_{11}$.

On treatment with warm dilute acid or with the enzyme invertase, sucrose is hydrolytically cleaved. In the hydrolysis, the elements of one molecule of water are taken up, and one molecule each of glucose and fructose are liberated.

Sucrose is not a reducing sugar, that is, it does not, for example, reduce cupric ions and hence does not possess either the free aldehydic group of glucose or the free ketonic group of fructose. It may be noted that whereas the glucose unit possesses a 6-membered pyranose ring structure, the fructose unit occurs as a 5-membered, or furanose ring.

The full name for sucrose is α-D- glucopyranosyl-β-D-fructofuranoside. Maltose is a reducing sugar (disaccharide-4-O-α-D-glucopyranosyl -β-D-glucopyranose) found in small amounts in many plants. Other oligosaccharides found in higher plants include trisaccharide raffinose; and the tetrasaccharide stachyose.

Polysaccharides

Polysaccharides are similar to oligosaccharides in that they are condensation polymers based on monosaccharides joined together by glycosidic linkages. They are, however of much higher molecular weight. Many polysaccharides, including the most abundant such as starch and cellulose, contain a single type of monosaccharide subunit and are called *homoglycans*. Those which contain more than one type of monosaccharide subunit are called *heteroglycans* and include many of the plant cell wall polysaccharides. The degree of polymerization, which is the number of monosaccharide submits present, is generally of the order of 80-100. Some cellulose molecules, however, may contain as many as 5000 residues.

Some Cell wall Polysaccharides

Polysaccharide	Main chain Monosaccharide	Side chain Monosaccharide
CELLULOSE	*D* - Glucose	—
MANNANS		
galactoglucomannas	*D* - Mannose	*D* - Galactose
	D - Glucose	
PECTIC SUBSTANCES		
arabinogalactans	*D* - Galactose	*L* - Arabinose
rhamnogalacturonans	*D* - Galacturonic acid	*D* - Xylose
		D - Galactose
	L - Rhamnose	*L* - Arabinose
		L - Fucose
XYLANS	*D* - Xylose	*L* - Arabinose
		4 - Methyl - α -
		D - glucuronic acid
		D - Glucuronic acid
		D - Xylose

SUCROSE METABOLISM

In most plants, sucrose is the form in which carbohydrate is transported from the sites of synthesis to the sites of unitization or storage. It is the most common disaccharide of plant tissues and is the precursor of most of the carbon-containing molecules found in plants. For example, it is the primary precursor of the lipid, protein, and carbohydrate stores of mature seeds ; it moves out from the stores during germination to provide energy and substrates for cell growth and differentiation in the absence of photosynthesis ; and it is presumably the immediate source of energy for maintenance of plant life during periods of darkness.

SUCROSE TRANSLOCATION

As the plant grows, material is exported from the leaves to the places of storage and growth. Those tissues where new organic materials are synthesized or where storage materials are mobilized are termed sources. These include the leaves as well as germinating seeds. The sites where the transported material is utilized are termed sinks and include meristems and cotyledons or endosperms of developing seeds. Material is transferred to sinks from sources by the process of translocation. In vascular plants this consists of two main pathways. The first is via the cytoplasm of the plant cell (symplast) and includes long-distance transport through the cells of the phloem as well as transfer over short distances via intercellular connections called plasmodesmata.

The main conducting unit of the phloem is the sieve element which includes specialized cells called sieve cells and sieve-tube elements. The sieve elements are elongated and arranged end-to-end in longitudinal rows to make sieve tubes. Associated with these are companion cells which may be concerned in the secretion of enzymes into the sieve elements. The direction of flux within the phloem is mainly from sources to sinks.

The second pathway is located in the extracellular tissues (apoplast) of the plant and serves to conduct water, ions, and some organic substances from the roots to the upper transpiring parts of the plant. These specialized tissues are called the xylem and consists of thick-walled conducting cells. These include vessels, which are large cells with no end walls that form tubes up the length of the stem, and tracheids, which are smaller cells with end walls. Carbohydrate is present only in trace amounts in xylem. The greater part of carbohydrate translocation thus takes place in the phloem. The loading of material into the sieve element companion cell complex thus generates the increased osmotic pressure to drive the mass flow of assimilates. Phloem loading is a selective process requiring metabolic energy and although the mechanism of sucrose uptake into the sieve elements is not yet clear, it is likely that it may be coupled to the co-transport of protons. The apoplast is relatively acidic (pH 5.5) and sieve tube elements are relatively alkaline (pH 8.5). Thus, the "downhill" transport of protons may be coupled to the secondary transport of sucrose.

Chapter 11

STAINING

INTRODUCTION

The subject of dyes or stains and methods of using them has become one of vast proportions. The number of dyes available is enormous, and ways of employing stains are almost as enormous as the workers using a particular dye or combination of stains. The preparation and use of stains has all become a science in itself, and there exists a most valuable journal, *Stain Technology,* devoted to these and related subjects. Experimenting with new dyes and new staining schedules is a most fascinating occupation, and there is a very great deal yet to be learned about stains and their utilization.

STAINS AND THEIR USE

Although stains were used as early as 1770 in the study of the structure of the wood, it was not until 1839 that Ehrenberg used them to study microbes. Their use depends upon the following properties :

1. The chemical must be *chromogenic*–in other words, it must possess colour – forming groups in its molecules.
2. The stain must exist as a salt in aqueous solutions–that is, it must break up (dissociate) into positively and negatively charged particles. These are called, respectively, cations and anions. For example, in the case of methylene blue, when placed in an aqueous medium the following chemical reaction occurs :

   Methylene blue chloride → methylene blue + chloride.

In this case the dye is cationic. However, sodium eosinate, a common form of the dye eosin, is anionic.

It is thought the proteins of cells react with the stains to produce the coloured products.

An *acid stain*, therefore, is a dye salt whose acid radical combines with the basic groups of tissue components staining them selectively. Conversely, a *basic stain* is a dye salt in which the basic radical combines with the acidic radicals of the tissue components stained by it. In a *double stain*, two dyes of contrasting colours are used either mixed together or successively. Commonly, an acid and a basic stain are combined. A stain produced by a chemical combination of an acid and a basic dye, and employed in alcoholic solution, is called a *neutral stain*, Eosin–methylene blue combinations are frequently so employed.

It is thought that the proteins of cells, react with the stains to produce the coloured products which we view under the microscope. A reason for this is that the acid dyes like eosin may combine with basic part of the protein through forces which resemble those that result in the attraction between the opposite poles of a magnet. Reciprocally, basic dyes like methylene blue will be attached to acid groups of proteins. It is likely that these facts explain many types of staining reactions.

Negative staining is a technique used in light microscopy and electron microscopy whereby the background is stained and organelles, bacteria, viruses, etc., are contrasted as unstained areas. In electron microscopy, solutions of heavy metal salts are used, most commonly potassium phototungstate. For example, Indian – ink staining method for *Spirochaets*, where they stand out white against a black background. Similarly, the Nigrosin method is used for cellular morphology, especially the *Diphtheria bacilli*. The organisms are again white against a black background.

Nuclear stains have greater affinity with chromatin of the nucleus. Such stains are usually basic dyes, such as haematoxylin or methylene blue. A dyestuff which has a special affinity for the cell cytoplasm (such as eosin) is known as a *plasmatic* or *plasmic stain*.

THE NATURAL DYES

The older botanical technicians placed much of their reliance upon natural or textile dyes. Very few of the natural, and practically none of the technical dyes are still in use (Johnson, 1940). Only three natural dyes (carmine, brazilin, and haematoxylin) are still used by botanical technicians. These dyes have not been manufactured synthetically.

Cochineal and Derivatives

Carmine is bright red coloured dye obtained from the female scale insect *Coccus cacti*, which lives on the cactus *Opuntia coccinellifera*. The bodies of the dried females make cochineal, and carminic acid (the actual colouring agent) is produced by extracting cochineal with boiling water, treating it with lead acetate, and treating the resultant lead carminate with sulphuric acid. The dye, carmine, is formed by mixing an alum with the carminic acid (Cohon, 1979). Carminic acid is only slightly soluble in water, and in 70% alcohol dissolves less than 3%.

Carmine is very widely employed by cytologists for studying, in fresh tissues, the chromosomes of both animals and plants as indicated by its extensive use in the iron – acetocarmine smear technique. One of the advantages of this stain is its combination with the fixing fluid. Thus only one step is necessary ; staining and fixing being carried out simultaneously. According to Johnson (1940), acetocarmine is prepared as follows :

Add 90 ml of glacial acetic acid to 110 ml of distilled water, heat to boiling and add 1 g of carmine dye. After cooling, add a few drops of 10% aqueous solution of ferric acetate till it acquires a wine–red colour. Allow the mixture to settle, and filter. Store the main bulk in a refrigerator, keeping a small quantity in a dropper bottle in the laboratory for immediate use (usually less than 0.5 g of carmine will dissolve in 100 ml of 45% acetic acid).

Brazilin

Brazilin is obtained principally from *Caesalpinia cristata* and *C. echinata*. However, it is neither so active nor so strong stain as haematoxylin.

Haematoxylin

Haematoxylin is one of the most important and useful biological stains. It is natural dye, extracted from logwood *Haematoxylum campechianum.* The product is purchased in the form of a pale yellow or brownish powder. Haematoxylin is not a dye in itself, but in the presence of certain alums (Aluminium or potassium phosphate), which serve as mordants, haematoxylin stains specific cell structures (mordants are salts, used to improve the selectivity or intensity of certain dyes).

The following formulae of haematoxylin are recommended by BDH (1972).

Delafield's Haematoxylin

Haematoxylin	3.2 g
Aluminium ammonium sulphate (ammonium alum)	30 g
Ethanol (absolute)	100 ml
Glycerol	80 ml
Distilled water	320 ml

Takes 6 to 8 weeks to ripen.

Ehrlich's Haematoxylin

Solution–A

Haematoxylin	2 g
Ethanol (absolute)	100 ml

Solution–B

Glycerol	100 ml
Distilled water	100 ml
Acetic acid (glacial)	10 ml
Aluminium potassium sulphate	3 g

Mix the two solutions and allow to ripen (unstoppered in bright daylight) for several weeks.

Harris's Haematoxylin

Solution–A

Haematoxylin	1 g
Ethanol (absolute)	6.7 ml

Solution–B

Aluminium potassium sulphate	13.3 g
Distilled water (warm)	133 ml

Mix the two solutions, bring quickly to the boil and then add 0.13 g of mercuric oxide. Cool rapidly. The solution is then ready for use. To improve nuclear selectivity, 535 ml of glacial acetic acid may be added to each 100 ml of stain.

Heidenhain's Haematoxylin

Haematoxylin	1 g
Ethanol (absolute)	10 ml
Distilled water	90 ml

Dissolve the haematoxylin in the ethanol and add the distilled water. Stopper the bottle and allow to ripen for a few days.

Weigert's Iron Haematoxylin

Solution–A

Haematoxylin	1 g
Ethanol, 95%	100 ml

Allow the solution to ripen in bright daylight for a few days.

Solution–B

Ferric chloride (30% in distilled water)	4 ml
Hydrochloric acid	1 ml
Distilled water	95 ml

Staining Solution : Mix equal parts of solutions A and B.

THE COAL–TAR DYES

All the dyes now used in biological technique are synthetic chemical compounds made from the substances found in coal tar.

OBJECTS OF STAINING

Most constituents of cells are, in their natural state, either colourless or only faintly coloured. Thus, they are only visible if their refractive indices differ from those of the media in which they are immersed. This is greatly facilitated by staining them either more deeply than, or of a different colour from their surroundings.

The chief object of histological staining is then, to cause certain constituents of the cells to take on a different intensity of tint from others. This may be done in various ways, but it is usual to distinguish two kinds of selective staining, *histological* and *cytological*. In the former, an entire tissue or group of tissue elements is prominently stained, the elements of other kinds present, remaining colourless or being differently stained. In the latter, the stain is taken up or retained by some constituent element of the cell, such as the chromatin of the nucleus or an element of the cytoplasm.

STAINING TECHNIQUE FOR LIVING AND FIXED CELLS

Vital dyes penetrate living cells and stain structures without serious injury. For example, Janus green B stains mitochondria selectively and methylene blue stains the Golgi complex. Vital dyes are not entirely harmless, but they kill only after long exposure. Although helpful, the vital dye technique has limited use because many cell organelles are not selectively stained or are too small to be seen with light microscopy.

Most early work on cell organelles was done with fixed and stained preparations. Fixing agents, such as formalin, alcohol, acids, salts of heavy metals or mixtures of these render proteins insoluble. Next, water is removed from the fixed tissues by *dehydrating agents,* such as alcohol, and the tissues are embedded in paraffin and sectioned with a microtome. The sections are affixed to the slides, the paraffin removed with xylol and the slides washed in xylol – alcohol. By washing in decreasing concentrations of alcohol the sections are partially hydrated and the proteinaceous matter of the cell is then differentially stained to distinguish the structures present. Natural dyes (*e.g.,* haematoxylin) or basic anilin dyes (*e.g.,* safranin and basic fuchsin) stain the nucleus selectively, and acid dyes (*e.g.,* orange G, eosin, and fast green) stain the cytoplasm. The sections are then dehydrated with alcohol. To reduce scattering of light, it is necessary to replace the alcoholic medium with a substance having the same refractive index as that of the protein particles. This is accomplished with *clearing agent,* such as xylol, which infiltrates among the protein particles. The preparation is then mounted in canada balsam (or DPX), which has a refractive index about equal to that of the cell proteins, making it possible to see the stained structures clearly.

Principal botanical uses for the common stains (Johnson, 1940, Sass, 1967).

Cellulose cell walls :

Delafield's Haematoxylin, congo red, fast green, light green, aniline blue, Bismarck brown, acid fuchsin.

Lignified cell walls :

Safranin, iodine green, methyl green, methylene green, crystal violet.

Cutinized cell walls :

Safranin, acid fuchsin, crystal or methyl violet, erythrosin, methyl green, methylene blue.

Suberized cell walls :

Safranin, Sudan III or IV.

Middle lamella :

Iron haematoxylin, ruthenium red.

Chitin :

Safranin.

Chromosomes :

Iron haematoxylin, safranin, crystal violet, carmine (for acetocarmine smears), ocrein.

Mitochondria :

Iron haematoxylin, acid fuchsin, crystal violet.

Cytoplasm :

Eosin, malachite green, erythrosin, methyl orange, fast green, light green, orange–G or gold orange.

Plastids :

Crystal or methyl violet, iron haematoxylin.

Simple Bacterial Stains :

Most routine stained preparations of bacteria are prepared by the use of a dye like methylene blue, gentian violet, or carbon fuchsin. To stain in this way, the dye is poured onto a slide containing a bacterial smear. The time of exposure depends on the stain used and sometimes on the bacterium. After the dye is washed off the slide, the smear is dried. The smear is covered with methylene blue stain for about five minutes. Subsequently, the stain is washed under tap water, drained, and dried.

THE GRAM STAIN

This is the most important stain as it serves as a basis for differentiating and classifying microorganisms. A Danish physician named Hans Christian Jaochim Gram (b – 1853) worked out its details when he noted that certain bacteria, after being stained with gentian violet, become decolourized when washed in an iodine solution followed by ethyl alcohol. These bacteria are referred to as being Gram – negative, while those that retain the stain under these conditions are called Gram – positive. The Gram stain is significant because many other characteristics appear to be correlated with this reaction, including types of flagellation, response to certain poisons, etc.

The procedure of Gram Stain is as follows :

1. Prepare a smear.
2. Cover the smear with ammonium oxalate crystal violet stain and allow to remain for one minute.
3. Drain the excess stain and gently wash the slide in tap water.
4. Flood with Gram's iodine solution and leave for one minute.
5. Rinse with 95% ethyl alcohol until no more dye flows away from the smear (about 30 seconds to one minute).
6. Wash in tap water.
7. Counterstain with safranin for 10 to 30 seconds.
8. Wash in tap water, drain off the excess water, blot, and allow to air – dry.

   Gram–positive organisms stain violet, and

   Gram–negative organisms stain pink.

A single stain is preferably used where tissue differentiation is not pronounced or a specific structure is to be distinguished. Filaments of algae, and sections of bryophytes are generally stained singly with iodine (because the cells contain plenty of starch), and mounted in glycerine. Iodine solution may be prepared by any of the following formulae.

Gram's Iodine

Iodine, resublimed	1 g
Potassium iodide	2 g
Distilled water	300 ml

Lugol's Iodine

Iodine, resublimed	1 g
Potassium iodide	2 g
distilled water	100 ml

NOTE Starch is stained blue or black with iodine solution, cytoplasm yellow brown and cellulose walls pale yellow.

In the case of plant materials where tissue differentiation (*i.e.*, lignified and nonlignified tissue) is very prounced, such as in pteridophyta, gymnosperms, and angiospesms, a combination of stains is used (called double staining). The following combination of an acidic and a basic dye are usually recommended.

Delafield's haematoxylin and safranin, safranin and fast green, or crystal violet and erythrosine.

FEULGEN STAINING

The Feulgen smear or squash technique to distinguish the particular type of nucleic acid in chromatin seems to be the most precise and the most satisfactory, for general cytological purposes. The method developed in 1924 depends upon the presence among the products resulting from a mild hydrolysis of nucleoproteins, of an aldehyde, and the brilliant colouration this gives with fuchsin – sulphurous acid. Generally, the *Feulgen–nucleal reaction* is specfic for DNA (Cohn, 1979). The cytoplasm and the RNA – containing portions of the nucleus are Feulgen–negative. The dye component used a chloride of pararosaniline is commonly known as basic fuchsin.

The staining solution is prepared as follows :

Dissolve 0.5 g basic fuchsin by pouring over in 100 ml boiling distilled water and shaking thoroughly. Cool to 50°C and filter. To the filtrate add 10 ml 1/N hydrochloric acid, then 0.5 g potassium metabisulphite. Sulphur dioxide is given off and the liquid slowly turns colourless or of a light straw – colour. Allow to stand for 23 hours before using, and keep the bottle well stoppered and in the dark. The clear liquid is otherwise known as Schiff reagent (1/N hydrochloric acid may be prepared by adding 82.5 ml of HCl to a litre of glass – distilled water). In the course of time the staining solution becomes coloured and then it should be discarded.

FUNGAL STAINING

Lactophenol blue method for mycelia – cotton blue is frequently used as a stain for fungal hyphae. Tease out a portion of the mycelial culture and mount wet in a drop of lactophenol blue (Phenol, melted 20 ml, lactic acid 20 ml, glycerol 40 ml, aniline blue, water soluble, 0.05 g). Place a cover slip over the preparation, press gently, and remove excess of the fluid.

Result : *Mycelial structures* – blue.

NOTE
1. As lactophenol blue in also a mountant, the preparations may be kept safely.
2. Staining is accelerated by gently warming the slide over a spirit flame.

REFERENCES

1. Alfred S. Sussman, 1964. *A Laboratory Block. Microbes : Their Growth, Nutrition and Interaction.* D. C. Heat & Co., Boston.
2. Arthur C. Giese, 1979. *Cell Physiology.* W. B. Saunders Company, Philadelphia.
3. *Biological Stains and Staining Methods*, 1972, 5 th. edition. BDH Chemicals Ltd. Poole, England.
4. D. A. Johnson, 1940. *Plant Microtechnique*, Mc Graw – Hill Book Company, Inc., New York.
5. J. Bronte Gatenby and Theophilus S. Painter (eds.), 1946. *The Microtomisr's Vade-Mecum.* J. & A. Churchill Ltd., London.
6. John E. Sass, 1967. *Botanical Microtechnique*, 3rd. edition. Oxford & IBH Publishing Co., Calcutta.
7. Macdonald Critchley (Editor-in-Chief), 1986. *Butterworths Medical Dictionary*, 2nd. edition. English Language Book Society, England.
8. Norman S. Cohn, 1979. *Elements of Cytology*, 2nd. edition. Freeman Book Co., Delhi.

Chapter 12

CHROMATOGRAPHY

INTRODUCTION

Chromatography is an analytical technique for separating two or more chemical compounds in solution by taking advantage of the fact that they are removed from solution at different rates when the latter is percolated down a column of a powdered adsorbent or passed across the surface of an absorbent paper. This is one of the most significant and reliable methods used in the fields of chemistry and biological sciences to obtain and identify substances (such as proteins, chlorophyll pigments, etc.), in a high state of purity.

Chromatography, a term derived from the Greek words *chroma*, colour, and *graphien*, to record, was introduced by the Russian botanist Michael Tswett in 1906. He described the separation of a mixture of leaf pigments on a column of calcium carbonate.

CHROMATOGRAPHIC METHODS

In all the chromatographic techniques, difference in affinity involves the process of adsorption, or partition. In adsorption, the binding of a compound, to the surface of the solid phase takes place. In partition the relative solubility of a compound in two phases, results in the partition of the compound in two phases.

Adsorption Chromatography

In the general chromatographic method as introduced into biology by Tswett in 1906, various adsorbants (*e.g.*, starch, aluminium oxide) are poured as a slurry into a tube. An unknown is poured on the widened top of the tube and then "developed" by the slow trickle of an appropriate solvent through the column. The rate at which any compound migrates through the column depends upon the balance between its affinity for the solvent and its adsorption on the particles of the column. This method of separating compounds is spoken of as *adsorption chromatography*.

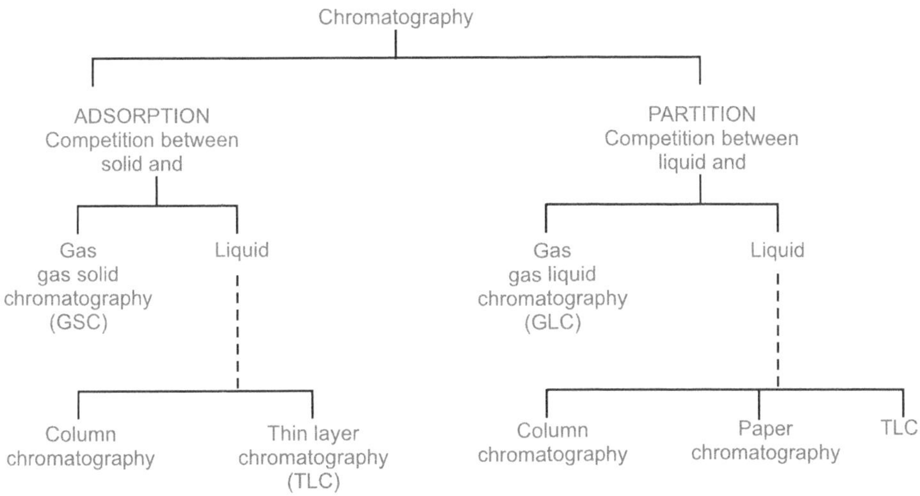

(Srivastava and Kishore, 1985)

Partition Chromatography

Another chromatographic method depends upon partition of a soluble between two solvents in which the solute has differential solubility. If a solute has greater solubility in one solvent than in another, the two solvents being immiscible, it is possible to remove and concentrate the solute in the solvent which it has greater solubility.

Something like this separation can be achieved on a column of some powdered substance if only one of the solvents wets the powder. The solvent wetting the powder will form a stationary solvent phase, and the powder in this case holds the solvent that wets it. In practice, the solute in the appropriate organic solvent is added to the top of the wet column and developed by washing with additional aliquots of the solvent. As the second solvent, immiscible with the first, moves through the column, the solute is partitioned between the two solvents, as if an infinite number of separatory funnels were being used in series to extract the solute. If instead of one solute several have been dissolved in the solution poured on the top of the column, they will separate as several bands on the column because of their differences in partition between the two solvents. The method has been used to separate protein hydrolysates (amino acids), and has tremendous applications in medical sciences.

Paper Chromatography

The technique of paper chromatography, as developed in 1941 by Martin and Synge, is a variation of Tswett's partition chromatographic method. It is a method based also on the differential solubility of compounds in solvents, and it achieves separation by partition of the solute between the two solvents. The paper acts as support for one of the solvents.

The solution to be tested is placed as a spot on a large sheet of filter paper. When the spot is dry paper is placed with one edge dipping into a tray containing the mixture of solvents, the choice of solvents depending upon the solubility of the solutes to be chromatographed. After the solvent has moved the desired distance beyond the sample spot, the solvent front is marked, and the paper is dried and sprayed with some agent to "develope" the spots of materials that have migrated from the original sample spot.

For example, a mixture of amino acid is spotted on filter paper, and the paper is dipped into a butyl alcohol–water mixture. Diverse amino acids are differentially soluble in these two solvents, some being more soluble in water, others more soluble in butyl alcohol. The alcohol – water mixture migrates along the filter paper fibres upward through the sample spot, and the various substances with differential solubility migrate to different distances from the original spot, those most soluble in butyl alcohol moving farthest. After solvent front has advanced past the spot to a distance considered from experience to be adequate, the line of the solvent front is marked and the paper is dried. For detection of amino acids the paper is then sprayed with ninhydrin, which forms a purplish colour on reacting with an amino acid. Control chromatograms, made from known mixtures of amino acids, are run in parallel with the unknown. From the positions of known amino acids the unknown acids can be determined. From the depth of the colour reaction quantitative data may be gathered.

The ratio of the distance traveled by any particular substance to the distance covered by the solvent is known as the *ratio of the fronts* or Rf. This ratio is usually constant for a given substance in a particular solvent and under given conditions (*e.g.*, temperature). Thus, the Rf values determined from controls of known amino acids under similar conditions may be used as standards for identification of unknown amino acids.

Two–dimensional Chromatography

When a "one – dimensional" chromatogram does not adequately separate the components of a mixture, the once – developed paper is placed, after it has dried, in a pan of another mixture of solvents in such a way that the movement of the second mixture of solvents occurs at right angles to the movement of the first, producing a "two – dimensional" chromatogram.

When the substances separated by paper chromatography are radioactive, they can be visualized by photography. The filter paper is pressed against a sensitive photographic film. The ionizing radiations from the radioactive material will so affect the film that on development the film will show darkening in proportion to the radiation, the amount of radiation depending upon the concentration of the radioactive substance. From the Rf values of controls and from the density of the darkened areas on film exposed to known concentrations of radioactive material, identification and quantitative analysis of an unknown material is possible, even when the unknown is present only in traces, as for example, after a few seconds of photosynthesis.

Gas Chromatography

A method of analysis by vaporization of volatile compounds through suitably packed columns, ending in a detector from which a recording is made. The substance to be analysed is partitioned between mobile phase (usually nitrogen and hydrogen), and a stationary phase (a liquid material coated on an inert solid medium).

Ion–exchange Chromatography

The separation of the components of a mixture (usually metallic salts) in solution by percolating the latter through a long column of ion – exchange material (usually synthetic ion – exchange resins). By subsequent elution with an acid the metal ions can be carried forward in bands which travel down the column and are re–absorbed at lower ends to from distinctive zones. The method has been applied to a wide variety of compounds (proteins, nucleic acids, vitamins, etc.).

Thin Layer Chromatography

A method by which dissolved chemical and drug substances can be identified by a chromatogram run out under electrical impulsion into a silica, or other "thin layer" gel spread on a glass plate.

REFERENCES

1. Arthur C. Giese, 1979. *Cell Physiology*. W.B. Saunders Company, Philadelphia.
2. H. S. Srivastava, 1991. *Elements of Biochemistry*, 3rd edition, Rastogi Publications, Meerut.
3. Macdonald Critchley (Editor – in – Chief), 1986. *Butterworths Medical Dictionary*, 2nd edition, English Language Book Society, England.
4. V. K. Srivastava and K. Kishore, 1985. *Introduction to Chromatography* – Theory and Practice. S. Chand & Company, New Delhi.

Chapter 13

TRACER TECHNIQUE—THE USE OF ISOTOPES AS TRACERS

INTRODUCTION

The atoms of most chemical elements exist in more than one variety. Each kind of a given element has a different atomic weight, but all of them carry the same nuclear charge. For example, there are three different kinds of magnesium atoms with atomic weights of 24, 25, and 26 respectively. Such different varieties of atoms of a given element are called *isotopes*. Differences in the chemical behaviour of two isotopes of the same element are so slight as to be barely detectable, and ordinarily they cannot be separated by chemical methods. The chemical properties of all isotopes of a given element are virtually identical, because all have the same electronic configuration ; one isotope differs from another only in the constitution of the atomic nucleus which is composed of protons and neutrons.

RADIOACTIVE ISOTOPES

In addition to the stable isotopes described above, there are many radioactive isotopes of certain heavy elements such as uranium and radium occur in nature. The atoms of many elements not normally radioactive may be made so by bombardment with various types of elementary particles such as neutrons, protons, deutrons, and alpha particles. Such bombardments are usually accomplished with the aid of a cyclotron or a uranium pile reactor.

"HALF-LIFE" OF ISOTOPES

Atoms of all radioactive elements gradually disintegrate, emitting charged particles and radiations in the process. For the artificially produced radioactive elements which are of greatest significance in biological sciences, the charged particles emitted are electrons and positrons. As a result of the loss of particles and radiations by radioactive isotopes one element is transmuted into another. Radioactive Na^{21}, for example, with the loss of an electron, is converted into the stable naturally occurring Mg^{24} isotope. The *half–life* of a radioactive isotope is the length of time required for half of any given weight of the isotope to be converted into the kind of atom into which it is degraded. The half – lives of radioactive isotopes ranges from less than a second to many hundreds of years.

The situation with regard to isotopes is well illustrated by the biologically important element carbon. Naturally occurring carbon consists almost entirely of stable C^{12} isotope. The C^{13} isotope is also stable and occurs in traces in nature. The atomic weight of naturally–occurring carbon is a little more than 12, because of the presence of a slight admixture of the C^{13} isotope. In addition, three radioactive isotopes can be made : C^{10}, with a half–life of 9 sec., C^{11}, with a half–life of 21 min., and C^{14}, with a half–life of about 5000 years.

A radioactive isotope betrays its presence wherever it may be by the continuous emission of charged particles which can be detected by an electroscope or Geiger counter. Similarly, the presence of any of the rarer stable isotopes in a system can be detected because of their unusual mass, by means of the instrument known as the mass spectrograph. Hence, if molecules of any compound are tagged, by incorporating into them atoms that are radioactive or atoms of the rarer kinds of stable isotopes, it is possible to trace such molecules through a system. After the absorption or ingestion of such molecules by an organism it is possible to determine the reactions in which they participate. This is not possible by ordinary methods of chemical analysis because by such methods it is impossible to distinguish between introduced molecules and others of the same species which were already present in the organism. Some of the more widely used isotopes in plant research are the stable isotopes of H^2, O^{18}, C^{13}, and N^{15}, and the radioactive isotopes Na^{24}, P^{32}, K^{42}, C^{11}, C^{14}, S^{35} and Ca^{45}.

If a compound containing tagged atoms is introduced through the roots of a plant, subsequent distribution of the labelled atoms in the plant can be followed. Certain examples of these investigations are given below.

1. Translocation of Carbohydrates

Rebideau and Burr (1945) labelled the carbohydrate formed in bean plants by allowing one leaf on the plant to photosynthesize in an atmosphere containing carbon dioxide made with the identifiable C^{13} isotope. They then showed that movement of carbohydrates occurred readily in the stem, both upward and downward, from the node at which the leaf petiole was attached. If short segments of the stem above and below the node at which the petiole was attached were killed by treatment with hot wax, however, no upward or downward movement of carbohydrates occurred. Phosphates, containing radioactive phosphorus (P^{32}), however, were readily conducted through such killed stem segments, indicating that their translocation occurred through the xylem.

2. The Entrance of Carbon Dioxide

Although the bulk of the carbon dioxide entering the leaves of land plants undoubtedly passes in through the stomata, substantial amounts may diffuse directly into the epidermal cells in some species. For example, a considerable portion of the carbon dioxide used in photosynthesis in certain species which have stomata only in the lower epidermis, such as coleus, begonia, etc., passes into the leaves through the upper epidermis. By the use of carbon dioxide made with the radioactive C^{14} isotope, it has been confirmed that this gas can pass into the leaves of coleus and hydrangea through their upper, unstomated surface.

3. Source of the Oxygen Released in Photosynthesis

A knowledge of the source of the oxygen released in photosynthesis is one of the keys to an understanding of the mechanism of the process. Many investigators of photosynthesis have assumed in the past that the oxygen released came from carbon dioxide. By using water and carbon dioxide made with heavy isotope of oxygen (O^{18}) as the raw materials of photosynthesis, however, Ruben *et. al.* (1941) have shown that the oxygen set free in the process came from water molecules and not from carbon dioxide molecules.

4. The Carbon Pathway in Photosynthesis

Considerable progress has been made in tracing the pathway of carbon through the photosynthetic mechanism by the use of radioactive isotopes. When a suspension of the alga *Chlorella* or of *Scenedesmus* was allowed to photosynthesize for the very short period of 5 sec. in the presence of carbon dioxide made with radioactive C^{14} isotope, a very large proportion (about 70%) of the labelled carbon was found by subsequent analysis of the cells to be present in the 3-carbon compound 3-phosphoglyceric acid (3-PGA). According to Calvin and his associates 3–PGA appears to be a key intermediate in photosynthesis.

Radioactive carbon appears in sucrose molecules after relatively short periods of photosynthesis. It is also found after very short periods of photosynthesis to be present in alanine, serine, and other amino acids. This is evidence in favour of the view that the synthesis of amino acids in photosynthetic tissues is closely associated with the intermediate steps in photosynthesis.

5. Permeability of Cell Membranes

Valuable quantitative data on permeability of ions such as sodium and potassium have been obtained by the use of radioactive tracers. Some ions (sodium for example) move in and out of cells so that little net change occurs in concentration of quantitative analysis. Such movement could be detected only by the tracer method. Some of the isotopes commonly used for such tracer studies are – H^3, C^{14}, Na^{24}, P^{32}, S^{35}, Cl^{36}. K^{42}, Ca^{45}, Fe^{59}, Cu^{64}, Zn^{65}, Br^{82}, Sr^{89} and I^{131}.

6. Measuring Growth Rates of Cells

Growth of certain cell constituents may be followed in living cells by the rate of tracer incorporation (*e.g.*, S^{35} incorporated into protein). Since protein increases continuously in a healthy, growing culture, the label concentration in the culture may be a satisfactory measure of the overall growth rate. C^{14} – thymidine has been used in a similar manner to label DNA.

7. Mycorrhizas

The roots of many species of plants are regularly infected with the mycelium of fungi. Such a root, together with its associated fungal hyphae, is called a mycorrhiza (literally "fungus – root"). As regards the significance of these root – fungus associations, it is believed that mycorrhizas may play a significant role in the absorption of mineral salts. Experiments with radioactive phosphorus have shown that mycorrhizal portions of the roots of two species of *Pinus* accumulate much larger quantities of phosphate than nonmycorrhizal portions.

8. Autoradiography

If a cell is suspended in a solution of radioactively labeled substance, it is possible to determine where in the cell incorporation occurs and to what extent. For example, ^{3}H–uridine may be presented to cells for a brief time and the cells are mounted on a slide and fixed. A layer of photographic emulsion is deposited over the preparation. The radioactive emission from the ^{3}H–uridine will reduce silver halide to silver in the emulsion. On development, the grains affected by the radiation (beta particles of low energy value) appear as black grains or traces over the structure into which the ^{3}H–uridine has been incorporated. If the grains appear over the nucleus we know that the nucleus incorporates uridine into ribonucleic acid (RNA). Similar experiments can be performed with ^{3}H– thymidine, which is selectively incorporated into deoxyribonucleic acid (DNA), and with ^{3}H – amino acids, which are incorporated into proteins. Autoradiography has been of immense value in localizing the sties of synthesis in the cell.

REFERENCES

1. Arthur C. Giese, 1979. *Cell Physiology,* W. B. Saunders Company, Philadelphia.
2. Bernard S. Meyer and Donald B. Anderson, 1958. *Plant Physiology,* 2nd edition, D. Van Nostrand Company, Princeton.

Chapter 14

SPECTROPHOTOMETRY

INTRODUCTION

Coloured substances absorb certain wavelengths of light and reflect or transmit rest of it. A material or a group which absorbs ultraviolet or visible light is known as a chromophore (coloured substance). Each chromophore is characterised by an *absorption maximum,* which is the radiation of a specific wavelength where the maximum excitation of a molecule takes place.

The colourless substances also have the property of absorbing certain wavelengths of invisible part of the spectrum. This principle of utilizing the ultraviolet, infrared, and fluorescent methods of analysis used in the detection and measurements of very small quantities of chemical substances is known as spectrophotometry. Absorption curves for most substances enable accurate quantitative and qualitative data to be obtained.

Spectrophotometry is a very useful device for determination of pigments, inorganic compounds (colorimetry), aromatic and heterocyclic organic compounds (ultraviolet spectrophotometry), and even gases (infrared spectrophotometry).

Spectrophotometer is an instrument used to compare the intensities of two differently coloured light sources by resolving their light into spectra and measuring the relative intensities of those spectra, wavelength by wavelength.

An instrument for the measurement of variation of the intensity with wavelength of the absorption spectrum produced by a substance in the infrared region is called an infrared absorption spectrophotometer :

A typical spectraphotometer consists of the following components :

1. Source of light
2. Monochromator unit
3. A detector, and
4. A meter

The most common source of light is an incandescent tungesten filament lamp. The light rays are rendered parallel by a collimating lens or mirror before passing to the monochromator unit. The unit is normally made up of a rotating prism and a filter with a slit. The prism resolves light into different wave bands and the filter allows certain wavelengths to pass through it. The aperture of the filter is adjustable. A sample holder is placed between the monochromator unit and the detector which is actually a photosensitive device. It is connected with a meter to show the exact amount of light passing through the sample.

APPLICATION OF SPECTROPHOTOMETRY

Some of the most important examples where spectrophotometry has been used in biochemical or cytochemical studies are given below.

1. Determination of Cell Constituents

In some cases the quantity of a cell constituent may be determined by the amount of light transmitted through a section of an organelle stained with a specific dye and comparing it with the light transmitted by an equral thickness of a standard solution of the substance. Ultraviolet spectrophotometry of nucleic acid has been made quantitative by comparing transmission of ultraviolet light through cell organelles and the same thickness of nucleic acid solutions.

2. Recombining of DNA Strands

The hydrogen bonds between the two DNA strands are broken by heat. When the strands separate from one another, they may fail to recombine in the same manner (renature) when the temperature is lowered. Because it is accompanied by progressively increased absorption at 260 n m, breaking the hydrogen bonds of DNA may be followed quantitatively by spectrophotometry.

3. Localization of DNA and RNA in Nuclei

Nucleic acids can be located in nuclei by ultraviolet photomicrography. This method is based on the fact that no other protoplasmic constituents absorb ultraviolet light significantly at the 260 n m absorption peak of nucleic acid. Therefore, the density of the image in photomicrographs taken at this wavelength measures the nucleic acid content in the cell.

Lambert's law states that the proportion of incident light absorbed by a stained specimen is directly related to its thickness, and Beer's law states that the proportion of light absorbed is directly related to the concentration of absorbing molecules (Beer–lambert law). Either ultraviolet or visible light may be utilized in cytopotometry, and their use depends on these relationships.

4. Estimation of Nucleic Acids

It is even possible to estimate the amount of nucleic acid present in a nucleus, for example, by comparing the optical density of the photograph through a nucleus with known nucleic acid concentrations. The special measuring equipment is called photoelectric microspectrophotometer. Because both DNA and RNA absorb ultraviolet radiation equally, it is necessary to use specific staining to localize each of the nucleic acids separately.

The Feulgen reaction detects DNA in cells. After fixation, tissue sections, or cells are hydrolyzed with hydrochloric acid to liberate the aldehyde group in the deoxyribose sugar of DNA and then treated with leucofuchsin (basic fuchsin in sulphurous acid). The aldehyde reacts with leucofuchsin to give a purplish colour, the depth of which can be estimated by spectrophotometry. Ribonucleic acid may be selectively stained with the dye pyronin, RNA content may then be estimated spectrophotometrically.

5. The absorption Spectra of Chlorophyll Pigments

The chlorophylls *a* and *b* exhibit different absorption spectra (an *absorption spectrum* is a measure of the extent to which a given substance absorbs the light of different colours or wavelengths). As determined with a spectrophotometer, both the chlorophylls show absorption maxima in the blue – violet region and orange – red region of the visible spectrum (minimum absorption of the green and yellow wavelengths, *i.e.,* 500 to 600 n m). The absorption spectrum of chlorophyll provides indirect evidence of the wavelengths of light that are absorbed for the process of photosynthesis. The absorption maxima of these photosynthetic pigments give an idea as to the quality of light most effective in the process of photosynthesis.

6. Measurement of CO_2—Evolution during Respiration

The rate of an enzymatic reaction is often measured spectrophotometrically because accurate measurements can be made quickly and continuously during the course of the reaction. Infrared spectrophotometry method can also be used for measuring and analyzing CO_2 evolved during respiration.

REFERENCES

1. David T. Plummer, 1988. *An Introduction to Practical Biochemistry.* 3rd edition, Tata McGraw-Hill Publishing Co., New Delhi.
2. H. S. Srivastava, 1991. *Elements of Biochemistry,* 3rd edition, Rastogi Publications, Meerut.
3. Macdonald Critchley (Editor – in Chief), 1986. *Butterworths Medical Dictionary,* 2nd edition, English Language Book Society, England.
4. Mammen Daniel, 1991. Methods in *Plant Chemistry and Economic Botany,* Kalyani Publishers, New Delhi.
5. Norman S. Cohn, 1979. *Elements of Cytology,* 2nd edition, Freeman Book Co., Delhi.

INDEX

A

B

C

D

E

F

G

H

I

K

L

M

N

O

P

R

S

T

V

W

Z